VIBRATIONAL SPECTROSCOPY OF MOLECULES AND MACROMOLECULES ON SURFACES

VIBRATIONAL SPECTROSCOPY OF MOLECULES AND MACROMOLECULES ON SURFACES

MAREK W. URBAN

Department of Polymers and Coatings
North Dakota State University
Fargo, North Dakota

A Wiley-Interscience Publication

JOHN WILEY & SONS, INC.

New York / Chichester / Brisbane / Toronto / Singapore

Library of Congress Cataloging in Publication Data:
Urban, Marek W., 1953–
      Vibrational spectroscopy of molecules and macromolecules on
   surfaces / Marek W. Urban.
          p.   cm.
      "A Wiley-Interscience publication."
      Includes bibliographical references and index.
      ISBN 0-471-52815-3 (acid-free paper)
      1. Vibrational spectra.   2. Surface chemistry.   I. Title.
   QD96.V53U73   1993
   541.3'3—dc20   93-18560

10 9 8 7 6 5 4 3 2

PREFACE

In the last 30 years experimental sciences have developed to a profound level of sophistication and modern vibrational spectroscopy is a vital example of the phenomenon where an instrument is often treated as a black box, and the output is supposed to solve all practical problems. Whereas new emerging technologies are expected to further advance the sensitivity and selectivity of existing instrumentation, let us not forget about the fundamental principles governing measured physical processes. Perhaps the most practical scientist was Albert Einstein, who, among many earth-shaking discoveries, related the simple feeling of wet and dry sand on his feet to the surface tension phenomenon. He loved sailing, and evidently spending time on a beach may have positively influenced his discoveries that later on provided a foundation for understanding many physicochemical processes.

Motivation for writing this book came about after studies on the vibrational spectroscopy of polymers, polymeric thin films, and coatings had proceeded for a few years in the author's laboratories. Graduate students, postdoctoral fellows, and research associates unanimously realized that there is a need for a book covering those aspects of vibrational spectroscopy that would bridge the gap between basic principles and practical aspects of vibrational spectroscopy and focus on structure–property relations on surfaces and interfaces. As a result, the first few chapters cover fundamental aspects of vibrational spectroscopy, including normal coordinate analysis with selective surface examples and the differences between bulk and surface spectroscopy. The remaining chapters provide selected band assignments, structural identification, and structure–property relations of monomers and polymers on surfaces. These sections are separated by the coverage of the fundamental principles of several surface techniques. The author hopes that such an approach will make this book not only a suitable source for those in need of theoretical background in surface vibrational spectroscopy but also a useful reference source. Ideally, the ultimate objective is to correlate surface and interface vibrational features with the structures that may or may not be responsible for certain properties. However, as one of the reviewers stated, "the topic is broad and task is difficult." Although efforts were made to provide a broad and balanced coverage of the field, it is simply not possible to cover all the frequent controversial findings in this limited space without omitting important studies. Hopefully the review articles and books that are abundantly quoted throughout this text will suffice for that.

Fargo, North Dakota   MAREK W. URBAN
November 1993

ACKNOWLEDGMENTS

The author would like to take this opportunity to thank all his graduate students, postdoctoral fellows, and other former or current associates who, through their involvement in the author's research and dedication, encouraged him to write this book. Learning spectroscopy is a complex process because it requires not only understanding of fundamental principles but also to be able to interpret the spectra. Special thanks are therefore extended to the author's outstanding spectroscopy teachers: Professor M. Handke, who was the first person to stimulate the author's interest in vibrational spectroscopy; and Professors K. Nakamoto, B. C. Cornilsen, and J. L. Koenig, the author's thesis and postdoctoral advisors. Writing a book is an effort that without secretarial and editorial efforts would be impossible to accomplish. The author sincerely thanks Mrs. Jeannette Lynch-Shaw for her endless efforts and help in the manuscript preparation. Incidently, this is her second spectroscopic effort as she was also involved in the preparation of the first edition of Prof. K. Nakamoto's book. Finally, the author is thankful to his family, his wife Kasia and daughter Anna, for encouragement and understanding while working on this and other manuscripts. Unfortunately, life is not always as we plan it to be, and the author's late mother, Wieslawa, will not be able to see this monograph in its final form. She is especially acknowledged for many years of inspiration.

MAREK W. URBAN

CONTENTS

5 ADSORPTION ON METAL OXIDES

6 VIBRATIONAL FEATURES OF INORGANIC MACROMOLECULES

7 BONDING TO POLYMERIC SURFACES

VIBRATIONAL SPECTROSCOPY OF MOLECULES AND MACROMOLECULES ON SURFACES

CHAPTER 1

FUNDAMENTALS OF VIBRATIONAL SPECTROSCOPY

1.1 INTRODUCTION

The primary motivation for identifying surface structures is to determine the relationship between surface species and their architecture and surface performance properties. The ability of surfaces and interfaces to participate in various chemical processes make the surface studies particularly useful not only from the fundamental point of view but also in practical applications. Ultimately, through the understanding of structures involved in various surface and interfacial chemical processes, one would like to control and tailor these processes in an effort to optimize the surface and interfacial properties. As a starting example, let us consider a polymeric mixture containing oligomer, crosslinker, and initiator, all in proper proportions dissolved in some sort of solvent. When such a mixture is deposited on a substrate, usually the evaporation of solvent molecules parallels the crosslinking reactions that are initiated by the initiator. As the reactions are almost completed, a solid film is formed that adheres to a substrate. The question now one needs to raise is what forces govern adhesion to the substrate. On a similar note, what surface properties make a polymeric film protective for many substrates? Of course, from the chemist's point of view, it is desirable to determine the nature of chemical bonds involved and, on that basis, predict the properties of the film–substrate and film–air interfaces. In a similar manner, an analogy would be gaseous species that are being attached to the surface and the surface reactions governing their bonding. This is the nature of the bonding that will determine the stability of the newly formed layers.

In view of these simplified but practical considerations, it is therefore the objective of most studies to develop the basic understanding necessary to optimize the structures to produce improved properties. This can be accomplished by determining the bonding and structures that develop on surfaces and/or interfaces. One important and useful technique often used is vibrational spectroscopy, which consists of two physically different yet conceptually complementary methods, infrared and Raman spectroscopies. Their increasing importance in the surface science and the surface modification analysis is because in recent years there have been tremendous instrumental advances in both types of spectroscopies, giving increased sensitivity and speed of measurements, and surface selectivity. Additionally, the availability of a dedicated computer enhances the utilization of data processing techniques, thus giving the spectroscopist several options to follow in analyzing surface or interfacial chemistry. Although the instrumental advances play a significant role, it is of great importance to understand the origin of physical processes governing spectral detection and interpretation of vibrational spectra.

1.2 ORIGIN OF MOLECULAR SPECTRA

One fundamental question confronting scientists is the problem of measuring and describing how atoms are held together to form molecules or macromolecules. While theoretical descriptions can be obtained by using quantum-mechanical methods, the measurements of forces and energies of bonding are achieved by spectroscopy, the science of interactions between electromagnetic radiation, and matter. Thus, in order to obtain spectroscopic, molecular-level information, one needs three components: a source of electromagnetic radiation, a sample that is the subject of study, and a detector. Light can, however, interact in different ways with matter. It can be absorbed, emitted, or it may be scattered. Although the processes governing such interactions are quite different, the aspects of light absorption and scattering are particularly important, because understanding these interactions will yield information about the molecular structures. Since each atom contains a positive core surrounded by negatively charged electrons that form chemical bonds with the neighboring atoms, each molecular system can be visualized as connected oscillating electrical dipoles. Using classical theories, it can be shown that an oscillating dipole of a molecular system can be either permanent or induced, and that dipoles can interact with electromagnetic radiation. A change in the dipole moment that is induced by vibrations are responsible for infrared activity. On the other hand, the movement of the cloud of electrons surrounding the atoms causes the polarizability changes that are responsible for Raman bands. Postponing temporarily the description of both effects until later, let us establish the general principles governing interactions of electromagnetic radiation with matter.

Because electromagnetic radiation is a form of energy, and its interactions with a molecular system are essentially the energy absorption processes, first let us establish what response one would expect from a molecular system exposed to electomagnetic radiation. The total energy of any multiatom configuration forming a molecule consists of the sum of four types of energy: rotational, vibrational, electronic, and translational.

$$E_{tot} = E_{rot} + E_{vibr} + E_{elect} + E_{trans} \tag{1.1}$$

Fortunately, because of the possibility of separating the total energy (E_{tot}) of a molecule or macromolecule into four separate components—namely, rotation of the molecule (E_{rot}), vibration of the constituent atoms (E_{vibr}), movement of electrons associated with the atoms (E_{elect}), and translation of a molecule as a whole (E_{trans})—each energy can be measured almost separately without interacting with the others. The basis for this separation is that the velocity of electrons is much greater than the vibrational velocity of nuclei. The latter is again greater than the velocity of molecular rotations. Because both electronic and translational energies are much greater than the vibrations and rotations of atoms, these forms of energy will be ignored in this context.

Electromagnetic radiation is characterized by wavelength λ, frequency v, and wavenumber $\bar{v}$. It is a common practice to express a given vibration in wavenumbers [reciprocal centimeters (cm^{-1})], which is related to the other units by

$$\bar{v} = \frac{1}{\lambda} \qquad \bar{v} = \frac{v}{c} \tag{1.2}$$

where c is the velocity of light (2.997925×10^{10} cm/sec), v is the frequency in cycles per second [reciprocal seconds (sec^{-1}) or hertz (Hz)], and λ is the wavelength in centimeters. Because wavenumbers (cm^{-1}) are energy units, they can be converted to the other energy units with the appropriate coefficients, which are provided in Table 1.1.

If electromagnetic radiation impinges at the molecule, only the waves of incident radiation corresponding to the energy of the vibrating atoms forming a bond will interact with the molecule. This is governed by the relationship given by Bohr:

$$\Delta E = E_1 - E_2 = hv \tag{1.3}$$

where ΔE represents the difference in energy between two quantized states E_1 and

TABLE 1.1 Conversion of Energy Units

Ergs/Molecule		eV	cm^{-1}	kJ/mol	MHz
			Multiply by		
Ergs/ molecule	·	6.242×10^{11}	5.035×10^{15}	6.023×10^{13}	1.509×10^{20}
eV	1.602×10^{-12}	·	8.067×10^3	9.649×10^1	2.418×10^8
cm^{-1}	1.986×10^{-16}	1.024×10^{-4}	·	1.196×10^{-2}	2.988×10^4
kJ/mol	1.660×10^{-14}	1.036×10^{-2}	8.359×10^1	·	2.506×10^6
MHz	6.626×10^{-21}	4.136×10^{-9}	3.336×10^{-5}	3.990×10^{-7}	·

E_2, h is Planck's constant (6.6256 × 10^{-27} ergs/sec or 6.6256 × 10^{-34} J/sec), and v is the oscillating frequency of the incident light.

The separation between the bands in rotational spectra occurs in the range between 1 cm^{-1} (10^4 μm) and 10^2 cm^{-1} (10^2 μm). However, this low separation of energy levels is not observed in the vibrational spectra, and pure vibrational spectral separation appears in the energy range between 10^2 cm^{-1} (10^2 μm) and 10^4 cm^{-1} (1 μm). The energy levels of electronic transitions are even further apart and are observed at much higher energies, typically between 10^4 cm^{-1} (1 μm) and 10^5 cm^{-1} (10^{-1} μm). On the basis of these energy differences between microwave and far-infrared, the infrared, and the visible and ultraviolet regions of the electromagnetic spectrum, the spectra have been classified as rotational, vibrational, and electronic. Figure 1.1 depicts three types of transitions described above. It should be noted, however, that this division is arbitrary to a certain extent and,

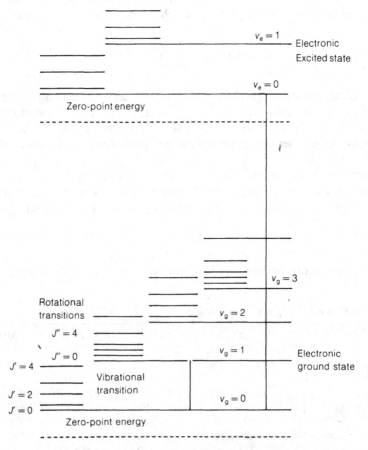

Figure 1.1 Electronic, vibrational, and rotational energy levels for a diatomic molecule.

because the major emphasis of this book is focused on surface vibrational spec-troscopy, no further description of rotational and electronic spectra will be given. The interested reader is referred to the appropriate literature.

1.3 SELECTION RULES IN VIBRATIONAL SPECTROSCOPY; DEGREES OF VIBRATIONAL FREEDOM

The excitation of a purely vibrational transition by infrared light obeys the selec-tion rules for electric dipole transitions and is described by classical electromag-netic theory; specifically an oscillating dipole moment can either absorb or emit radiation. As a result, the periodic dipole moment changes of the vibrating molecule will cause the absorption or emission of radiation of the same frequency as that of the oscillating dipole moment. Following the principles of electromagnetic theory, the intensity of the absorption or emission is proportional to the square root of the dipole moment variations. Consequently, for a kth transition, such as depicted in Fig. 1.1, the intensity will be

$$I_k = A \left(\frac{\partial \mu_d}{\partial x} \right)^2 \tag{1.4}$$

where μ_d is the dipole moment and x is the displacement from equilibrium. This fundamental requirement for the dipole moment changes during the vibration can be simply illustrated by many examples of homodiatomic molecules such as O_2, H_2, N_2, or halogen gases that exhibit no infrared activity. While other examples of the infrared inactive, totally symmetric vibrations will be considered later, the principal rule is that the symmetry of the molecule is maintained during the particular vibrational mode and there is no net change in dipole moment. Conversely, for asymmetric vibrations, the symmetry of the molecule is per-turbed. This results in the net dipole moment, which leads to infrared absorption.

According to the rules of quantum mechanics, the probability of vibrational infrared transition is proportional to the square of the dipole moment, and the condition for the vibration to be infrared-active is that the integral, often called the *transition moment integral*, is not equal to zero:

$$[\mu]_{i,j} = \int \mu_{df}(Q_i) \mu_{di}(Q_f) dQ_k \tag{1.5}$$

where $[\mu_d]_{i,j}$ is the dipole moment of the molecule and Q_i and Q_f are the vibrational eigenfunctions of the initial and final states of the kth vibrational mode described by the normal coordinate Q_k. Thus, for a given vibration to be infrared-active, the integral (1.5) will not be equal to zero, which, in turn, guarantees that the dipole moment changes also are not equal to zero during the vibration.

The selection rules for the Raman spectrum impose different requirements. Although the Raman effect will be described in greater depth in Chapter 3,

Raman activity arises not from the dipole moment changes during vibrations, but from interactions of the electric field of the electromagnetic radiation with electrons. These subsequently induce a temporary dipole moment by a change of the polarizability of the molecule. Such polarizability changes may be visualized as shape changes of the electron clouds surrounding the molecule, which are caused by the oscillating incident radiation. This radiation will distort the electron cloud of the molecule and transmit energy to it. Again, using classical theory, one can depict this process as the generation of a variable dipole moment in the molecule by the oscillating electric vector of the electromagnetic radiation. As a result of these interactions, the dipole moment oscillates with the frequency of the electric vector of the incident radiation. Therefore, it emits radiation in all directions. The intensity of the emitted radiation can be expressed as shown below, and is proportional to the fourth power of the incident frequency:

$$I = \frac{16\Pi^4 v^4}{3c^2} P^2 \tag{1.6}$$

where P is the induced dipole moment given by

$$P = \alpha E \tag{1.7}$$

and α is the molecular polarizability tensor and E is the electric vector of the incident electromagnetic radiation. Since

$$E = E_0 \cos(\omega) = E_0 \cos(2\Pi v_0 t) \tag{1.8}$$

the induced dipole moment becomes

$$P = \alpha E_0 \cos(2\Pi v_0 t) \tag{1.9}$$

When a molecule exhibits a normal vibrational mode of frequency v_k, the polarizability of this mode can be expressed in a form of periodic motion

$$\alpha = \alpha_0 + \alpha_k \cos(2\Pi v_k t + \Phi) \tag{1.10}$$

where α_0 is a constant, α_k is the maximum change in α, and Φ is a phase change.
   Combining Eqs. 1.10, 1.9, and 1.8 into 1.11 gives the expression for the induced dipole moment changes:

$$P = [\alpha_0 + \alpha_k \cos(2\Pi v_k t + \Phi)][E_0 \cos(2\Pi v_0 t)] \tag{1.11}$$

After rearrangements, Eq. 1.11 becomes

$$P = A + \tfrac{1}{2}\alpha_k E_0 \{\cos[2\Pi(v_0 + v_k)t + \Phi] + \cos[2\Pi(v_0 - v_k)t - \Phi]\} \tag{1.12}$$

where $A = \alpha_0 E_0 \cos(2\Pi v_0 t)$.

Equation 1.12 clearly illustrates that when one of the normal vibrational frequencies of the molecule is designated as v_k, the induced dipole is induced not only with the incident frequency v_0, but also with frequencies $v_0 + v_k$ and $v_0 - v_k$. While the v_0 is referred to as *elastic* or *Rayleigh scattering* with no change of energy on collision with the molecule, the terms $(v_0 + v_k)$ and $(v_0 - v_k)$ are the inelastic anti-Stokes and Stokes lines of the Raman spectrum. Thus, the Raman bands are symmetrically distributed on both sides of the excitation v_0 band. Figure 1.2 illustrates the energy diagrams for the Rayleigh and Raman scattering processes.

Having the choice of recording either the Stokes or anti-Stokes portions of the Raman spectrum, it is important to consider which "half" of the Raman spectrum would give stronger bands $(v_0 + v_k)$ or $(v_0 - v_k)$ and, therefore, greater sensitivity. In view of the classical theory described above, it should not make any difference because substituting either portion of the induced dipole moment from Eq. 1.10 into Eq. 1.12 will give the same intensity. However, Fig. 1.3 illustrates the entire Raman spectrum of sulfur S_8 with both Stokes and anti-Stokes wings, with an extremely intense Rayleigh band in the middle. It is quite apparent that the intensities of the Stokes lines are greater. This inadequacy of the classical theory can be corrected by using quantum-mechanical description of the Raman effect. The quantum theory is based on a second-order perturbation theory of Raman intensities. Quantum mechanics assumes that the molecule with discrete energy levels interacts with photons and gives elastic (Rayleigh) and inelastic (Raman) scatterings. These are the inelastic collisions that lead to either loss or gain of vibrational quanta. This process, schematically illustrated in Fig. 1.2 indicates that if the populations of the first (or ground $(v_k = 0)$ and the second $(v_k = 1)$ vibrational states are equal, the Stokes and anti-Stokes intensities would have appeared to be the same. In most cases, however, the Raman spectra are collected at room or near room temperature. Therefore, the majority of the molecules are in the ground vibrational state $(v_k = 0)$, with only a small fraction in higher vibrational energy

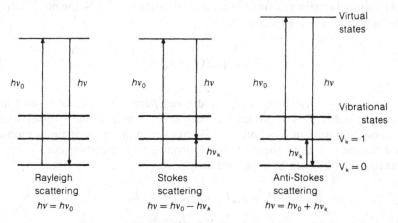

Figure 1.2   Rayleigh and Raman scattering processes.

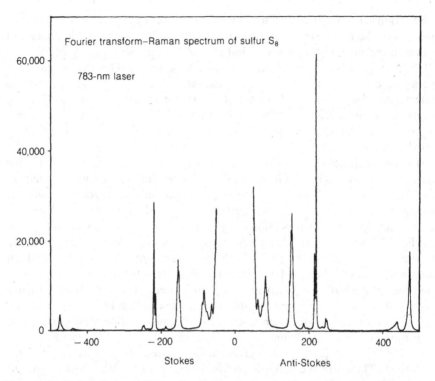

Figure 1.3   Stokes and anti-Stokes Raman spectrum of sulfur S_8 (courtesy of B. Chase).

levels being occupied. As a result of a smaller population in the higher levels, the anti-Stokes scattering process is much weaker. Conversely, because the ground state is highly populated, the efficiency of scattering is high, giving rise to much greater intensities of the Stokes portion of the Raman spectrum.

Using the quantum-mechanical description similar to that introduced for the dipole moment changes during the atomic vibrations (Eq. 1.5), the polarizability of the molecule can be expressed as

$$[\alpha]_{i,j} = \int \alpha_f(Q_k)\alpha_i(Q_k)dQ_k \tag{1.13}$$

where α is the polarizability tensor and i and j represent the initial and final vibrational states. This approach has been developed from the symmetry considerations and Raman selection rules. Because both the induced dipole moment P and the electric component of the electromagnetic radiation are the tensor quantities, $\alpha_{i,j}$, they can be expressed as

$$P_x = \alpha_{xx}E_x + \alpha_{xy}E_y + \alpha_{xz}E_z$$
$$P_y = \alpha_{yx}E_x + \alpha_{yy}E_y + \alpha_{yz}E_z$$
$$P_z = \alpha_{zx}E_x + \alpha_{zy}E_y + \alpha_{zz}E_z$$

or using matrix notation

$$
\mathbb{P} = \begin{vmatrix} P_x \\ P_y \\ P_z \end{vmatrix} = \begin{vmatrix} \alpha_{xx} & \alpha_{xy} & \alpha_{xz} \\ \alpha_{yx} & \alpha_{yy} & \alpha_{yz} \\ \alpha_{zx} & \alpha_{zy} & \alpha_{zz} \end{vmatrix} \begin{vmatrix} E_x \\ E_y \\ E_z \end{vmatrix}
\tag{1.14}
$$

The $\mathbb{P}$ matrix is the symmetric polarizability tensor with $\alpha_{xy} = \alpha_{yx}$, $\alpha_{yz} = \alpha_{zy}$ and $\alpha_{xz} = \alpha_{zx}$. Following the quantum-mechanics rules, the vibration is Raman-active if one of the six polarizability components changes during the vibration or, equivalently, if one of the integrals in Eq. 1.13 is nonzero. Similarly, the vibration is infrared-active if one of the three integrals in Eq. 1.5 of the dipole moment μ_x, μ_y, or μ_z changes in the three directions is not equal to zero during the vibration.

Having established these very simplified principles of the infrared and Raman activities, let us go through a numerical example and calculate the vibrational energy of two atoms attached by a chemical bond. For that purpose, we will consider two vibrating atoms with masses m_1 and m_2 bonded to each other; the internuclear forces holding the molecule together are massless springs. As the atoms vibrate, the springs try to restore the bond lengths or bond angles to a certain equilibrium position. Let the displacement of atoms m_1 and m_2 from equilibrium along the bond be x_1 and x_2. The average displacement from the equilibrium will then be equal to $x = x_2 - x_1$. If the bond between atoms obeys Hook's law, the motion of each atom can be expressed by Newton's law (since force = mass × accelaration) as follows:

Atom 1:
$$
m_1 \frac{d^2 x_1}{dt^2} = -k(x_2 - x_1)
\tag{1.15}
$$

and

Atom 2:
$$
m_2 \frac{d^2 x_2}{dt^2} = -k(x_2 - x_1)
\tag{1.16}
$$

Classical mechanics teaches us, however, that the two vibrating atoms in a diatomic molecular configuration can be reduced to the motion of a particle with mass μ and displacement x from its equilibrium position. The mass μ, called the *reduced mass*, is expressed by the following equation:

$$
\frac{1}{\mu} = \sum_{i=1}^{n} \frac{1}{m_i}
\tag{1.17}
$$

where n is the number of atoms with a mass m_i. For a diatomic molecule, the reduced mass is simply described by

$$
\frac{1}{\mu} = \frac{1}{m_1} + \frac{1}{m_2}
\tag{1.18}
$$

where m_1 and m_2 are the masses of the two nuclei.

The total energy of a molecular system (E) consists of the kinetic (T) and potential energy (V). The kinetic energy of the diatomic molecule is

$$T = \tfrac{1}{2}\mu\dot{x}^2 = \frac{1}{2\mu}p^2 \tag{1.19}$$

where p is the conjugate momentum, $\mu\dot{p}$. With a simple parabolic potential function, this diatomic system will behave as a harmonic oscillator and its potential energy will be represented by

$$V = \tfrac{1}{2}kx^2 \tag{1.20}$$

where k is the force constant of the vibration. Thus, assuming that Φ is a single-value, finite, continuous function, and vanishes at infinity, we can solve the Schrödinger wave equation

$$\frac{d^2\Phi}{dx^2} + \frac{8\Pi^2\mu}{h^2}(E - \tfrac{1}{2}kx^2)\Phi = 0 \tag{1.21}$$

for E but only when E values are

$$E_v = h\nu(v + \tfrac{1}{2}) = h\bar{\nu}c(v + \tfrac{1}{2}) \tag{1.22}$$

with the vibrational frequency equal to

$$\nu = \frac{1}{2\Pi}\sqrt{\frac{k}{\mu}} \quad \text{or} \quad \bar{\nu} = \frac{1}{2\Pi c}\sqrt{\frac{k}{\mu}} \tag{1.23}$$

where v is the vibrational quantum number, and may take the values 0, 1, 2, 3, and so on.

According to Eq. 1.23, the frequency of the vibration in a diatomic molecule is proportional to the $\sqrt{\frac{k}{\mu}}$. If the constant k is approximately the same for a series of diatomic molecules, the vibrational frequency is inversely proportional to the square root of μ. Similarly, if μ is approximately the same for a series of diatomic molecules, the vibrational frequency is proportional to the square root of k. This simplified discussion indicates that the bond becomes stronger as the force constant becomes larger. It should be remembered, however, that even for the most ideal cases such as diatomic molecules, this general theoretical relationship between the dissociation energy and the force constant is difficult to derive. This is because, according to the force constant definition, its magnitude represents the curvature of the potential energy well (second derivative) near the equilibrium position:

$$k = \left(\frac{d^2V}{dx^2}\right)_{x \Rightarrow 0} \tag{1.24}$$

Although various potential functions will be discussed in later chapters, here it should be remembered that the dissociation energy D_e is given by the depth of the potential-energy curve. In essence, a large force constant is indicative of sharp curvature of the potential well near the bottom but does not necessarily mean a deep potential well (large D_e). However, for the same type of molecule, a larger force constant is an indication of a stronger bond. The situation becomes more complex when the molecules are adsorbed on surfaces because surface energy may result in surprizing energy potential changes.

1.4   SYMMETRY AND INFRARED AND RAMAN SELECTION RULES

As was discussed earlier, the dipole moment and the polarizability tensor changes are immediate factors affecting infrared and Raman activity. However, both quantities are not always available, and a spectroscopist still would at least like to make a guess and predict vibrational activity of a given species. Because the dipole moment and polarizability tensor changes are directly related to the symmetry properties and chemical formula and structure are usually known, such predictions can be done.

Perhaps the most common approach in estimating vibrational activity is to go back to "pencil chemistry" and dream up on a piece of paper all possible structures and vibrational features associated with them. Such predictions can then be taken to the lab and compared with the real spectra.

The issue of symmetry is particularly important because in order to understand the origin of infrared and Raman activity and their selection rules on surfaces, a clear understanding of the molecules in an unperturbed state is required. Let us first consider the normal vibrational modes of CO, CO_2, and H_2O molecules such as that illustrated in Fig. 1.4. A motion of an individual nucleus, indicated by the arrow, is simply a harmonic motion and all nuclei of a particular species in phase motion with the same oscillating frequency. This is the frequency of the normal vibration. The lengths of the arrows illustrate approximate values of the relative displacements and amplitudes of each nuclei. It should be noted that in CO_2 there are two v_{2a} and v_{2b} vibrations having the same oscillating frequency. The only difference is the direction of oscillation. Apparently there is an infinite number of such motions, so that normal vibrations can be reduced into two that are perpendicular to each other. Such vibrations as v_2 in CO_2 molecule are called *doubly degenerate vibrations*. Doubly degenerate vibrations occur only in the species with an axis higher than twofold. Triply degenerate vibrations are also seen in molecules with more than one C_3 axis. Somewhat more complex but useful examples in surface science are normal vibrations and selection rules for NH_3, $H_2C\!=\!CH_2$, and $H_3C\!-\!CH_3$ molecules. Similar to the cases discussed previously, Fig. 1.5 illustrates normal vibrational modes of these species. A common approach in predicting normal vibrations is to identify a point group and based on its diameter table determine normal vibrations. Appendix A provides principles leading to this determination.

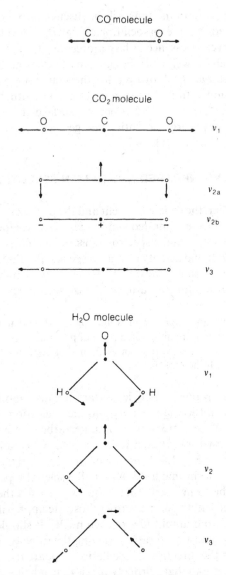

Figure 1.4 Normal vibrations of CO, CO_2, and H_2O molecules.

Because IR (infrared) and Raman activity of a given vibration depend on symmetry of the species, we will try to develop a methodology and discuss the previously introduced kinetic and potential energy of the system, and at the same time, keeping in mind that for an n-atomic molecule, we will see that a total number of the normal vibrations is equal to $3n - 6$. The $3n$ of total degrees of freedom is reduced by six since six coordinates are required to describe translational and rotational motion of the molecule (three for each). For linear molecules the total number of the vibrational degrees of freedom is $3n - 5$. These molecules exhibit almost no or only marginal rotational freedom around the molecular axis; thus one rotational degree of freedom can be neglected. With these fundamental principles in mind, one can develop a method that allows one to calculate and establish IR and Raman activity; this is called *normal-coordinate analysis* and is briefly discussed below.

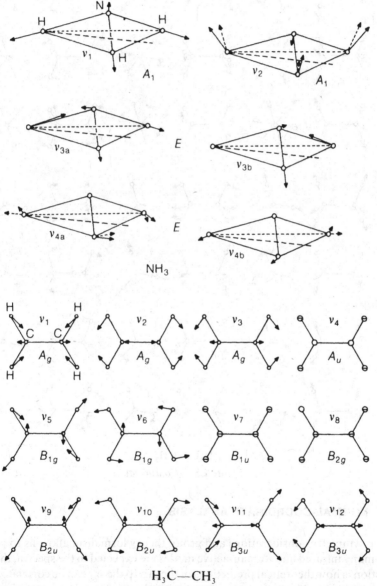

Figure 1.5 Normal vibrations of NH_3, $H_2C=CH_2$, and H_3C-CH_3 species.

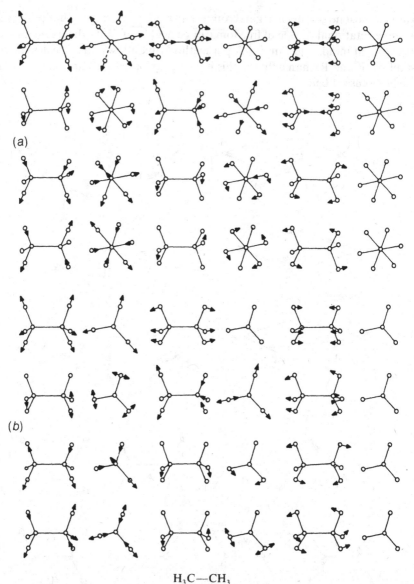

$$H_3C—CH_3$$

Figure 1.5 (*Continued*)

1.5 NORMAL-COORDINATE ANALYSIS

While symmetry considerations and group theory formalism allow us to predict how many infrared and Raman-active modes are expected in the spectra, the next question is how theoretical predictions of the group theory can be correlated with the experiment. After all, we are faced with the problem of band assignment, which, in turn, allows us to determine structural features of a given species. A common method used to do this is normal-coordinate analysis. This method is based on the fact that the displacement of each vibrating atom can be expressed in terms of a rectangular coordinate system with the origin of each atomic coordinate system at the equilibrium position of each atomic nucleus. For such defined coordinates for n atoms, the kinetic vibrational energy of n-atom-containing molecule can be expressed as

$$T = \frac{1}{2} \sum_n m_n \left[\left(\frac{d\,\Delta x_n}{dt} \right)^2 + \left(\frac{d\,\Delta y_n}{dt} \right)^2 + \left(\frac{d\,\Delta z_n}{dt} \right)^2 \right] \tag{1.25}$$

With the internal coordinates described for each atom such as

$$q_1 = \sqrt{m_1}\,\Delta x_1, \quad q_2 = \sqrt{m_1}\,\Delta y_1, \quad q_3 = \sqrt{m_1}\,\Delta z_1 \qquad \text{for atom 1} \quad (1.26)$$

$$q_4 = \sqrt{m_2}\,\Delta x_2, \quad q_5 = \sqrt{m_2}\,\Delta y_2, \quad q_6 = \sqrt{m_2}\,\Delta z_2 \qquad \text{for atom 2} \quad (1.27)$$

and so on. The kinetic energy can be expressed as

$$T = \frac{1}{2} \sum_i^{3n} \dot{q}_i^2 \tag{1.28}$$

Of course, the total energy of each system consists of kinetic and potential energy. Because the potential energy of all atoms involved is a complex function involving all coordinates, it is usually expressed as a Taylor series

$$V(q_1, q_2, \ldots, q_{3n}) = V_0 + \sum_i^{3n} \left(\frac{\partial V}{\partial q_i} \right)_0 q_i + \frac{1}{2} \sum_{i,j}^{3n} \left(\frac{\partial^2 V}{\partial q_i \partial q_j} \right)_0 q_i q_j + \cdots \tag{1.29}$$

with the derivatives being calculated at the equilibrium position ($q_i = 0$). Let us now evaluate the potential-energy terms in an effort to define which term gives the most important contribution. If the potential energy at $q_i = 0$ is taken as a standard, the constant term V_0 will also be zero. Since V must be a minimum at $q_i = 0$, the $(\partial V/\partial q_i)_0$ terms become zero as well. With these assumptions and neglecting higher-order terms, we can simplify V to one term:

$$V = \frac{1}{2} \sum_{i,j}^{3n} \left(\frac{\partial^2 V}{\partial q_i \partial q_j} \right)_0 q_i q_j = \frac{1}{2} \sum_{i,j}^{3n} b_{ij} q_i q_j \tag{1.30}$$

This equation contains cross products $q_i q_j$. If this were not the case, the potential-energy problem could be solved combining the V and T terms into the Newton's equation of motion:

$$\frac{d}{dt} \left(\frac{\partial T}{\partial \dot{q}_i} \right) + \frac{\partial V}{\partial q_i} = 0 \qquad i = 1, 2, \ldots, 3n \tag{1.31}$$

Substituting for T from Eq. 1.28 and for V from Eq. 1.30, Eq. 1.31 becomes

$$\ddot{q}_i + \sum_j b_{ij} q_j = 0 \qquad j = 1, 2, \ldots, 3n \tag{1.32}$$

For the case when $b_{ij} = 0$ for $i \neq j$, Eq. 1.33 can be written as

$$\ddot{q}_i + b_{ii} q_i = 0 \tag{1.33}$$

Solving this equation with respect to q_i gives

$$q_i = q_i^\circ \sin(\sqrt{b_{ii}} t + \delta_i) \tag{1.34}$$

where q_i° and δ_i are respectively the amplitude and the phase of the motion.

However, such a simplified approach cannot be applied because we would lose cross product for $i \neq j$. One approach is that the previously introduced coordinates q_i must be transformed into a matrix of new coordinates Q_i through the following relationship:

$$q_1 = \sum_i B_{1i} Q_i$$

$$q_2 = \sum_i B_{2i} Q_i \tag{1.35}$$

$$\vdots$$

$$q_k = \sum B_{ki} Q_i$$

By doing so, we introduced new coordinates, Q_i, which are commonly called *normal coordinates*, and by choosing proper B_{ki} coefficients, both the potential and the kinetic energy of the system containing n atoms can be expressed as

$$T = \frac{1}{2} \sum_i \dot{Q}_i^2 \tag{1.36}$$

$$V = \frac{1}{2} \sum_i \lambda_i Q_i^2 \tag{1.37}$$

Such an approach allows one to express both energies of the system without any complications related to the cross products. Another advantage is that the preceding expressions can now be substituted into Newton's equation (Eq. 1.31), resulting in

$$\ddot{Q}_i + \lambda_i Q_i = 0 \tag{1.38}$$

with the solution

$$Q_i = Q_i^\circ \sin(\sqrt{\lambda} t + \delta_i) \tag{1.39}$$

which gives the frequency of oscillation

$$v_i = \frac{1}{2\Pi}\sqrt{\lambda_i} \tag{1.40}$$

This vibration is called a *normal vibration* since it was derived using normal coordinates and represents the displacement coordinate from the origin related to the normal coordinates through Eq. 1.35.

Let us establish the physical meaning of the normal vibrations by taking into account only one single vibration of the system. In that case, $Q_1^\circ \neq 0$, whereas all other Q values are equal to zero. By combining Eq. 1.35 and 1.39, we obtain

$$q_k = B_{k1}Q_1 = B_{k1}Q_1^\circ \sin(\sqrt{\lambda_1}t + \delta_1) = A_{k1}\sin(\sqrt{\lambda_1}t + \delta_1) \tag{1.41}$$

This relationship indicates that the excitation of one normal vibration within the system will induce vibrations of all the atoms in the system in which all atoms vibrate with the same frequency and phase, although they may vary in direction. Now, we can combine Newton's equation of motion (Eq. 1.31) with Eq. 1.41 rewritten in the more general form such as

$$q_k = A_k \sin(\sqrt{\lambda}t + \delta) \tag{1.42}$$

and obtain

$$-\lambda A_k + \sum_j b_{kj}A_j = 0 \tag{1.43}$$

With the use of a matrix notation, Eq. 1.42 allows us to find a solution with respect to A. For all the A values to be nonzero, the following $3n$ order secular equation must be solved:

$$\begin{vmatrix} b_{11}-\lambda & b_{12} & b_{13} & b_{14} & \cdots \\ b_{21} & b_{22}-\lambda & b_{23} & b_{24} & \cdots \\ b_{31} & b_{32} & b_{33}-\lambda & b_{34} & \cdots \\ b_{41} & b_{42} & b_{43} & b_{44}-\lambda & \cdots \end{vmatrix} = 0 \tag{1.44}$$

To solve this equation it is necessary to find one λ_1 and then insert it into Eq. 1.43. This way we will obtain $A_{k1}, A_{k2}, A_{k3}, \ldots$. In the same manner other λ values can be obtained to give a general solution for all normal vibrations

$$q_k = \sum_i B_{k1}Q_1^\circ \sin(\sqrt{\lambda_1}t + \delta_1) \tag{1.45}$$

To make this general discussion more applicable to the surface spectroscopy, let

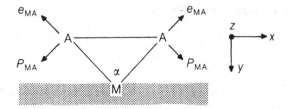

Figure 1.6 A normal vibration of an A_2 molecule attached to the surface through the atom M.

us consider an example of a diatomic molecule attached to the surface having the configuration of coordinates, such as those shown in Fig. 1.6. For the system defined as above, the potential energy will be given by

$$V = \tfrac{1}{2}k\{(\Delta x_1 - \Delta x_2)\}^2 + (\Delta x_2 - \Delta x_3)^2\} \tag{1.46}$$

In this case, it is also assumed that the surface atoms are bonded equivalently to the atom M, which implies that both M—A force constants are the same. Because the masses of atoms A are equal ($m_A = m_B$), it can be easily shown that the kinetic energy of such a system can be expressed by

$$T = \tfrac{1}{2}m_A(\Delta \dot{x}_1^2 + \Delta \dot{x}_3^2) + \tfrac{1}{2}m_M \Delta \dot{x}_2^2 \tag{1.47}$$

Introducing the general coordinates q_n, such as in Eqs. 1.34 and 1.35, we can present both energies as

$$2V = k\left\{\left(\frac{q_A}{\sqrt{m_A}} - \frac{q_M}{\sqrt{m_M}}\right)^2 + \left(\frac{q_M}{\sqrt{m_M}} - \frac{q_A}{\sqrt{m_A}}\right)^2\right\} \tag{1.48}$$

$$2T = \sum_i \dot{q}_i^2 \tag{1.49}$$

If we combine Eqs. 1.48 and 1.49, we obtain

$$b_{11} = \frac{k}{m_A} \quad \text{and} \quad b_{22} = \frac{2k}{m_2}$$

$$b_{12} = b_{21} = \frac{-k}{\sqrt{m_A m_M}} \quad \text{and} \quad b_{23} = b_{32} = \frac{-k}{\sqrt{m_A m_M}} \tag{1.50}$$

$$b_{13} = b_{31} = 0 \quad \text{and} \quad b_{33} = \frac{k}{m_A}$$

Now all b_{ij} terms can be inserted into matrix 1.44 and as a result, we obtain the following:

$$\begin{vmatrix} \dfrac{k}{m_A} - \lambda & \dfrac{-k}{\sqrt{m_A m_M}} & 0 \\[2mm] \dfrac{-k}{\sqrt{m_A m_M}} & \dfrac{2k}{m_M} - \lambda & \dfrac{-k}{\sqrt{m_A m_M}} \\[2mm] 0 & \dfrac{-k}{\sqrt{m_A m_M}} & \dfrac{k}{m_A} - \lambda \end{vmatrix} = 0 \qquad (1.51)$$

The solution of this secular equation leads to the three roots:

$$\lambda_1 = \frac{k}{m_A} \qquad \lambda_2 = k\mu \qquad \lambda_3 = 0 \qquad \mu = \frac{2m_A + m_M}{m_A m_M}$$

where μ is the reduced mass of m_A and m_M.

For this particular triatomic system, Eq. 1.51 gives a set of three equations:

$$-\lambda A_1 + b_{11}A_1 + b_{12}A_2 + b_{13}A_3 = 0$$
$$-\lambda A_2 + b_{21}A_1 + b_{22}A_2 + b_{23}A_3 = 0 \qquad (1.52)$$
$$-\lambda A_3 + b_{31}A_1 + b_{32}A_2 + b_{33}A_3 = 0$$

which can be also expressed in terms of normal coordinates q

$$(b_{11} - \lambda)q_1 + b_{12}q_2 + b_{13}q_3 = 0$$
$$b_{21}q_1 + (b_{22} - \lambda)q_2 + b_{23}q_3 = 0 \qquad (1.53)$$
$$b_{31}q_1 + b_{32}q_2 + (b_{33} - \lambda)q_3 = 0$$

Since $\lambda_1 = k/m_A$, the above equations become

$$q_1 = -q_3 \qquad \text{and} \qquad q_2 = 0$$

and similar rearrangements leads to

$$q_1 = q_3 \quad \text{and} \quad q_2 = -2\sqrt{m_A/m_M}\,q_1 \quad \text{when} \quad \lambda_2 = k\mu$$
$$q_1 = q_3 \quad \text{and} \quad q_2 = \sqrt{m_M/m_A}\,q_1 \quad \text{when} \quad \lambda_3 = 0 \qquad (1.54)$$

Now we are in a position to depict the displacement of atoms, such as that depicted in Fig. 1.6. The physical meaning of λ_3 being zero is such that it represents the translational mode where all atomic displacements move in one direction; that is, $\Delta x_1 = \Delta x_2 = \Delta x_3$. However, we purposely included λ_3 to illustrate its meaning. One can exclude it and simplify the calculations by assuming that the center of gravity of the A—M—A molecule does not move. In this case $m_A (\Delta x_1 + \Delta x_3) + m_A \Delta x_2 = 0$.

As mentioned earlier, polyatomic molecules have a maximum of $3n - 6$ and, if linear, $3n - 5$ vibrational degrees of freedom or normal vibrations. However, not all of them are necessarily infrared or Raman-active. Whether a given vibration is infrared or Raman-active is dictated by what we often call *selection rules*. These, in turn, are ruled by the symmetry of the molecule. For example, the normal O—O stretching vibration in diatomic O_2 molecule will be Raman-active because of the polarizability tensor changes during the vibration. In contrast, the same vibration is infrared inactive since there is no change in the dipole moment as the oxygen atoms vibrate.

In view of the preceding discussion, let us consider the symmetry properties of the A—M—A molecule, assuming that the A—A is bonded such as shown above, and neglecting the effect of M bonding to the surface. Using the group theory formalism, we can immediately recognize that such species are assigned to the C_{2v} group symmetry. The interested reader is referred to the group theory literature for further considerations and practical applications in vibrational spectroscopy.[1][5] Here, only Table 1.2 is presented to identify symmetry elements and operations pertinent to the C_{2v} system.

As a useful exercise, let us consider two vibrations from the character table (Table 1.2), A_1 and B_2. All character tables are arranged so that $+1$ and -1 for a given normal vibration denote symmetric and antisymmetric modes, respectively. The analysis of the first row for A_1 indicates that all symmetry properties are maintained during the vibration. This is manifested by the appearance of $+1$ values for all symmetry operators. Hence, such vibrations are fully symmetric. To emphasize their symmetric character, they are often called *totally symmetric vibrations*. On the other hand, the analysis of B_2 normal vibration indicates that such symmetry elements as C_2 (rotation by 180° about the z axis) and $\sigma_v(xz)$ (xz plane of symmetry) are lost. This is manifested by the presence of -1 values. In this case, such vibrations are nonsymmetric.

To understand how symmetry of a molecule may affect the selection rules and therefore vibrational spectra, consider again a diatomic O_2 molecule chemically bonded to the surface. Free O_2 belongs to the $D_{\infty h}$ group symmetry with one totally symmetric A_1 Raman-active mode at 1620 cm^{-1}. However, when such species become bonded to the surface, the selection rules change because the

TABLE 1.2   Character Table for C_{2v} Symmetry Elements and Operations

C_{2v}	E	$C_2(z)$	$\sigma_v(xz)^a$	$\sigma_v(yz)^a$		
A_1	1	1	1	1	T_z	$\alpha_{xx}, \alpha_{yy}, \alpha_{zz}$
A_2	1	1	-1	-1	R_z	α_{xy}
B_1	1	-1	1	-1	T_x, R_y	α_{xz}
B_2	1	-1	-1	1	T_y, R_x	α_{yz}

^{a} σ_v is a vertical plane of symmetry.

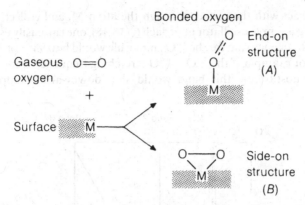

Figure 1.7   Side-on (*A*) and end-on (*B*) structures of diatomic oxygen molecule adsorbed on a metal surface.

symmetry of the molecule changes. Figure 1.7 schematically illustrates the various possible structures of the O_2 molecule. Although the stability of each structure may raise some uncertainties, such structures are known to exist at low temperatures. Now, it is our task to predict the vibrational spectra of the species depicted above or, conversely, on the basis of vibrational spectra, to establish which of these structures are actually present. Appendix A provides complete at point groups for various symmetries.

Having the two physically different techniques in hand, infrared and Raman spectroscopy, let us first establish what response one would anticipate from the infrared measurements. As mentioned earlier, there is no dipole moment change during its vibration and therefore, the O—O stretching mode is infrared-inactive. Once the molecule becomes attached to the surface, the selection rules change in relation to the symmetry changes. This results in a nonuniform distribution of electrons, which, in turn, leads to dipole moment changes. As a consequence, the O—O stretching mode becomes infrared-active. However, depending on the bonding characteristics of the surface, various structures can exist. In the case of end-on structures (Fig. 1.7*A*) one would expect to observe the O—O stretching vibration in the infrared spectrum, but its frequency would be somewhat below $1620\,cm^{-1}$, depending on the strength of the M—O bond and its symmetry. Similarly, the side-on structure would also be infrared-active with the one infrared band due to the O—O stretching normal mode. Although its frequency would be expected to be below that for the end-on species, the exact determination of either structure based on vibrational energy differences becomes difficult and the assignment is highly speculative. A similar problem may be encountered in the case of Raman spectra.

At this point one needs to raise the question of how to make unambiguous structural determination. One of the most powerful spectroscopic approaches, which unfortunately is often neglected, is the use of isotopic substitutions. In the case of the O_2 molecule, we will purposely introduce its isotopic counterpart,

$^{18}O_2$, let it react with the surface through the atom M, and collect vibrational spectra. Using a simple oscillator approach ($\sqrt{16/18}$), one can easily estimate that the expected frequency shift for the $^{18}O_2$ molecule would be 0.9428 of that of $^{16}O_2$ frequency. For example, if the ^{16}O — ^{16}O stretching frequency were 1620 cm^{-1}, on isotopic substitution, this band would shift downward to approximately

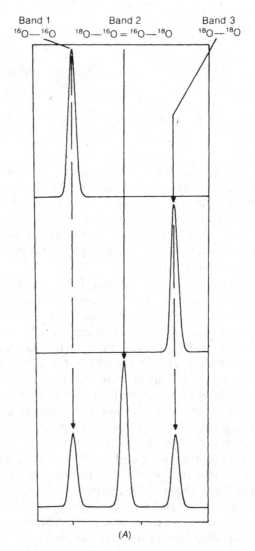

Figure 1.8 (*A*) 1 — Raman spectrum of $^{16}O_2$; 2 — Raman spectrum of $^{18}O_2$; 3 — Raman spectrum of the mixture containing $^{16}O_2$, ^{16}O — ^{18}O, and $^{18}O_2$ mixture in the ratio of 1:2:1; (*B*) side-on structure of the molecular oxygen attached to the surface and the corresponding infrared spectra; (*C*) end-on structure of the molecular oxygen attached to the surface and the corresponding IR spectra.

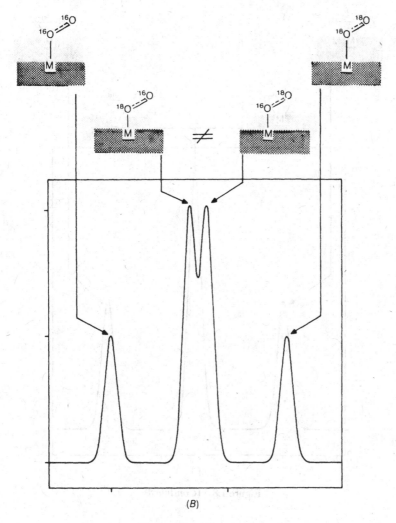

Figure 1.8 (*Continued*)

1461 cm^{-1}. Although such an experiment will provide a clearcut band assignment (namely, the 1620 cm^{-1} band is due to O—O stretch), our goal is to establish bonding characteristics. While the isotopic substitutions are extremely useful in making appropriate band assignments, in this case, to determine the structural features of the oxygen molecule on the surface, it is necessary to attach the ^{18}O—^{16}O isotope to the surface and analyze vibrational spectra. Let us consider how this approach may help answer the question posed at the beginning of this section.

First, if only ^{16}O—^{16}O molecules were present, only one ^{16}O—^{16}O stretching Raman band would be detected, such as that depicted in Fig. 1.8*A*, trace 1.

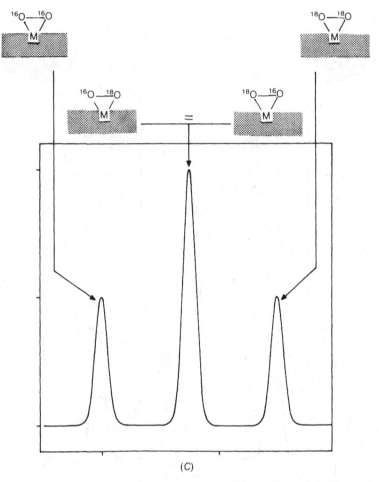

(C)

Figure 1.8 (*Continued*)

Similarly, if only the ^{18}O — ^{18}O molecules are present, the same vibration will be Raman-active, but its energy will be lower, such as shown in Fig. 1.8A, trace 2. However, the presence of the mixture containing ^{16}O — ^{16}O, ^{16}O — ^{18}O, and ^{18}O — ^{18}O species will generate three bands in the Raman spectra. If the initial mixture of the oxygen gases (before scrambling) consists of a 50/50 mixture of $^{16}O_2$ and $^{18}O_2$, the three Raman bands will have the intensities 1:2:1, such as shown in Fig. 1.8A, trace 3. In all three cases, the infrared spectra will remain "silent" with no activity. Only the masses of atoms have changed without disturbing the electronic structure, and the dipole moment changes have been left intact and equal to zero. Once the molecule is bonded to the surface, the situation will be substantially different.

It should be kept in mind that the presence of isotopic substitutions does not affect the strength of the chemical bonding but may significantly influence reactivity. The effect of isotopic substitution on reactivity has been known and studied by many scientists.

Let us take advantage of the fact that once the molecule becomes attached to the surface, the O—O stretching will be infrared-active. Therefore, using IR and isotopic scrambled substitutions, such as that presented in Fig. 1.8A, we expect to observe the same infrared activity as in the naturally abundant case, except the number of bands will change. In the case of the end-on structure (Fig. 1.8B), there will be two spectroscopically different species because the nonbonded oxygen atoms have different atomic weights (16 and 18) and the lighter one (^{16}O) will have slightly higher vibrational energy. Therefore, these two structures will generate two bands separated by approximately 15 cm^{-1}. Of course, the magnitude of splitting may be different depending on the nature of the system and the strength of the O_2-surface bonding. Thus, the infrared spectrum will exhibit a total of four bands, the two discussed above and the two due to the ^{16}O—^{16}O and ^{18}O—^{18}O species.

In the case of the side-on structures (Fig. 1.8C), both oxygen atoms are bonded to the same surface atom, and as long as the strength of the two M—O bonds is the same, both structures are equivalent. This will give us one ^{16}O—^{18}O stretching mode and two modes due to ^{16}O—^{16}O and ^{18}O—^{18}O species. To conclude the preceding discussion, let us compare the normal vibrational modes of the side-on A_2 M (or O_2 M) species depicted in Fig. 1.6 with that for water. Note that both structures exhibit the same C_{2v} symmetry and, therefore, spectral activity (number of infrared and Raman bands) and the direction of atomic displacements will be the same. Obviously, the magnitude of the displacements will change, which is reflected in the force constant differences. Such analogies, combined with the powerful isotopic substitutions leading to the determination of structural features on the surface, are helpful because they may allow us to predict surface properties. Although this approach should be considered as a rather idealized one, it is always useful to have a fundamental understanding of the processes responsible for reactions. As will be seen in later chapters, there are other complications which may lead to various discouraging difficulties. So for now before things become a bit more complex, let us enjoy what we understand and build further on these blocks of knowledge.

1.6   STRENGTH OF BONDING AND BAND ASSIGNMENTS

In the previous sections it was stated that using normal-coordinate analysis, the frequency of the normal vibration can be calculated from the kinetic- and potential-energy considerations. Kinetic energy is determined by the oscillating masses of individual atoms and their geometric arrangements. Potential energy is related by the force constants with the strength resulting from bonding interactions between atoms. Thus, knowledge of the potential energy of the system is important because it provides valuable information about the nature of interatomic forces. Examination of Eq. 1.23 shows us that if we know k and μ, we can calculate vibrational frequency. The question is, though, how to determine force constants.

Based on the knowledge of vibrational frequencies, a commonly used method for calculating the force constants is the GF matrix method. Assuming a set of the force constants for a given system and modifying it during calculations, one can match the calculated and observed frequencies. The GF matrix has been extensively presented and explored in the literature; thus only a brief, general description will be given here. It will parallel the example calculation for the diatomic molecule attached to the surface through M atom (Fig. 1.6), which has been considered previously.

The first step in the GF matrix method is to express both the kinetic and potential energies in terms of cartesian or internal coordinates. Although it is desirable in terms of defining the kinetic energy to represent the system using cartesian coordinates, it is easier to picture the physical meaning of the force constants and exclude from calculations translational and rotational degrees of freedom with internal coordinates. This is why they are most commonly used

The potential energy of the system can be expressed using internal coordinates as

$$2V = \tilde{\mathbb{R}}\mathbb{F}\mathbb{R} \tag{1.55}$$

which for the bent A_1—M—A_2 molecule with the internal coordinates defined in Fig. 1.6 is a column matrix

$$\mathbb{R} = \begin{vmatrix} \Delta r_1 \\ \Delta r_2 \\ \Delta\omega \end{vmatrix} \tag{1.56}$$

and $\tilde{\mathbb{R}}$ is its transpose matrix

$$\tilde{\mathbb{R}} = |\Delta r_1, \quad \Delta r_2, \quad \Delta\omega| \tag{1.57}$$

and $\mathbb{F}$ is the matrix representing a set of the force constants defined as below

$$\mathbb{F} = \begin{vmatrix} f_{11} & f_{12} & r_1 f_{13} \\ f_{21} & f_{22} & r_2 f_{23} \\ r_1 f_{31} & r_2 f_{32} & r_1 r_2 f_{33} \end{vmatrix} \tag{1.58}$$

where r_1 and r_2 represent the equilibrium .lengths of the A_1—M and M—A_2 bonds. To avoid confusion, we labeled chemically equivalent atoms A. The $\mathbb{F}$

matrix can be redefined as

$$
\mathbb{F} = \begin{vmatrix} F_{11} & F_{12} & F_{13} \\ F_{21} & F_{22} & F_{23} \\ F_{31} & F_{32} & F_{33} \end{vmatrix}
\tag{1.59}
$$

Because the expression for the kinetic energy cannot be that easily defined using internal coordinates, Wilson[1] has derived an expression that relates the kinetic energy of the system and $\mathbb{R}$ matrix.

$$
2T = \tilde{\dot{\mathbb{R}}} \mathbb{G}^{-1} \dot{\mathbb{R}}
\tag{1.60}
$$

Postponing temporarily a definition of the $\mathbb{G}$ matrix, but combining Eqs. 1.59 and 1.56 and Newton's equation expressed by

$$
\frac{d}{dt}\left(\frac{\partial T}{\partial R_k}\right) + \frac{\partial V}{\partial R_k} = 0
\tag{1.61}
$$

We can derive the following matrix relationship:

$$
\begin{vmatrix} F_{11} - (G^{-1})_{11}\lambda & F_{12} - (G^{-1})_{12}\lambda & \cdots \\ F_{21} - (G^{-1})_{21}\lambda & F_{22} - (G^{-1})_{22}\lambda & \cdots \\ \vdots & \vdots & \end{vmatrix} \equiv |\mathbb{F} - \mathbb{G}^{-1}\lambda| = 0
\tag{1.62}
$$

Multiplying the left-hand side of the Eq. 1.61 by the determinant of $\mathbb{G}$, we obtain

$$
\begin{vmatrix} G_{11} & G_{12} & \cdots \\ G_{21} & G_{22} & \cdots \\ \vdots & \vdots & \end{vmatrix} \equiv |\mathbb{G}|
\tag{1.63}
$$

and thus

$$
|\mathbb{G}| \equiv \begin{vmatrix} \sum G_{1t}F_{t1} - \lambda & \sum G_{1t}F_{t2} & \cdots \\ \sum G_{2t}F_{t1} & \sum G_{2t}F_{t2} - \lambda & \cdots \\ \vdots & \vdots & \end{vmatrix} \equiv |\mathbb{G}\mathbb{F} - \mathbb{E}\lambda| = 0
\tag{1.64}
$$

where $\mathbb{E}$ is the unit matrix and $\lambda = 4\Pi^2 c^2 \nu^2$, and the dimension of the matrix is equal to the number of internal coordinates used in a particular problem. The F matrix is the force constant matrix, and usually it is assumed. In order to solve Eq. (1.64) one needs to construct the $\mathbb{G}$ matrix in the following way:

$$
\mathbb{G} = \mathbb{B}\mathbb{M}^{-1}\hat{\mathbb{B}}
\tag{1.65}
$$

where $\mathbb{M}^{-1}$ is a diagonal matrix composed of reciprocal masses of the nth atoms. For the diatomic A_2 molecule attached to the surface through atom M, the matrix is

$$
\mathbb{M}^{-1} = \begin{vmatrix} \mu_A & & & & \\ & \mu_A & & & \\ & & \mu_A & & \\ & & & \ddots & \\ & & & & \mu_M \end{vmatrix}
\qquad (1.66)
$$

where μ_A and μ_M are the reciprocal masses of the A and M atoms. The $\mathbb{B}$ matrix is defined by the matrices of internal ($\mathbb{R}$) and cartesian ($\mathbb{X}$) coordinates as

$$
\mathbb{R} = \mathbb{B}\mathbb{X} \qquad (1.67)
$$

and for MA_2 molecule, Eq. 1.67 can be presented as

$$
\begin{vmatrix} \Delta r_1 \\ \Delta r_2 \\ \Delta \omega \end{vmatrix} =
$$

$$
\begin{vmatrix}
-\sin\!\left(\dfrac{\omega}{2}\right) & -\cos\!\left(\dfrac{\omega}{2}\right) & 0 & 0 & 0 & \sin\!\left(\dfrac{\omega}{2}\right) & \cos\!\left(\dfrac{\omega}{2}\right) & 0 \\
0 & 0 & 0 & \sin\!\left(\dfrac{\omega}{2}\right) & -\cos\!\left(\dfrac{\omega}{2}\right) & -\sin\!\left(\dfrac{\omega}{2}\right) & \cos\!\left(\dfrac{\omega}{2}\right) & 0 \\
\dfrac{-\cos(\omega/2)}{r} & \dfrac{\sin(\omega/2)}{r} & 0 & \dfrac{\cos(\omega/2)}{r} & \dfrac{\sin(\omega/2)}{r} & 0 & \dfrac{-2\sin(\omega/2)}{r} & 0
\end{vmatrix}
\begin{vmatrix} \Delta x_1 \\ \Delta y_1 \\ \Delta z_1 \\ \vdots \\ \Delta x_2 \\ \Delta y_2 \\ \Delta z_2 \\ \vdots \\ \Delta x_3 \\ \Delta y_3 \\ \Delta z_3 \end{vmatrix}
$$

$$(1.68)$$

where r is the equilibrium distance between the M and A surface atoms, and xyz coordinates and the vector notation, such as shown in Fig. 1.6. One can take advantage of the notation depicted in this figure and describe the atom displacement in terms of vectors. If δ_1 and δ_2 are the displacement vectors of the atoms A, and δ_3 is the displacement of M, the $\mathbb{R}$ matrix can be represented as a scalar product of the two vectors

$$
\mathbb{R} = \mathbb{S}\delta \qquad (1.69)
$$

where the $\mathbb{S}$ vector represents the direction of the atom displacement in which a given displacement gives the greatest increase of Δr or $\Delta \omega$. For example, Δr_1, which corresponds to the displacement of the A and M atoms or the A—M stretching will be represented by

$$\Delta r_1 = \Delta r_{A\ M} = e_{A\ M} \cdot \delta_A - e_{A\ M} \cdot \delta_M$$

Similarly, the angle $\Delta \omega$, or the bending mode of vibration, will be represented by

$$\Delta \omega = \Delta \omega_{A\ M\ A} = \frac{\delta_{M\ A} \cdot \delta_A + \delta_{M\ M} \cdot \delta_M - (\delta_{A\ M} + \delta_{M\ A}) \cdot \delta_M}{r}$$

With the use of the $\mathbb{S}$ matrix, the $\mathbb{G}$ matrix (Eq. I.4.11) can be expressed as

$$\mathbb{G} = \mathbb{S}\,\mathbf{m}^{-1}\hat{\mathbb{S}} \tag{1.70}$$

and for the A—M—A structure presented in Fig. 1.6, becomes

$$\mathbb{G} = \begin{array}{|ccc|} e_{M\ A} & 0 & -e_{M\ A} \\ 0 & e_{M\ A} & -e_{M\ A} \\ \dfrac{p_{M\ A}}{r} & \dfrac{p_{M\ A}}{r} & -\dfrac{(p_{M\ A} + p_{M\ A})}{r} \end{array} \begin{array}{|ccc|} \mu_A & 0 & 0 \\ 0 & \mu_A & 0 \\ 0 & 0 & \mu_M \end{array}$$

$$X \begin{array}{|ccc|} e_{M\ A} & 0 & \dfrac{p_{M\ A}}{r} \\ 0 & e_{M\ A} & \dfrac{p_{M\ A}}{r} \\ -e_{M\ A} & -e_{M\ A} & -\left(\dfrac{p_{M\ A} + p_{M\ A}}{r}\right) \end{array} \tag{1.71}$$

and considering that the atoms A are equivalent

$$e_{M\ A_1} \cdot e_{M\ A_1} = e_{M\ A_2} \cdot e_{M\ A_2} = p_{M\ A_2} \cdot p_{M\ A_2} = 1,$$

$$e_{M\ A_1} \cdot p_{M\ A_1} = e_{M\ A_2} \cdot p_{M\ A_2} = 0,$$

$$e_{M\ A_1} \cdot e_{M\ A_2} = \cos \omega, \qquad e_{M\ A_1} \cdot p_{M\ A_2} = e_{M\ A_2} \cdot p_{M\ A_1} = -\sin \omega$$

and

$$(p_{M\ A_1} + p_{M\ A_2})^2 = 2(1 - \cos \omega)$$

the $\mathbb{G}$ matrix can be expressed as

$$
\mathbb{G} = \begin{vmatrix}
\mu_M + \mu_{A_1} & \mu_M \cos \omega & -\dfrac{\lambda_M}{r} \sin \omega \\[2ex]
\mu_M \cos \omega & \mu_M + \mu_{A_1} & -\dfrac{\mu_M}{r} \sin \omega \\[2ex]
-\dfrac{\mu_M}{r} \sin \omega & -\dfrac{\mu_M}{r} \sin \omega & \dfrac{2\mu_{A_1}}{r^2} + \dfrac{2\mu_M}{r^2}(1 - \cos \omega)
\end{vmatrix} \qquad (1.72)
$$

At this point we can use the symmetry properties of the molecule attached to the surface. First, we should recognize the equivalence of the two M—A bonds of a bent MA_2 species; and second, we can calculate the force constants. The force constants are presented by the $\mathbb{F}$ matrix (Eq. 1.59), and then Eq. 1.64 can be solved. Of course, in order to solve this equation one needs to assume a type of force field which holds atoms together. These force fields are described in the following sections.

1.7   MOLECULAR FORCE FIELD POTENTIALS AND FORCE CONSTANTS

A primary purpose of normal-coordinate analysis is the determination of appropriate force constants from the observed frequencies. As was discussed earlier, such analysis requires the definition of the force field, which may create various difficulties. For example, using Eqs. 1.62 and 1.63, we can write the potential energy of a MA_2 molecule as

$$
2V = f_{A_1 A_1}(\Delta r_{A_1})^2 + f_{11}(\Delta r_{A_2})^2 + f_{MM} r^2 (\Delta \omega)^2 + 2 f_{A_1 A_2}(\Delta r_{A_1})(\Delta r_{A_2})
$$
$$
+ 2 f_{A_1 M} r (\Delta r_{A_1})(\Delta \omega) + 2 f_{A_1 M} r (\Delta r_{A_2})(\Delta \omega) \qquad (1.73)
$$

This is often called a *generalized valence force* (GVF) *field* or, when only stretching or bending force constants are included in the calculations, a *simple valence force field*. In our case, in addition to the stretching and bending forces, the interaction force constants between stretching or bending forces are also included.

Shimanouchi (6) has proposed the Urey–Bradley force field, which, in addition to the stretching and bending forces, utilizes the repulsive forces between nonbonded atoms. The basis for this force field is the electronic structure of atoms forming the molecules. Specifically in addition to the attractive energy associated with the valence electrons, the repulsive energy, which is associated with the electrons localized at the nuclei, contributes to the total energy.

The general expression for the potential energy can be written

$$
V = V_R(r) - V_A(r) \qquad (1.74)
$$

where $V_R(r)$ and $V_A(r)$ are respectively the repulsive and attractive components of the potential energy and are graphically illustrated in Fig. 1.9. At the equilibrium distance r_e, the repulsive and attractive components are at equilibrium:

$$\left(\frac{\partial V_R}{\partial r}\right) = \left(\frac{\partial V_A}{\partial r}\right) \tag{1.75}$$

As the distance between atoms becomes greater than r_e, the $\partial V_R/\partial r$ decreases more slowly than $\partial V_A/\partial r$ because the attractive forces dominate the equilibrium. Conversely, as the distance becomes smaller than r_e, the repulsive term dominates. Since we are concerned with the force constants represented as a second derivative of the potential energy, it follows that $\partial^2 V_R/\partial r^2$ is larger than $\partial^2 V_A/\partial r^2$ at equilibrium. The quadratic force constant is dependent on the repulsive term. In general, assuming the harmonic approximation of the potential function, the potential energy can be expressed as

$$2V = \sum_{i,j=1}^{3n} f_{ij}q_iq_j \tag{1.76}$$

This equation would imply a tremendous number of force constants, but the force

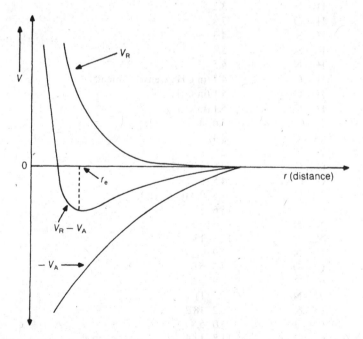

Figure 1.9 A plot of attractive and repulsive energy profiles giving rise to the potential-energy curve $V_R - V_A$ with a minimum at the equilibrium distance between two atoms.

constant matrix can be reduced because of a maximum $(3N)^2$ terms allowed and the fact that $f_{ij} = f_{ji}$.

One principle that should be kept in mind is that the more force constants we use, the easier it is to match the calculated and observed frequencies. As a matter of fact, if one uses too many force constants, one can fit anything he or she wants to, even the spectrum of the sample other than the object of interest. Therefore, the number of force constants should be reduced to the minimum, always keeping in mind their physical significance.

The number of force constants in the general quadratic potential function is equal to the elements on or above the diagonal of the force constant matrix having $3N - 6$ independent elements (without redundanceies) on each side. this number

TABLE 1.3   Force Constants for Selected Common Bonds

Bond	k (mdyn/ Å)
H—F	9.7
H—Cl	5.2
H—Br	4.1
H—I	3.2
H—O	7.8
H—S	4.3
H—Se	3.3
H—N	6.5
H—C	4.7 (in CH_3X environment[a])
H—C	5.1 (in C_2H_4)
H—C	5.1 (in C_6H_6)
F—C	5.6
Cl—C	3.4
Br—C	2.8
C=C	7.6 (in C_6H_6)
C—C	4.5–5.6
C=C	9.5–9.9 (in C_2H_4)
C≡C	15.6–17
N—N	3.5–5.5
N=N	13.0–13.5
N≡N	22.9 (in N_2)
O—O	3.5–5.0
C—N	4.9–5.6
C=N	10–11
C≡N	16.2–18.2
C—O	5.0–5.8
C=O	11.8–13.4

[a] X = Cl, F.

can be expressed as

$$1 + 2 + 3 + \cdots + (3N - 7) + (3N - 6) = \tfrac{1}{2}(3N - 6)(3N - 5) \qquad (1.77)$$

This equation is particularly useful for the surface analysis because it is valid for the species with no symmetry, such as the end-on structures shown in Fig 1.7A. On the other hand, if the molecule has C_{2v} symmetry, such as the side-on structure MA_2, the force constant matrix will take the following form:

$$2V = \begin{vmatrix} f_{11} & f_{12} & f_{13} \\ & f_{22} & f_{23} \\ & & f_{33} \end{vmatrix} \qquad (1.78)$$

where f_{11} and f_{12} are the stretching force constants and f_{33} is the bending force constant. However, because $f_{11} = f_{22}$ and $f_{13} = f_{23}$ are equivalent, the maximum number of independent force constants needed is four instead of six.

The force constants for a particular type of bond or bond order for different monomeric molecules does not change much. As a result, the vibrational frequency difference for a given bond in various species does change significantly. This observation has provided the basis for allowing the transferring of bond vibrational energies (or vibrational frequencies) from one molecule to another. This practice, although convenient, should be applied very carefully since the vibrations of not only polymeric molecules but even polyatomic monomers involve complex displacements of all constituent atoms. Hence, they are sensitive to the bond environment. Table 1.3 summarizes selected force constants for the most commonly occurring characteristic bonds.

1.8 SUMMARY

Although we have tried to demonstrate, in as a simple form as possible, how to utilize the symmetry and selection rules to establish infrared and Raman activities, certain complexities arise from the frequent use of complex mathematics. However, let us not get discouraged by this because things will get much better in the later chapters. Remember, only direct molecule-level information obtained with truly spectroscopic probes (among which IR and Raman have and will play a key role) can provide an understanding the chemistry, reactions, and other features of surface structures. The role of IR and Raman spectroscopy will become even more important in view of the new instrumental advances and the specific information content, because we will need to determine not only the structures but also their typically small quantities. Since surfaces of materials are always complex mixtures of molecules, macromolecules, and their various structural arrangements, vibrational spectroscopy does meet these requirement criteria. Therefore, it is one of the most versatile spectroscopic methods of surface analysis. With proper instrumentation, surface and nonsurface vibrational spectroscopy can be as nondestructive to the sample as one can desire. In contrast, such spectroscopic techniques as UV–VIS (ultraviolet–visible) or mass spectroscopy do not fit into this contest. Therefore, stick around and explore further the vibrational features of materials surfaces.

Many scientists claim that in order to understand certain phenomena, they must deal with what they often call, "high-purity samples" or "highly controllable conditions." In actual life, there are only a few cases were the sample has not been dropped on the floor and therefore does not contain a controllable amount of dirt. Therefore, it is considered to be of a "high purity." For that matter, all polymers and surfaces (especially polymers) are dirty because they do not exist in a pure form. May be this is why they are so useful, and polymer and surface scientists are still in demand.

REFERENCES

1. E. W. Wilson, J. C. Decius, and P. C. Cross, *Molecular Vibrations*, McGraw-Hill, New York, 1955.
2. G. Hertzberg, *Molecular Spectra and Molecular Structure*, Parts I–III, 2nd edn., Van Nostrand, New York, 1950.
3. D. Steele, *Theory of Vibrational Spectroscopy*, Saunders, Philadelphia, 1971.
4. L. A. Woodward, *Introduction to the Theory of Molecular Vibrations and Vibrational Spectroscopy*, Oxford (Clarendon Press), New York, 1972.
5. F. A. Cotton, *Chemical Applications of Group Theory*, Wiley-Interscience, New York, 1972.
6. T. Shimanouchi, in *Physical Chemistry: An Advanced Treatise*, Vol. 4, H. Eyring, D. Henderson, and W. Jost, Eds., Academic Press, New York, 1971, Chapter 6.
7. K. Nakamoto, *Infrared and Raman Spectra of Inorganic and Coordination Compounds*, Wiley Interscience, New York, 1992.

CHAPTER 2

VIBRATIONAL FEATURES OF MOLECULES ADSORBED ON SURFACES

2.1 THEORY OF SURFACE VIBRATIONS

As was discussed earlier, the number of dipole active vibrations of a molecule is determined by degrees of vibrational freedom and symmetry. Since the freedom of the molecule depends on the molecule's physical state, the degree of freedom is subsequently related to whether the molecule is in a gas, liquid, solid, or adsorbed state. For example, let us consider a molecule freely travelling through a gas phase with $3n$ degrees of freedom (n — number of atoms in the molecule). When the molecule becomes attached to the surface, this process, often called *adsorption*, is virtually a transition from the gas phase to the adsorbed state on a solid surface. This is because the previously freely moving gaseous molecule, on being attached with one or both ends to the surface, has a restricted motion. While translational motion or translational degrees of freedom in three (x, y, z) directions is completely eliminated, rotations around the "surface-bonded atom" bond may exist, depending on the nature of the bonding. For example, if one end of the molecule is not attached to the surface and the bonding with the surface has a single-bond character, the molecule may be able to rotate along this bond. This, however, is not the case for vibrational energy because as a result of bonding, the vibrational degrees of freedom will change in such a way that the number of expected vibrations will depend on the symmetry changes resulting from the new surface structures imposing new vibrational selection rules.

To illustrate what can happen when gaseous species adsorb on the surface, let us consider the carbon monoxide molecule that becomes attached to the surface of a metal oxide substrate, such as depicted in Fig. 2.1. Before the surface reaction occurs, a linear $C{\equiv}O$ molecule has $3n - 5$ degrees of vibrational freedom

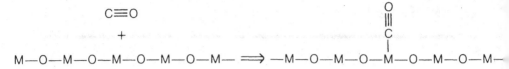

Figure 2.1   Carbon monoxide reaction with a metal surface.

because it is a linear molecule and one rotational degree of freedom along the molecular axis can be neglected. When $C\equiv O$ becomes attached to the surface, which usually occurs through the carbon atom, a new geometry of such species is established. As a result of new atomic environments, new vibrational selection rules are created. In addition, the strength of bonding with the surface will determine whether physical or chemical adsorption has occurred. Physical adsorption is usually a reversible process, and volatile adsorbers can be easily removed from the surface by evacuating the system at elevated temperatures. On the other hand, chemisorption is believed to involve stronger bonding and is usually irreversible at moderate temperatures. It should be kept in mind, however, that these definitions are fairly arbitrary and should not be considered as definite criteria.

2.1.1 Vibrational Spectra of Adsorbed Species

In addition to vibrational degrees of freedom changes, the most evident effect of adsorption is a loss of rotational freedom. Figure 2.2A illustrates this effect on carbon monoxide.[1] The fine structure in the gas-phase spectrum corresponds to transitions between various rotational energy levels of the lower and upper vibrational energy levels. When the molecule is adsorbed on the surface, it does not have full ability to rotate. This results in the appearance of one vibrational band due to the $C\equiv O$ stretching vibration. As illustrated in Fig. 2.2B, a single band of the Loretzian, Gaussian shape, or their mixture is observed, and the degree of broadening may indicate the extent of restricted rotation. The spectral changes produced by physical adsorption of gas molecules onto the solid surface often is similar to the changes that occur on dissolution of molecules in solvents. Rotational structures of adsorption bands are lost, and in most cases, a single adsorption band with the vibrational frequency lower than that of the gaseous molecule is observed. To get a better feeling of how environment may affect rotational energy and how the physical state of a molecule may influence certain characteristic vibrational modes, Table 2.1 is provided. It lists vibrational frequencies of selected organic molecules in gaseous, solution, and adsorbed phases.

Another comparison indicating the effect of physical state and environment can be made for the extensively studied ammonia molecule adsorbed on the silica surface. Table 2.2 illustrates these results and indicates that the vibrational frequencies of the surface hydrogen-bonded molecule, such as depicted in Fig. 2.3, are between the pure liquid and solid ammonia frequencies and those obtained in

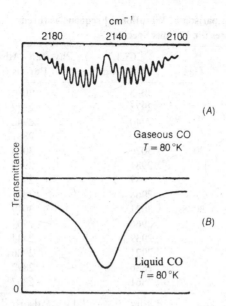

Figure 2.2 Rotational and vibrational IR spectra of carbon monoxide: (*A*) in a gas phase; (*B*) after absorption to the surface. (Adopted from ref. 1.)

a dilute carbon tetrachloride solution.

The presence of an asymmetric surface force field produces another effect on the adsorbed molecules. In solution, the molecule is surrounded by solvent molecules forming a more or less uniform electronic environment. Whereas on the surface, usually one side of the molecule interacts with the environment. Such an asymmetric environment generates a distortion in the symmetry of the molecule, giving rise to infrared vibrations that were previously not allowed because of the symmetry and freedom of movement. A good example is adsorption of the ethylene molecule on silica. The adsorption of the ethylene molecule will be discussed in greater detail later on. Here let us analyze the normal C—H stretching vibrations of isolated ethylene. As was described in Chapter 1, among four IR and Raman-active stretching modes, the symmetric nature of the v_1 and v_5 normal vibrations makes them Raman-active. On the other hand, nonsymmetric vibrational modes v_9 and v_{11} are infrared-active because of dipole moment changes during vibrations. All four normal vibrational modes are illustrated in Fig. 2.4. When the ethylene molecule was adsorbed on the silica glass surface, instead of the two IR-active bands, two previously inactive vibrations are detected at 3007 and 3070 cm^{-1} and are attributed to the Raman-active bands that also become IR-active as a result of the symmetry changes on adsorption and nonuniform surface environments.

TABLE 2.1   Comparison of Vibrational Frequencies (in cm^{-1}) in Gas, Liquid, and Adsorbed States for Various Species

Molecule	Gas	CCl$_4$ Solution	Physically Adsorbed on Porous Glass	Ref.
1-Butene	3094	3085	3080	2
		2975	2972	2
		2940	2942	2
		2880	2881	2
cis-2-Butene	3034	3026	3023	90
		2981	2983	90
		2937, 2926	2931	90
		2968	2871	90
trans-2-Butene	3028	3021	3019	2
		2966	2969	2
		2939	2944	2
		2921	2926	2
		2886	2892	2
		2861	2867	2

	Gas	Liquid	Solid	Adsorbed	
Methane	3018	3018		3006	3
	2916	—		2899	4
	2823	2820		2819	4
Ethylene	3105	3105	3089	3100	4
	3019	—	—	3010	4
	2989	2980	2973	2980	4
Acetylene	3287	—	—	3240 (90 K)	4
				3255 (300 K)	4
Hydrogen	4160	—	—	4131	4

TABLE 2.2   Infrared Bands of Ammonia in the 3600–2800-cm^{-1} Region

Phase	Frequency (cm^{-1})			
	N—H Stretching Mode		Overtone or Combination Mode	Ref.
	Symmetric	Asymmetric		
Gas	3444	3326		5,6
Liquid	3375	3285	3220	7
Solid	3378	3323	3297	8
Dissolved in CCl$_4$	3417	3315	3297	9
In solid N$_2$	3440	3332		10
Porous SiO$_2$	3400	3320		11

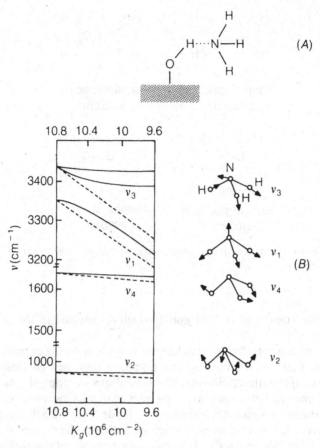

Figure 2.3 (*A*) Bonding of ammonia to the surface hydroxyl groups on silica and ammonia normal vibrational modes; (*B*) normal vibrations in NH_3 molecule.

A similar effect can be observed for the totally symmetric H_2 molecule that has only one Raman-active H—H stretching vibration. On adsorption on the surface, a new band at 4131 cm^{-1} is detected in the infrared spectrum. Its presence is attributed to the distortion of the electronic cloud, which, in the unadsorbed state, is totally symmetric. Of course, in the nonadsorbed state, the H—H stretch is only Raman-active.

There are numerous other examples, such as acetylene adsorbed on alumina surface, which was found by Yates et al.[12] by the appearance of a new infrared (IR)-active carbon–carbon stretching vibration, which, under normal conditions, is Raman-active. On the basis of analysis of infrared spectra the authors attributed the presence of two bands at 1950 and 2007 cm^{-1} to two forms of acetylene present on the surface: one weakly bonded (1950-cm^{-1} band) and the other one strongly attached to the surface (2007 cm^{-1}). Both bands correspond to

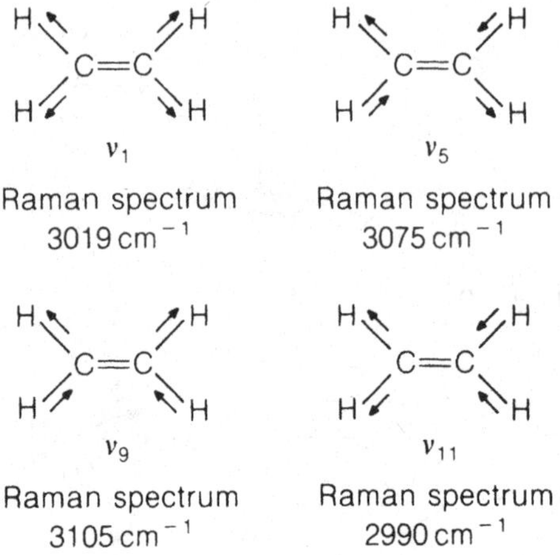

Figure 2.4 Four symmetric and nonsymmetric stretching vibrations of an isolated ethylene molecule.

Raman-active vibrations at 1974 and 1961 cm^{-1} detected in the unperturbed state.

Because of molecular symmetry changes, when a molecule or macromolecule undergoes a transition from the gas to solid phase, these new environments may also lead to conformational changes. Vibrational spectroscopy plays a key role in monitoring chemical processes involving such changes. One useful and practical example is the autooxidative crosslinking of alkyds. Figure 2.5 illustrates a series of spectra recorded as a function of time and indicates the disappearance of the 3015-cm^{-1} band due to the C—H stretching vibrations of the allyl group with hydrogens being in a cis position.[13] As a result of the reaction, this molecular segment goes to trans conformation; hence, the C—H stretching band becomes Raman-active and IR-inactive. This is demonstrated in trace E, where the 3015-cm^{-1} band disappears and OH stretching at 3430 cm^{-1} forms. Alkyls are discussed further in Chapter 7.

The selected examples provide a strong experimental basis that symmetry changes are essential factors in surface or nonsurface reactions. It is equally important to have a theoretical foundation for these changes to account for and predict surface modifications leading to the desired properties. Therefore, in the next sections, we will focus on outlining several approaches to the substrate–adsorbate interactions, which may or may not account for the observed spectral features.

2.1.2 Degrees of Freedom and a Physical State

As was stated earlier, the number of dipole active vibrations of the molecule on the surface is determined by its degree of freedom. Since the molecular freedom

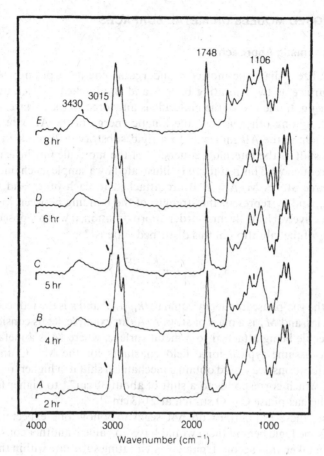

Figure 2.5 A series of photoacoustic FT–IR spectra of autooxidative curing of alkyd coatings recorded as a function of time. (Reprinted with permission from ref. 13.)

depends on the molecule's physical state, the degrees of freedom are ultimately related to whether the molecule is in the gas, liquid, solid, or adsorbed state. One can picture the process of adsorption of a gas as a transition from the gas phase to the adsorbed state on a solid surface such as a condensation-like mechanism. If a gas-phase linear molecule is adsorbed on the surface, the number of vibrational degrees of freedom changes—that is, from $3n-5$ (linear) or $3n-6$ (nonlinear) to somewhat different. What effectively happens, is that a molecule attaches to at least one end to the surface. This process changes the overall degrees of freedom because surface atoms are now bonded to the adsorbed molecule. Such a process will lead to the mechanical effect, which may become an important issue in the consideration of vibrational frequency shift due to mechanical constraints.

2.2 PROPOSED MODELS ON METAL SURFACES

2.2.1 Mechanistic Approach

Let us visualize a diatomic molecule undergoing the adsorption process to a substrate surface as two vibrating balls A and B, attached to each other by a massless spring. If one end of the molecule is anchored to the surface, it is quite certain that, among other factors, the kinetic energy of the AB vibration will change. Attaching the AB molecule to a rigid substrate will result in a purely mechanical shift of the internal frequencies of the molecule (in this case A—B stretching frequency). This result can be illustrated by a simple mechanical model in which three atoms M, A and B are attached to each other and the force constants k_0 and k_1 represent the strength of the bonding between M—A and A—B, respectively. Using the first-order approximation, it was proposed that the frequency v_a of the vibration in this disturbed state is[14]

$$v_a = v_0 \left\{ 1 + \frac{k_1}{k_0} \frac{\mu^2}{2M_A^2} \right\} \tag{2.1}$$

where v_0 is the gas-phase frequency equal to $(k_0/\mu)^{1/2}$ and μ is the reduced mass of the AB species, and M_A is a mass of atom A. As an example, let us consider again $C\equiv O$ molecule being attached to a metal surface, where M is a metal surface atom. If we assume typical force field constants for the M—C and $C\equiv O$ stretching vibrations, we would obtain a mechanical shift to a higher frequency of about 1.4% which corresponds to a shift of about 30 cm^{-1} to higher frequency values for the gas-phase C—O stretch at 2143 cm^{-1}.

Of course, these estimations do not seem to follow our expectations, but nevertheless the logic behind them would have no value if another consideration has not been taken into account: namely, a vibrating substrate within the surface lattice environments. Although the lattice vibration—which are in the range of 500–50 cm^{-1}—will mainly affect the M—C stretching vibrations instead of the $C\equiv O$ mode at about 2000 cm^{-1}, they should not be neglected. Calculations by Richardson et al.[15] indicated that when Ni_5CO and Ni_6CO clusters were compared to the heavy, indefinite mass metal atoms, the frequency shifts of about 90 cm^{-1} for the M—C stretching vibrations and only 1 cm^{-1} for $C\equiv O$ would occur. For the case of CO adsorbed on a metal surface, and accounting for the significant influence of the lattice vibrations, only an expected shift to higher frequencies for the metal–carbon vibrations is anticipated which may not always be experimentally confirmed. Thus, the mechanical model predicts upward frequency shift on the adsorption of C—O from the gas phase. Therefore, in light of the experimental results, where in most cases downward shifts are observed, a purely mechanistic approach is simply not adequate to explain many effects.

2.2.2 Self-Image of the Isolated Molecule

The self-image model of an isolated molecule is based on the principle that the induced dipole moment in the adsorbed molecule has its own image on the metal surface with which it can interact. The induced dipole moment P_{dyn}, excited by a local electric field E_{loc} on the surface, may interact with an adsorbed molecule. Its magnitude is expressed by the following equation:

$$P_d = \alpha(v)E_{loc}(\omega) \tag{2.2}$$

where $\alpha(v)$ is the dynamic polarizability of the adsorbed molecule. The local electric field is expressed by

$$E_{loc}(\omega) = (1 + r_p)E_0 + \frac{P_{dyn}}{4d^3} \tag{2.3}$$

where the $(1 + r_p)E_0$ component is the surface electric field from the incident and reflected light with the frequency v.[16] According to the dielectric theory, the electric field E_0 at the surface is a superposition of the incident and reflected light, and its magnitude will depend on the angle of incidence and the optical properties of a substrate. The second term is due to the field of the image created by the vibrating dipole and d is the distance between the image and the dipole moment. The dynamic polarizability can be expressed as a sum of an electronic and ionic part described mathematically as

$$\alpha(v) = \alpha_e + \frac{\alpha_v}{1 - v^2/v_0^2 + i\gamma(v/v_0)^2} \tag{2.4}$$

Because the electronic part α_e can be independent on the spectral range, a magnitude of the vibrational or ionic part is related to the vibrations of atoms at the normal mode frequency v_0 associated by the A—B stretch with the assumption that the lineshape is of Lorentzian type with the width of γv_0.[17] Following Lucas et al.,[18] and solving (2.2) through (2.4) for v it has been showed that

$$\left(\frac{v}{v_0}\right)^2 = 1 - \frac{\alpha_v/(4d^3)}{1 - \alpha_e/(4d^3)} \tag{2.5}$$

According to Eq. 2.5, the effect of the own image tends to lower the frequency of the isolated A—B molecule adsorbed on a metal surface. For example, assuming typical values for CO adsorbed on Pt(111) surface with values $\alpha_v = 0.057 \text{ Å}^3$, $\alpha_e = 1.89 \text{ Å}^3$, and $d \cong 1 \text{ Å}$, the shift by 30–45 cm^{-1} to lower wavenumber for the C—O stretch was predicted.[19] It should be mentioned that the d value, that is, the spacing between dipole moments, may significantly affect the calculated shift, and its precise determination may be troublesome. This is because the dipole moment of a diatomic molecule, and for that matter any other molecule, cannot be presented as a single point in space, but it is expended to be spaced above the surface. Of course, the orientation with respect to the surface further complicates the issue.

Using simple mathematics, it can be shown that the dipole moment distribution will vary depending on the adsorbate orientation with respect to surface. As a matter of fact, one drawback of many surface models is simply lack of the orientation considerations. It is quite apparent that depending upon the orientation of the surface species, such quantities as surface coverage and dipolar and vibrational decouplings will change.

Nevertheless, the mechanical model has proven at least some conceptual and educational values in giving a semiquantitative explanation of experimentally determined image-induced frequency shifts. A derivative of the model described above has also been developed[20,21] and includes extended dipole considerations. For the same CO molecule attached to the surface, the downward shifts in the range of $20-30\,cm^{-1}$ were predicted, but the same problem in locating the image plane was experienced. None of the above outlined models included the effect of a chemical bond formation, a key feature is surface stability and reactivity.

2.2.3   Chemical and Electrostatic Shifts

During the chemical bond formation, there is a substantial change of potential energy of the system forming new bonds due to the changes of electronic energies within the system and new configurations of electronic levels. These changes will be reflected in the structural changes such as bond lengths, symmetry, and molecular geometry. If a given molecule becomes chemically bonded to the surface, the vibrational energy is expected to change; this is also referred to as chemical shifts with respect to the nonbonded molecule. These changes are also related to the chemical interactions between the adsorbate molecule and the substrate. One of the most important consequences is that, depending on surface coverage, this effect is responsible for the appearance of absorption bands in the infrared spectrum. For example, when carbon monoxide is adsorbed on different sites of supported transition metals such as nickel or palladium, depending on the surface characteristics, various spectral responses can be attained. In this particular case, a theoretical consideration has been given by Blyholder[22,23] who assumed that the chemical bond between CO and metal is formed by charge transfer from the $5s$ MO of CO into metal and by "backdonation" from metal d bands into the unoccupied $2\Pi^*$ MO. This is shown in Fig. 2.6. Because the $5s$-MO is only weakly bonding and the $2\Pi^*$ bonding has an antibonding character, the $C{\equiv}O$ bond is weakened by a bond formation due to the chemisorption process and the amount of backdonation into the $2\Pi^*$ is drastically affected by a lowering of the $C{\equiv}O$ stretching frequency. A quantitative relationship between $C{\equiv}O$ stretching frequency and $2\Pi^*$ occupation has been also predicted.[24] In this case, $2\Pi^*$ populations obtained from LCAO–HF (linear combination of atomic orbitals/Hartree–Fock) calculations with experimental results of $C—O$ stretch frequencies for a series of $Ni(CO)_4$ complexes were compared. The relationship, shown in Fig. 2.7, indicates a linear behavior, although it cannot be applied without restrictions. Such effects as electrodynamic and mechanical reorganizations may also affect the $C{\equiv}O$ stretching frequencies. Nevertheless, this model agrees with experimental frequencies obtained for $C{\equiv}O$ adsorbed on palladium[25,26] and many other systems.

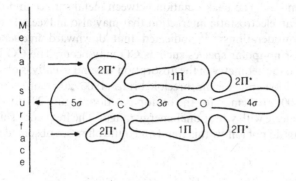

Figure 2.6 Schematic representation of backdonation to a metal surface.

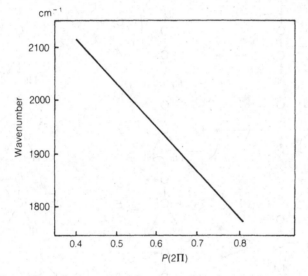

Figure 2.7 Experimentally determined C—C stretching frequencies plotted as a function of calculated $2\Pi^*$ CO population for metal carbonyl (from ref. 14).

Additional supporting evidence for backdonation can be obtained from coadsorption experiments. If C≡O is coadsorbed with hydrocarbons, shifts to lower frequency up to $100\,\text{cm}^{-1}$ were observed.[27,28] Hydrocarbons are electron donors, and hence provide a source of more extensive donation. On the other hand, the presence of molecular oxygen, which is considered to be an electron acceptor, reduces the backdonation to the CO molecule, and a shift to higher frequency is observed.[29] Perhaps the lowest observed frequency shift was detected for CO adsorbed on Pt(111) surface in the presence of a coadsorbed strong electron donor such as potassium. The observed CO stretching frequency was as low as $1400 \, cm^{-1}$.[30] The backdonation between metal surface and the molecule also results in an electrostatic interaction that may also induce a frequency shift. Theoretical considerations[31,32] indicated that downward frequency shifts of $10–20 \, cm^{-1}$ for nonpolar species such as CO may occur. The CO frequency is also sensitive to the geometry of the absorption site and usually is observed in the range of $1800–1900 \, cm^{-1}$ for the threefold bridge, $1900–2000 \, cm^{-1}$ for the twofold bridge, and $2000–2100 \, cm^{-1}$ for the on-top position. Figure 2.8 illustrates possible structures along with vibrational-energy ranges. The frequency shifts observed on nobel metals do not follow this trend and quite often are observed much above $2000 \, cm^{-1}$.

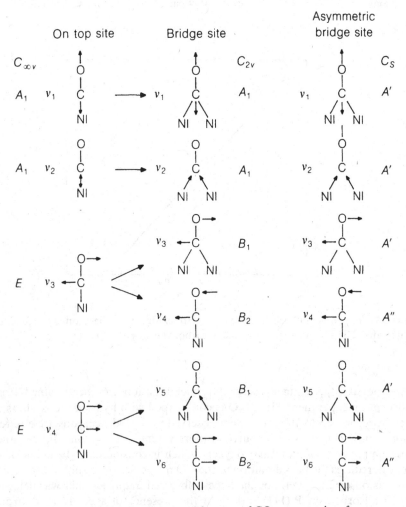

Figure 2.8 Possible structural features of CO on a metal surface.

2.3 FREQUENCY SHIFTS WITH THE SURFACE COVERAGE

In addition to the effects discussed above, there are the frequency shifts related to the surface coverage. These effects are perhaps the most documented in the literature. The experimental observation that the heat of chemisorption decreases with the increasing surface coverage has been the subject of substantial research. This effect has been attributed to the surface heterogeneities; specifically it is related to a nonuniform distribution of the surface active sites. These are the sites that cause greater local concentrations of the adsorbed species and create strong interactions between the adsorbate molecules. Usually, the repulsive interactions give raise to higher frequencies, which increase with the increasing population of adsorbate molecules on the surface. As an example IR spectra of the CO species adsorbed on the silica-supported Pd are presented in Fig. 2.9 and exhibit a shift of the CO stretching frequency by approximately $90\,\mathrm{cm^{-1}}$.[33] The frequency shift parallels the increasing intensity of the $2060\text{-}\mathrm{cm^{-1}}$ band with the increasing surface coverage.

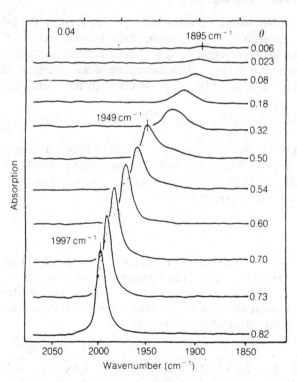

Figure 2.9 Infrared spectra of CO species adsorbed on Pd(100) at 100 K as a function of increasing surface coverage. (Reprinted with permission from ref. 33.)

2.3.1 Theoretical Approaches

A shift to the higher CO stretching frequency due to a surface coverage has been reported in numerous experiments[34] and attributed to increasing dipole–dipole interactions among neighboring adsorbate molecules.[35] Both theoretical[36] [38] as well as experimental[39,40] considerations have shown that the frequency shift is, indeed, related to the surface coverage. Using the isotopic substitutions, it is possible to distinguish between the frequency shifts caused by vibrational coupling and chemical effects such as backdonation.

One can draw an analogy that the frequency shift due to the surface coverage changes may be similar to that observed when a molecule in a gas phase becomes adsorbed on the surface. As a matter of fact, the physical principles underlying both processes are similar; the only difference is that in the adsorption processes we are dealing with the gas phase–surface interactions, whereas the increasing surface coverage adds one more player, that is, more adsorbate molecules. And this is the main reason why considerations of vibrational coupling are so essential for the surface interactions.

2.3.2 Vibrational Coupling

Vibrational coupling within the adsorbed layer can occur in the form of spacial dipole–dipole interaction[41] or via coupling through metal electrons.[42] The dipole–dipole coupling between neighboring adsorbed molecules can be considered a self-image shift; that is, it is vibrational coupling of the molecular dipole with its own image. This can be described mathematically by the following relationships:

$$E_{loc} = (1 + r_p)E_0 + E_{\text{own image}} + E_{\text{other dipoles}} + E_{\text{other images}} \qquad (2.6)$$

where the first two terms of the equation have already been recognized in Eq. 2.3 as contributions from the incident light and the image induced by the vibrating dipole, respectively. If an adsorption site L_j contributes to the surrounding dipoles, its effect can be expressed by[43]

$$E_{\text{other}} = -\sum_j \frac{P_j}{|L_i - L_j|^3} \qquad (2.7)$$

where the lattice sum includes all occupied adsorption sites L_j. The contributions from other images of the neighboring dipoles represented by $E_{\text{other images}}$ in Eq. 2.6 contain lattice sums. If one assumes a perfectly ordered layer of identical dipoles adsorbed on identical adsorption sites, the sum of all lattice sites is a function of the surface-coverage-dependent function $S(\theta)$:

$$S(\theta) = \sum_{\substack{K \\ k \neq 1}} \frac{1}{|L_1 - L_k|^3} + \frac{1}{(|L_1 - L_k|^2 + 4d^2)^{3/2}} - \frac{12d^2}{(|L_1 - L_k|^2 + 4d^2)^{5/2}} \qquad (2.8)$$

So the sum $E_{\text{other dipoles}} + E_{\text{other images}}$ is equal to "$-P_{\text{dyn}}S(\theta)$".

Now, by insertion of Eq. 2.6 into Eq. 2.2, the expression for dynamic dipole moment will be

$$P_{dyn} = \frac{\alpha(v)(1 + r_p)E_0}{1 + \alpha(v)[S(\theta) - (1/4d^3)]} \tag{2.9}$$

According to this equation, the interactions with the dipole image term (d^3) increase with the dipole moment, whereas when the image term decreases, the effect of other dipoles is diminished. The same relationships have been provided and hold for the frequency shifts as a function of the surface coverage:[44]

$$\left(\frac{v}{v_0}\right)^2 = 1 + \frac{a_v[S(\theta) - (1/4d^3)]}{1 + \alpha_e[S(\theta) - (1/4d^3)]} \tag{2.10}$$

It should be realized that the shift to lower frequency on adsorption due to the interaction of the dipole moment own image is highly affected by a shift to higher frequencies. This is due to the increase of surface coverage and resulting dipole interactions with the surface coverage between neighboring dipoles and the images of these and other dipoles. For example, knowing the appropriate values of polarizabilities α_v and α_e, and the dipole image distance d, it was shown[45] that the shift of an approximately $50 \, cm^{-1}$ at saturation coverages was obtained for CO/Pt[100]. Similar to the isolated molecule model, this model, lacks the ability to determine the image–dipole distance. As indicated in the previous studies,[46 48] the importance of an extended dipole, a correct image formula, the effect of the electrostatic interactions on the equilibrium position of the atoms and the charge transfer between the adsorbed molecule and a metal surface, all play a significant role in the theoretical predictions. For example, if the self-image is not included in the lattice sum, but is incorporated into the molecular polarizability, the vibrational polarizability $\alpha_v = 0.27$ Å for CO on Cu[100] was found to be about 4 times greater than the corresponding gas value ($\alpha_v = 0.057$ Å). Using the above α_v value, a good agreement between theory and experiment for the coverage de- pendence of the vibrational coupling shifts of CO adsorbed on Cu(100) and Ru(001) was obtained.[49]

The interactions between the dipoles of adsorbates through metal substrate electrons have also been proposed[50] and, to some extent, considered to be partially responsible for the observed frequency shifts with the surface coverage. Although the mathematical treatment of the vibrational coupling can be done in a way similar to that for dipole coupling, the physical concepts in both phenomena are quite different. The dipolar coupling comes from interactions between the neighboring dipoles and is related to through-space electromechanical interactions. On the other hand, vibrational coupling through the electrons of a metal substrate comes from the overlapping molecular orbitals of two neighboring adsorbed molecules. This difference will be reflected in the fact that a magnitude of the bond strength in one molecule will affect that in the other.

2.3.3 Chemical and Electrostatic Shifts

Figure 2.6 depicted the basis for backdonation responsible for the chemical shift. When electrons occupy the antibonding molecular orbitals, the strength of the chemical bonding between two atoms decreases, leading to a shift to frequencies lower than that observed for a free molecule (gas phase). It was proposed[51,52] that the shifts to higher frequencies with the increasing surface coverage could be attributed to competition among adsorbed molecules for backdonation electrons. As a result of the backdonation ability into the antibonding molecular orbitals decreases, causing a shift to higher frequency. In spite of rather extensive and qualitative successful use of this approach for explaining various experimental facts, there is no quantitative model. This is perhaps related to many complexities, for example, electrostatic shifts may cause charge transfers between the metal substrate and adsorbed molecule and, similar to other effects, the electrostatic shift will also affect interactions between neighboring molecules, which may give rise to upward frequency shifts.[53,54]

In spite of all these theoretical difficulties, intermolecular repulsions between adsorbate molecules have been accounted for frequency shifts at higher surface coverages.[55,56] At smaller intermolecular distances, the overlap between the wavefunctions of adjacent molecules adsorbed on the surface becomes more pronounced, because the occupation of the $2\Pi^*$ MO is significantly reduced. Another approach to this problem was proposed.[57] Intermolecular repulsions are responsible for the observed decrease in binding energy for CO on Pd[100] at higher surface coverages. Measurements of the M—C stretching frequency by electron energy-loss spectroscopy (EELS) confirmed these changes.[58] In this particular case, the M—C stretching frequency at $340 \, cm^{-1}$ remains unchanged up $\theta = 0.5$ surface coverage, and further increase in the surface coverage results in a downward shift to $290 \, cm^{-1}$. This observation demonstrates that the M—C bond weakening is indeed surface-coverage-dependent and may be attributed to intermolecular repulsions of the adsorbate molecules. The effect of the M—C stretch shift with surface coverage seems to affect the C≡O frequency, such as that depicted in Fig. 2.9. Because of the complexity of the electronic effects in CO, such as backdonation and other effects, quantitative relationships that relate M—C and C≡O stretching frequencies are yet to be resolved.

2.4 EXPERIMENTAL APPROACHES TO FREQUENCY SHIFT DETERMINATION: ISOTOPIC MIXTURES OF ¹²CO AND ¹³CO

Apparently, it is possible to design an experiment that may allow one to distinguish between the vibrational coupling and chemical frequency shifts. On the basis of theoretical considerations,[59] it was suggested that only molecules with the same vibrational frequency may strongly couple. Therefore, using isotopic mixtures with a sufficient frequency difference due to the mass differences such as that for ¹²CO and ¹³CO, it is possible to attain vibrational dicoupling of the

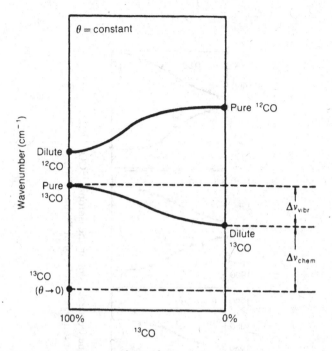

Figure 2.10 The effect of isotopic ^{12}CO/^{13}CO exchange on the surface on the frequency shift allowing the determination of contributions from vibrational coupling or chemical shifts. (Adopted from F. M. Hoffmann, *Surf. Sci. Rep.*, 1983, 3(2–3), 109.)

neighboring molecules adsorbed on a substrate. Using this approach, King et al.[60] have demonstrated that the frequency shift for CO on Pt(111) is caused by dipole coupling. The concept behind this is to exchange a mole fraction of ^{13}CO in ^{12}CO while maintaining the same total surface coverage. The results of these experiments are shown Fig. 2.10. As the isotopic ratio changes, the frequency of the ^{13}CO band decreases. This frequency shift is attributed to vibrational coupling because its origin comes from a gradual vibrational decoupling of the ^{13}CO molecules of the ^{12}CO layer due to frequency mismatch of the two isotopes. On dilution, although the ^{13}CO molecule is vibrationally dicoupled, it still feels the chemical interaction of the surrounding ^{12}CO layer, which, in this case, is not affected by isotopic substitution. Thus, the difference between the frequency value of ^{12}CO dilution and at no ^{13}CO coverage is related to the chemical shift.

It should be kept in mind that the CO/Pd system is a friendly one because there is no spectral disturbance from other features. In reality, this is often not the case and the frequency shifts are also detected. Since only one band due to CO stretch makes life a whole lot easier, it is relatively easy to perform a quantitative analysis, in this case, the frequency shift as a function of surface coverage.

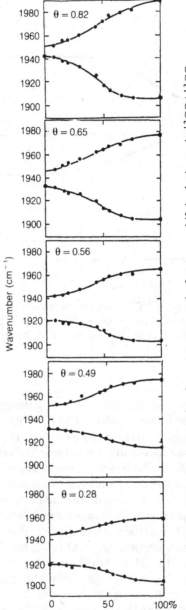

Figure 2.11 The effect of surface coverage on the frequency shift in the isotopic $^{12}CO/^{13}CO$ exchange. (Courtesy of A. Ortega, Ph.D. Dissertation, Freïe Universität Berlin, 1980.)

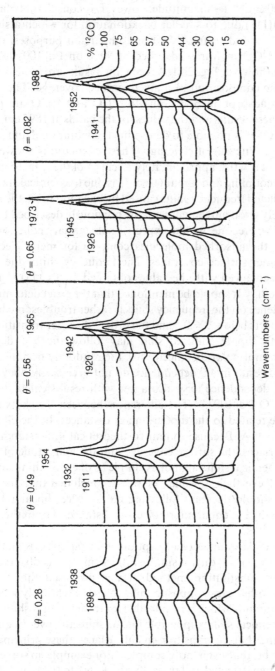

Figure 2.12 Infrared spectra for the $^{12}CO—^{13}CO$ surface exchange at constant surface coverages for CO/Pd(110). (Courtesy of A. Ortega, Ph.D. Dissertation, Freie Universität Berlin, 1980.)

As demonstrated in Fig. 2.9, an increase of the surface coverage produces a shift of approximately 100 cm^{-1}. Its magnitude, however, is considerably larger than that observed on Pt(111) due to smaller backdonation for which a shift of approximately 35 cm^{-1} was detected.[61,62] For comparison purposes, the results obtained for ^{12}CO/^{13}CO mixtures adsorbed at 90 K on Pd[100] for selected surface coverage are shown in Fig. 2.11. As seen, the results are quite different on changing the isotopic composition at a given surface coverage. The frequency shifts resulting from adsorption bands due to ^{12}CO and ^{13}CO are plotted in Fig. 2.12. Based on these results and the values for the bands at 1895 cm^{-1} (^{12}CO) and 1853 cm^{-1} (^{13}CO), it is possible to determine the vibrational and static part of the total shift as a function of coverage. These estimations showed that at saturation coverage of $\theta = 0.8$, approximately 35 cm^{-1} of the total 95 cm^{-1} shift is due to vibrational coupling and about 60 cm^{-1} is due to chemical static effects. As we recall the predicted frequency shifts caused by backdonation, the preceding results are in relatively good agreement with the previously described backdonation representation where one would expect a substantial reduction in $2\Pi^*$ occupation because the adsorbed molecules compete for metal electrons for backdonation at higher surface coverages. The frequency shifts due to dipole coupling (35 cm^{-1}) agree well with the values obtained for Pt(111),[63] Pt(001),[64] Cu(111),[65] and Cu(100).[66] It should be mentioned that the exact determination of the coverage dependence on the frequency shifts is rather troublesome because of so-called ordering effects below $\theta = 0.5$ and the order–disorder transitions above $\theta = 0.5$. Although the dipole–induced shift may follow the $\theta^{3/2}$ dependence indicating equal intermolecular distance [140], the randomly occupied adsorption sites that should result in a θ dependence do not necessarily obey this relationship.[67] The $\theta^{3/2}$ dependence may also agree with results calculated for the dipole shifts in for CO/Pd[100].[68] The differences between the shifts at $\theta = 0.5$ were suggested to be related to the dipole–image distances. In the cited case, it was assumed to be 0.96 Å. Because of limited theoretical understanding of the backdonation on surfaces, the surface coverage and the magnitude of donation dependence on the chemical shifts are not as simple as one may think and are not clearly understood. Nevertheless, it may be safety concluded that, for the most part, backdonation should decrease following more or less linear relationship with the surface coverage at equal intermolecular distances within the absorbed layer.

Let us consider several examples of adsorption on metal surfaces. In the case of CO adsorbed on Cu(111),[69,70] only a small shift (8 cm^{-1}) was observed. Using the familiar isotopic substitution method, it was proposed that two effects working against each other cause opposite shifts that give relatively small overall shift: a vibrational coupling, which is responsible for the frequency shifts to higher frequencies, and a chemical shift responsible for the shifts to lower frequencies. It should be kept in mind that for transition-metal surfaces, these principal sources of the frequency shift become even more complex. For example, in some cases the upward shifts were often detected and partially attributed to the s- and p-bonding characteristics between the metal and the adsorbate.[71]

Although discussion of experiments and theoretical considerations may—and for that matter usually do—create some controversies and disagreements, in practice surfaces are not uniform, and even for highly cleaned and polished metal surfaces, material defects such as dislocations and vacancies, may lead to disagreement between the experiment and theory. Moreover, it can be easily visualized that as the CO molecules are being deposited on the surface, initially the molecules are randomly distributed and disordered. With the increasing surface coverage, they become more packed and ordered, forming islands with more ordered domains or domains themselves. One may take advantage of the sensitivity of CO stretching vibrations to chemical effects and intermolecular distances, and monitor the formation of domains as they undergo order–disorder transitions. Such transitions can usually be monitored by changing temperatures while maintaining the same surface coverage, or conversely, changing the pressure while maintaining constant temperature.

A useful demonstration of this type of phenomenon was provided for the CO/Pd(100) system where such a transition occurs at the surface coverages $\theta \geqslant 0.5$.[33,72,73] Figure 2.13 shows that using IR, one can monitor the order–disorder transitions. In this particular case, the IR spectra were recorded at equili-

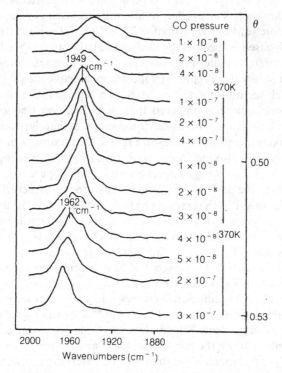

Figure 2.13   A series of IR spectra for CO/Pd(100) system recorded at 370 and 330 K with various CO partial pressures. (Reproduced with permission from ref. 33.)

brium pressures and temperatures of 330 and 370 K because the mobility of the molecules at these temperatures is higher and the CO coverage can be more accurately controlled and estimated by comparison with the isosteric data.[74] It is interesting to note that the 1949 cm^{-1} band (370 K, 4 × 10^{-7} torr) becomes sharper and its diminished halfwidth indicates a higher homogeneity of the surface sides. Above $\theta = 0.5$, however, the band does not broaden, as one would expect of both codomains which may undergo order–disorder transitions. Instead, a new band at 1962 cm^{-1} is detected. By examination of a halfwidth of this band, one may see that ordered domains are rather homogeneous in relation to adsorption sites and intermolecular distances. Another conclusion is that the transition occurs in a very narrow range of surface coverages, namely, $\theta = 0.5$ to 0.52. This demonstrates a high sensitivity of the method.

2.4.1   Ordered Islands of Carbon Monoxide on Surfaces of Ru(001), Pt(001), and Pt(111)

It is known that surfaces are not homogenous, and the spatial differences in the surface morphologies may lead to the areas with an excess or deficit of surface energy. A simple macroscopic example that has a molecular origin is the effect of surface tension. Specifically, any area on the surface with an excess of energy may likely adsorb other species in an effort to minimize this energy excess. Such situations may therefore lead to the formation of preferential surface areas, so-called islands, at which gaseous species may likely adsorb. Because of the high sensitivity of the C—O stretch to intermolecular distances, one can monitor a formation of islands on a substrate. Figure 2.14 illustrates a sequence of IR spectra of the C—O stretch region obtained at 200 K as a function of increasing surface coverage.[75] A single band is observed that shifts to lower frequency with the increasing surface coverage. This observation suggests that only one adsorption site, and most likely a top-on (or side-on Fig. 1.7B) structure; is present. A close examination of spectra also indicates that a considerable narrowing of the band occurs when the surface coverage approaches $\theta = 0.33$. Apparently, this observation corresponds to the presence of geometrically ordered structures.[76] Remembering that the bandwidth decreases as the ordering increases, the distribution of molecules with respect to adsite (adsorption site) and intermolecular distances that the islands are formed, the coverage increases by adding new molecules at the border of the island. Therefore, the intermolecular distances and resulting IR frequency remain constant. The absence of island formation at lower temperatures is expected because molecules have relatively low mobility. As the temperature is increased, surface diffusion increases and molecules may arrange into energetically stable island formation. Pfnür et al.[44] have calculated the IR bands for different island sizes. Using Scheffler's model[77] and assuming the presence of dipole–dipole interactions, the band splittings of about 8 cm^{-1} was determined. Its values are in general much lower than the one observes experimentally (about 20 cm^{-1}). Also, according to this model, the two bands shift with increasing island size, whereas the low frequency does not shift with the increasing surface

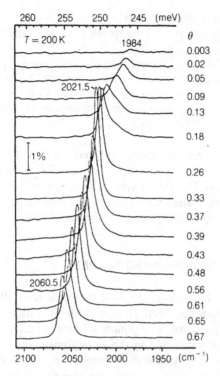

Figure 2.14 A series of IR spectra for the CO/Ru(001) system for various surface coverages. (Reproduced with permission from ref. 44.)

coverage. On that basis, it was assigned to the disordered phase, for which molecular intermolecular spacing is the same.

One important quantity to be determined from infrared data is the size of the islands. Although for this purpose it is necessary to know the distribution of the island size alone, quite often it is assumed to be a homogeneous one. With this assumption, the width of the band and the theoretical predictions provide only a rough estimate of the average at 80 K. In the case of CO, an average estimated island size of 30–50 Å was found. Using measurements of the change of the beam profiles, several other models of island formation based on dipole–dipole coupling have been proposed for CO on Cu(111)[78] and Pt(001)/Pt(111) surfaces.[64]

It is apparent that in many situations, especially where it is necessary to utilize a quantitative approach, such spectral properties as band intensity, shape, and resolution of the bands of interest are the most important criteria. These are the features that are utilized in making or establishing fundamental principles or experimental observations. Therefore, it is necessary to establish reliable methodologies that allow not only enhancement of the spectral information but also increased confidence. This issue is of particular importance in surface spectroscopy because the surface concentration levels are usually low, and numerical procedures can open new vistas in the spectral interpretation and further information.

There are spectroscopic situations where we are really trying to get the answer that we have asked, or worse yet, your boss has asked for. Perhaps one illustrative example is trying to interpret spectral changes within a 0.01-cm^{-1} range, whereas the experiment was conducted at a 2-cm^{-1} resolution. Unfortunately, there are numerous literature examples on that.

This is why, when we use what we often call "enhancement methods", it is necessary to first choose the proper method and, second, follow all precautions and be aware of potential traps. This can be achieved only if we have a full understanding of the source of various band shapes and their intensities and the experimental circumstances. The next section will deal with these issues.

2.5   BAND INTENSITIES AND BAND SHAPES

When an oscillating dipole moment in a molecule interacts with the electric vector of electromagnetic radiation, an absorption or scattering band is produced only if there are dipole moment or polarizability changes during the vibration. The absorption bands appearing in the IR region of the spectrum arise from rotational and vibrational transitions in the molecule. If the changes in the dipole moment during a vibration are strong, the intensity of the absorption bands increases. The intensity of the absorption bands for the IR-active vibrations is expressed by the formula

$$I = \frac{N}{3\mathbb{C}^2}\left[\frac{\partial\mu_d}{\delta Q}\right]^2 \tag{2.11}$$

where N is a constant, $\mathbb{C}$ is the velocity of light, μ_d is the dipole moment, and Q is the normal coordinate associated with the particular vibrational mode. For a stretching mode, the normal coordinate is a distance between two vibrating atoms, whereas for the bending mode, it is an angle between two bonds of three atoms. Usually the shape of the absorption bands for molecules in liquid and solid states is fit by a Lorentzian shape curve, which is given by the following equation:

$$\ln\left(\frac{I_0}{I}\right) = \frac{a}{(v - v_m)^2 + b^2} \tag{2.12}$$

where I_0 and I are the intensities of the incident and transmitted light at a given frequency v, respectively; v_m is the frequency of the band at maximum; and a and b are constants. This formula allows one to fit experimental data on either side of the band maximum. In actual absorbancy measurements, it is convenient to use integrated intensities, that is, the area under the band of interest. A relationship between the band intensity and concentration of species is related by the commonly known Beer–Lambert law. This is shown by the following formula:

$$A_s = \frac{\varepsilon}{cl} = \int_{band} \ln\left(\frac{I_0}{I}\right) dv \qquad (2.13)$$

where A_s is absorbance, c is concentration, l is a pathlength, and ε is the extinction coefficient.

Equation 2.13 really reflects the integrated absorbance, that is the area under the band. It is interesting to note, however, that both the band intensity and integrated absorbance are in use. As a matter of fact, both are used interchangeably. The reader will be faced with the choice because, according to the literature, both approaches can be used for quantitative purposes. One issue to keep in mind is that the spectra must be recorded in, or converted to a transmission mode.

In order to obtain a proper area under the band, it is necessary to precisely determine the integration boundary. In a quantitative analysis, where the band in several spectra is integrated, it is first necessary to establish the boundary of integration and then to use the same region for all the bands of interest. This procedure diminishes errors and makes the quantitative analysis more reliable.

Before we continue the discussion, let us go back to Eq. 2.13 and note that the concentration units are millimoles per cubic centimeter ($mmol/cm^3$), with the pathlength in centimeters (cm) and dv in reciprocal centimeters (cm^{-1}). Using these units, the intensity difference between strong and weak bands will be anywhere between 1000 and 0.1, respectively. When a molecule absorbs infrared radiation, the intensity of the band will depend on the movement of the electronic charges during the movement of vibrating atoms. Hence, the IR intensities provide information about the distribution or redistribution of charges during molecular vibrations. By using a simple expression for the electric dipole moment for a molecule,

$$\mu_0 = \sum_i e_i r_i \qquad (2.14)$$

where e_i is the charge on the ith particle and r_i is the vector from the particle to the origin of the coordinate system, one can express, using the first two terms of the power series, the electric dipole moment for the vibrating surface species in the normal coordinates:

$$\mu_d = \mu_0 + \sum \left[\frac{\partial \mu}{\partial Q_s}\right]_0 Q_s + \cdots \qquad (2.15)$$

where the subscript zero denotes the equilibrium position value and μ_0 the permanent dipole moment.

Using Eqs. 2.15 and 2.13, it is possible to determine A_s[79]

$$A_s = \left[\frac{8\Pi^2 N_{Avo}}{3hc}\right] v_s \left|\frac{\partial \mu}{\partial Q_s}\right|^2 \tag{2.16}$$

where N_{Avo} is Avogadro's number, h and c are respectively Planck's constant and the speed of light, and v_s is the center of the band of interest in cm^{-1}. Of course, one can express the preceding relationship as follows:

$$\left|\frac{\partial \mu_d}{\partial Q_s}\right| = 0.032(A_s)^{1/2} \tag{2.17}$$

which represents the relationship between the dipole moment changes and the normal coordinates Q_s.

2.5.1 Lineshapes

Customarily, spectroscopists describe bands as sharp and intense or broad and weak or any combination of these. It is, therefore, important to have a clear picture as to the origin of the band's shapes. Since we already established the origin of band intensities, the next step is to identify the origin of the bandwidth which, in turn, determines its shape. For that reason let us go back again to the fundamental principles of a spectroscopic experiment and remind ourselves that in order for an absorption process to occur, the Bohr condition for absorption from the initial to final state must be fulfilled:

$$E_f - E_i = hcv \tag{2.18}$$

One conclusion that follows from this condition is that only a relatively small frequency range dv_{fi} is effective in producing a spectral transition. This frequency range usually defines the absorption bandwidth and depends on the population of the transition, the wavelength region, and experimental conditions. In other words, if we were to divide Δv_{fi} into sections so that each section would correspond to a spectral resolution of the experiment, and plot a population of the states within each element, we would find that the distribution obtained would have the shape corresponding to either Gaussian[80] or Lorentzian[81] curves. These are the most frequently used lineshapes. The equations representing them are given by

$$\text{Gaussian lineshape:} \quad \gamma = \gamma_0 \left[-\left(\frac{2(v-v_0)}{\delta v}\right)^2 \ln 2\right]$$

$$\text{Lorentzian lineshape:} \quad \gamma = \frac{\gamma_0}{1 + [2(v-v_0)/\rho]^2}$$

where γ_0 = band maximum, v_0 = band position, ρ = bandwidth at half maximum, δv = full bandwidth at half maximum.

It should be kept in mind, however, that although the detailed theoretical understanding of the factors governing linewidth and line broadening are particularly important in surface spectroscopy, experimental verification of the sources and contributions to line broadening are often very difficult to separate.

In essence, the difficulty is most often related to the problem of distinguishing between homogenous and nonhomogenous broadening. Before we elaborate further on these two aspects of band broadening in surface vibrational spectroscopy, let us visualize one more time a vibrating molecule attached to a metal surface. As a result of bonding, the vibration will experience damping due to interactions of the molecule with the surface. Such an effect will interapt nondisturbed vibrational energy leading to a broader distribution of vibrational states of this particular absorption. Such broadening is called *homogenous broadening*. On the other hand, when the adsorbed surface molecules are inhomogeneously distributed as a result of inhomogeneous distribution of the surface absorption sites, interactions of the oscillating dipole moments or intermolecular interactions will lead to so-called inhomogeneous or resonance broadening.

2.5.2   Homogeneous Broadening

When a molecule is adsorbed on a metal surface, the elastic waves propagating through the metal, called *phonons*, may couple with the vibrational modes of the adsorber. In other words, atoms are displaced from their positions at different times, and different amplitudes, forming a wave. As a result of bonding, damping of motion may occur because there are many atom sites giving different vibrational energies. Such mechanical damping of adsorbate oscillations will be reflected in the band broadening. This phenomenon was experimentally observed for atomic hydrogen adsorbed on tungsten[82] and theoretically treated as well.[83,84] The theoretical studies were concerned mostly with the effect of mechanical coupling on the broadening of C—O stretching frequency. It was concluded that the effect is negligible because the highest phonon frequencies were much lower than the C—O stretching vibrations, and this separation would not result in broadening. If these studies are indeed valid, it seems that such effects should be observable for the M—C stretching modes.

Another effect that contributes to homogenous line broadening is the creation of an electron–hole pair. This mechanism involves the transition to the excited state and subsequent decay either electronic through or vibronic (phonon) relaxations.[85] Apparently, if the molecule adsorbed to the surface oscillates, the oscillating dipole moment can excite the electron–hole pair, which leads to vibrational damping. However, depending on the nature of the adsorbate–surface bonding and molecular structures on the surface, there are short- or long-range interactions. While short-range interactions can be described as periodic charge fluctuations caused by a vibrating molecule, long-range interactions appear to have a negligible effect on the band broadening. Let us further consider the short-range interactions, which can be envisioned as follows. At this point it is helpful to go back to Fig. 2.6 and consider coadsorption of a molecule on a metal surface. In that case, an empty or sometimes partially filled 2Π* antibonding orbital of the CO adsorbate is significantly lowered to the Fermi level. As a result, and in the simultaneous presence of freely moving electrons on a metal surface, it may become partially filled. During the oscillating vibrational motion of the carbon and oxygen atoms, the CO bond stretches. This results in energy changes of the $2\Pi^*$ orbital. When energy is lowered, a partial filling of the orbital occurs. In contrast, when energy is higher, the electrons flow away from the molecular $2\Pi^*$ orbital. This process may continue, and because the molecule is in an adsorbed state, the effect may become even more pronounced as compared to a nonadsorbed molecule flow of electrons because of increased dipole moment changes of the molecule. This will certainly cause vibrational dumping along with the shorter lifetimes of the excited states. For example, the lifetimes estimated for CO on Cu(100) are in the range of 2×10^{-8} sec. This gives a halfwidth of about $7-8$ cm^{-1}. In general, one would expect that for metals exhibiting a higher degree of the backdonation, the charge distribution changes and, therefore lifetime broadening, are expected to be greater, giving rise to broader bands. The bandwidth will also increase with a smaller surface coverage or, conversely, the bands will become narrower with greater surface coverage due to the increased tendency and competition for backdonation.

2.5.3 Nonhomogenous Broadening

As it was pointed out earlier, one of the factors determining the motion or frequency of the oscillating dipole moments of the adsorber molecules is the distribution of individual oscillators on the surface. If the oscillating molecules are being adsorbed on the certain privileged surface areas, homogeneity of intermolecular distances along with the vibrational coupling between individual oscillators may inherently affect the lineshapes. Let us consider several analytic situations for the CO molecule and first assume that the surface coverage is low. With this in mind, let us try to predict the spectral response resulting from differences in surface ordering.

In the case of a completely ordered surface coverage, even distribution of adsorbate molecules, and a low surface coverage, one would expect to detect a perfectly symmetric and relatively sharp band. If this highly ordered surface is disturbed by a more random orientation of adsorbed molecules, more randomly oscillating species will exist with a broader distribution of dipole moment frequencies. As a result, the band intensity will become weaker at the expense of the band broadening at the low frequency. This is because a certain fraction of the previously ordered oscillating dipoles became disordered and, as a result of interactions with other dipoles, a population with the lower oscillating energy is increased. The low-frequency absorption tail is characteristic of vibrational coupling of random molecular distribution. Such predictions have been theoretically established.[86-88] However, when the adsorbed surface layer mobility is enhanced, most likely repulsive forces will increase, giving rise to weaker but highly symmetric band since the molecules try to maximize their intermolecular distances right above the surface; this is so-called lattice gas. The presence of islands is often manifested by attractive forces that cause the appearance of an additional band since now there is an equilibrium between the island species and lattice gas. This will occur in the spectrum as an additional band with a broadening at a lower frequency site. The magnitude of the band intensity ratio will depend on the relative ratio of the species occupying the islands and lattice gas, and again, the band separation is often necessary for further enhancement of the spectral information.

2.5.4   Band Separation and Resolution

Keeping in mind homogeneous and nonhomogenous lineshapes, let us return to the issue of bond resolution and separation. This is particularly important because often a spectroscopist, and especially a surface spectroscopist, is confronted with the problem of identifying an actual number of overlaping bands and, moreover, must distinguish between broadening and intensity changes resulting from structural changes and unexpected experimental factors. Obviously, the experimental conditions should be the first criterion used in this evaluation. Once experimental data are sorted out, it is found that there are several overlapping bands, and the spectroscopist faces another problem: how to isolate these hidden spectral features.

Let us consider the example presented in Fig. 2.15, which illustrates three spectra of poly(dimethylsiloxane) (PDMS) oligomers[89] with the repeating unit x ranging from 4 to 20, such as shown below.

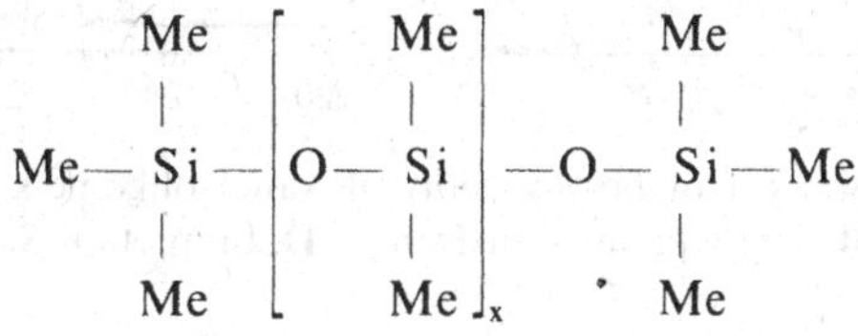

$$
\text{Me}-\underset{\underset{\text{Me}}{|}}{\overset{\overset{\text{Me}}{|}}{\text{Si}}}-\left[\text{O}-\underset{\underset{\text{Me}}{|}}{\overset{\overset{\text{Me}}{|}}{\text{Si}}}\right]_x-\text{O}-\underset{\underset{\text{Me}}{|}}{\overset{\overset{\text{Me}}{|}}{\text{Si}}}-\text{Me}
$$

(B)   (A)

It is quite apparent that the main band at $1260\,\text{cm}^{-1}$, due to symmetric deformation modes of methylene groups (A), exhibits a shoulder due to terminal methylene groups (B) of the oligomer backbone. It is also noted that the shoulder increases as the number of repeating units decreases. In an effort to resolve these two bands, Fourier self-deconvolution procedure (discussed later in this chapter) has been applied. The spectra presented on the right-hand side are the results of this operation. At first glance, it is seen that the band that appeared as a shoulder is now clearly resolved, which might have been an indication of a better resolution. Of course, the spectral resolution as such is determined by the experimental conditions and connot be changed. What really allows us to resolve these bands is a narrower width of the bands and a relative increase with respect to the baseline of the band intensities due to deconvolution. Because deconvolution methods have received so much attention in the literature, they will be discussed below. It

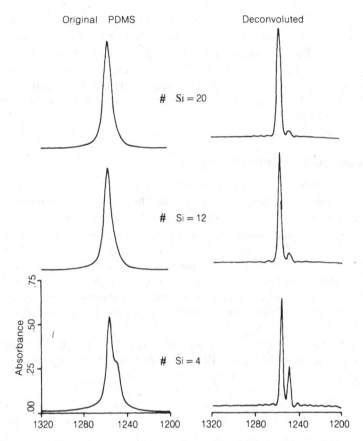

Figure 2.15   FT–IR spectra of poly(dimethylsiloxane) oligomers with 4, 10, and 20 rupiting units. [Reprinted with permission, from E. D. Lipp, *Appl. Spectrosc.*, 1986, **40** (7), 1009.]

should be kept in mind, however, that these techniques have nothing to do with surfaces. The advice is to convert all spectra to transmission or adsorption and, necessary, use deconvolution methods.

2.6   DECONVOLUTION METHODS

The main purpose of deconvoluting the spectra is to increase the level of chemical and structural knowledge of a given system. While the example presented in the previous section demonstrates how one can determine the length of the polymer backbone, in many situations one would like to determine the number of structural components and separate them. For that reason several techniques have been developed.

2.6.1 Subtraction and Ratio Method

This is the simplest technique of isolating spectral components in which two spectra, $A_1(v)$ and $A_2(v)$, are subtracted from each other in order to find the spectral features representing new structures. Mathematically, this can be expressed by a simple relationship:

$$A_1(v) - kA_2(v) = 0 \qquad (2.19)$$

It should be kept in mind, however, that the following assumptions and conditions should be fulfilled: the band shape and intensities do not change with the optical thickness and individual components do not interact with each other differently at different concentrations. This simple relationship demonstrates that if the result of subtraction produces a flat line with no negative or positive features, both $A_1(v)$ and $A_2(v)$ spectra therefore represent the same surface structures. The situation will be quite different, however, for a heterogeneous system consisting of two interacting phases. Under such circumstances, the composite spectrum will consist of not only the two phases but also the third component resulting from the interactions between the two phases. As an example, let us assume that the total surface spectrum is $A_t(v)$ and represents a sum of three components:

$$A_t(v) = A_b(v) + A_s(v) + A_i(v) \qquad (2.20)$$

where $A_b(v)$ and $A_s(v)$ are the pure absorbance spectra of the bulk and surface, respectively, and $A_i(v)$ is the spectrum resulting from interactions between the two components. Assuming that the $A_t(v)$ was obtained with a given defined surface coverage, the next step is to change it, and obtain new composite spectrum $A_{1t}(v)$:

$$A_{1t}(v) = A_{1b}(v) + A_{1s}(v) + A_{1i}(v) \qquad (2.21)$$

This composite spectrum represents a new relationship with the increased surface coverage or concentration on the surface. Subtracting each side of Eq. 2.20 from each side of Eq. 2.21 gives

$$(A_{1t} - kA_t) = (A_{1b} - kA_b) + (A_{1s} - kA_s) + (A_{1i} - kA_i) \qquad (2.22)$$

After choosing k such that $(A_{1b} - kA_b)$ is equal to zero, we obtain the interaction spectrum. It should be remembered that the following assumptions are made: the absorbance and shape of the bands do not change with the surface coverage and the components do not interact with each other differently at different relative concentrations or surface coverages. These assumptions impose rather drastic limitations on the surface subtractions because, as we have seen before, surface coverages may have a significant effect on the band shape and other spectral features. However, if for a given system these assumptions are valid, the subtraction can be conveniently used in determining the interaction spectrum.

In 1976, Hirschfeld[90] presented the ratio method that, in a nutshell, can be described as follows. If the spectrum is given by

$$A(v) = \sum_{i=1}^{n} a_i(v) \tag{2.23}$$

where $a_i(v)$ is the spectrum of the pure component i, and there is a change in relative concentration, the new spectrum becomes

$$A_i(v) = \sum_{i=1}^{n} k_i a_i(v) \tag{2.24}$$

Taking the ratio

$$R(v) = \frac{A_i(v)}{A(v)} = \frac{k_1 a_1(v) + k_2 a_2(v) + \cdots + k_n a_n(v)}{a_1(v) + a_2(v) + \cdots + a_n(v)} \tag{2.25}$$

and finding the k_n coefficients in the nonabsorbing areas allows us to determine the spectra of pure components.

2.6.2 Factor Analysis

Although subtraction or ratio methods can be successful when dealing with the relatively simple systems, the presence of several spectral components may be troublesome. Under such circumstances, it is necessary to use more advanced computational techniques. The first one is known as *factor analysis* or *principal component analysis* because it finds a set of abstract spectra called factors that can be combined to reproduce the original spectra. Each spectrum consists of a number of intensity values, and the frequencies at which these intensities are measured are the same for all spectra. As we have seen in Chapter 1, such data can be treated as a matrix. Starting with the Beer–Lambert law for a number of components j we obtain

$$A_i = \sum k_j c_{ij} \tag{2.26}$$

where A_i is the absorbance spectrum of mixture i, k_j is the absorptivity for the j th component, and c_{ij} is the concentration of component j in mixture i. The same equation can be rewritten in a form of matrix notation

$$\mathbb{A} = \mathbb{K}\mathbb{C} \tag{2.27}$$

where $\mathbb{A}$ is a normalized rectangular absorbance matrix containing absorbances at each wavenumeber in columns and the number of mixtures under investigation in rows. $\mathbb{K}$ is the molar absorption coefficient matrix with the number of rows corresponding to the number of mixtures (the same as in the $\mathbb{A}$ matrix). The $\mathbb{C}$

matrix is the concentration matrix with the dimensions of the number of components times the number of mixtures being studied.

Our goal is to determine the number of individual components contributing to an experimentally observed spectrum. Two assupmtions are used in factor analysis: (1) individual spectra of each pure component are not linear combinations of the other components, and (2) the concentration of species cannot be represented as a ratio of other species. With this in mind, the first step is to establish the rank of matrix A, as this rank is equal to the number of absorbing components, or in other words, determine the number of components required to reproduce the data matrix A (absorbance as a function of wavenumber).

As a first step in finding the rank of A, the matrix $M = A'A$ is created, where A' is the transpose of A and is often called the *covariance matrix*. One important property of this matrix is a square dimension, which represents the number of mixtures being examined. The rank of A is equal to the number of absorbing components. One method of finding the rank of M (or estimating k) is to plot the values of the eigenvalues, as a function of the number of eigenvalue and look for singular values. Assuming that the signal-to-noise ratio is high, and this is always helpful, a large drop between $k + 1$ and the subsequent values would be expected. When dealing with the high noise level and with a larger number of pure component spectra, things become rather hopeless.

In fact, it should be kept in mind that when you put garbage in, you will get garbage out, or using the causality principle, if there is no input, there will be no output. Let's face it, the numerical procedures can enhance spectral information as long as the signal-to-noise ratio is at the acceptable level, and with the availability of proper software, the applications of factor analysis are possible and have been shown by at least one research group.

Knowing the number of components n in a mixture, we know the dimension of the vector space and can obtain a set of mutually orthogonal abstract eigenspectra. The spectra define an n-dimensional coordinate system within each data spectrum, and each pure component spectrum is represented by points in space. At this point, the abstract spectra can be rotated to physically real components using the restriction that each pure component contains a peak in a region where the other pure components do not absorb. Although factor analysis can be quite useful in many situations,[91][93] the high signal-to-noise ratio is one of the essential factors leading to successful solutions. Morever, the problem of determining higher-dimension vector spaces becomes increasingly difficult with dimensions.

2.6.3 Nonlinear Methods

Following a confrontation with the problem of identifying unknown components of a composite spectrum, a spectroscopist would like to "purify" the spectra, that is, to isolate pure components from the mixture. One of the numerous approaches for extraction of pure component spectra and subsequent quantitative analysis is a nonlinear method developed by Koenig et al.[94] No a priori knowledge of the spectra of the components and their concentrations is required in this approach. These nonlinear algorithms were applied to poly(ethylene terethalate) and extraction of the "pure" spectra of gauche, the crystalline trans, and the amorphous trans isomers.

When information about only one component is available, the method of rank annihilation has proved to be successful.[95] In the case of multicomponent system, the rank of the correlation matrix M is set to be equal to the number of components. With the knowledge of the components N, by subtracting correct N from M, we will end up with the rank of M equal to unity. Under such circumstances, we will obtain the eigenvalue of M corresponding to N equal to zero. Knowing a known pure component in the mixture, we can express the process of rank annihilation can be mathematically by the residual matrix E as

$$E = M - \beta N \tag{2.28}$$

where M is the correlation matrix equal to $A^T A$. When the correct amount β of the known component N is subtracted, the rank of the matrix E will decrease by one.[96] This can be monitored by changing β and watching the changes of the eigenvalues of $E^T E$.

If one needs to analyze a complex and unknown mixture when one component is known, a helpful method is the cross-correlation method.[97] This method evaluates the similarity between the spectra of the two different systems, such as spectrum of a sample and a reference. The approach is essentially used to generate "reference spectra" from the knowledge of the sample's composition instead of the approach of using multilinear regression to generate a linear combination of the compositions of the mixture. The technique can be used in such circumstances where background fluctuations exceed the differences due to the compositional changes in the sample as well as to reconstruct the spectra of the pure components from the mixture spectra when the pure component spectra are not available.[98]

2.6.4 Maximum Likelihood Entropy Method

Although several spectral deconvolution procedures have been developed, many of them require prior knowledge of some spectral feature. The primary functions of deconvolutions are to enhance the intensities of overlapping bands. This selective enhancement of intensities with respect to the spectral baseline leaves us with a visual illusion that the spectral resolution has been improved. This is not, however the case. The restored bands in the spectrum may appear to be better resolved, but in reality, because their intensities are greater, the definition of individual spectral features is enhanced. With this in mind, let us briefly review the principles that govern the maximum likelihood restoration approach.[99]

The maximum likelihood (ML) criterion is often based on a Bayesian approach in which all data and prior information are included in the estimation of the true spectrum. Philosophically, it is like maxima–a posteriori estimation, which uses the maximum probability given the data and prior information. In spectroscopy, prior knowledge includes physical knowledge about the spectrum and its statistical nature of having an unknown number of overlapping bands. Usually, the prior knowledge is built into deconvolution algorithms such that one needs to provide a number of bands in the spectra and their definitions. However, in estimating unknown spectra, all possible physical and statistical prior knowledge should be utilized. Moreover, the approach should be physical. In the case of photons being absorbed by a sample, the Bose–Einstein approximation is usually the most suitable one. Such criteria will give the estimated spectrum the following property: if all known deterministic and statistical conditions for obtaining a given spectrum are repeated many times, the spectrum that occurred the largest number of times would coincide with the ML estimation. Depending on prior knowledge, the ML estimator can take various forms.

Because each recorded spectrum is blurred by various instrumental problems such as signal-to-noise ratio, the ultimate goal is to deconvolute and restore the spectral image consisting of O_m; $m = 1 \ldots M$ or $\{O_m\}$ intensities. Each of these intensities is at wavenumber $\{v_m\}$. Since $\{v_m\}$ can be subdivided depending on instrumental resolution, each v_m is centered around a resolution element of the instrument

$$v_m = m\Delta v_m \tag{2.29}$$

At a given v_m, the energy per photon is hv_m. If n_m denotes the number of photons radiating from the mth element, O_m obeys

$$O_m = n_m h v_m \tag{2.30}$$

If N photons contribute to the intensity of particular band, then

$$\sum_{m=1}^{M} n_m = N \tag{2.31}$$

Now we must estimate $\{n_m\}$ so that it will obey normalization conditions 2.31 and will resemble the physical appearance of recorded spectrum. Once the $\{n_m\}$ are known, Eq. 2.30 can be utilized to estimate $\{O_m\}$. Obtained in such a way, the deconvoluted band or bands in a spectrum are the most probable number of sets $\{n_m\}$. This is called the *maximum likelihood* (ML) estimate of the object. It obeys the rule that $P(n, \ldots, n_m) = $ maximum, where P is the probability of occurrence.

Under prior knowledge conditions, and assuming certain noise level, it is possible to determine the probability of the spectrum occurrence. The final expression that leads to restoration of a given spectrum is as follows:

$$\sum_m n_m \ln\left(\frac{Q_m}{n_m}\right) - \sum_m \frac{\epsilon_m^2}{2e_m^2} + \sum_m \lambda_m\left(i_m - h\sum_n n_n v_n S_{nm} - \epsilon_m\right) + \mu\left(\sum_m n_m - N\right)$$

$$= \text{maximum} \tag{2.32}$$

While the first sum estimates the spectrum utilizing the principle of maximum cross entropy, the second term estimates the noise by least-squares analysis. The solution $\{n_m\}$ and $\{\epsilon\}$ to Eq. 2.32 is obtained by setting $\partial/\partial\epsilon_m$ equal to zero with all $\{\epsilon_m\}$ fixed and by setting $\partial/\partial\epsilon_m$ equal to zero with all n_m fixed. As a result

$$n_m = Q_m \exp\left(-1 - \mu - hv_m \sum_n \lambda_n s_{nm}\right) \tag{2.33}$$

and

$$\epsilon_m = -\lambda_m e_m^2 \tag{2.34}$$

The two unknown parameters, μ and $\{\lambda_m\}$, can be found by substituting Eqs. 2.33 and 2.34 for $\{n_m\}$ and $\{\epsilon_m\}$ into the constraint equations. This results in the following relationships:

$$i_m = h \sum_n v_v s_{mn} Q_n \exp\left(-1 - \mu - hv_n \sum_k \lambda_k s_{kn}\right) - \lambda_m e_m^2 \tag{2.35}$$

$$N = \sum_n Q_n \exp\left(-1 - \mu - hv_n \sum_k \lambda_k s_{kn}\right) \tag{2.36}$$

Under these circumstances, the system of $M + 1$ equations and $M + 1$ unknowns μ and $\{\lambda_m\}$ may be solved by utilizing the Newton–Raphson iteration method. By setting initial values of $\lambda_m = O$ and $\mu = N$, the solution is usually obtained within 10–12 iterations. At this point one needs to use prior knowledge about the spectrum (Q_m) and the noise (e_m). By making all Q_m equal, we set up the spectrum to be a flat line, which implies that we have a maximum conviction case, and the prior knowledge should exert a smoothing influence on the solution. Thus, the spectrum is restored by the principle of maximum entropy.

In practice, the ML approach requires the band shape definition, which can be either Lorentizian or Gaussian or taken directly from the spectrum. In many cases, when heavily overlapping bands make the definition of the entire band impossible, one can generate the band from its halfwidth by calculating the second half. Defined in such a way, one can use the band to restore a spectrum of interest, without defining the number of expected bands. It should be noted that the spectrum of interest needs to be baseline corrected to a zero position since any nonzero intensity baseline will be treated as prior knowledge. This again reinforces the statement that the resolution increase is an illusion obtained by the enhancement of intensities with the use of physical and prior knowledge. A few examples utilizing the ML approach demonstrated that there are numerous areas of applications where signal-to-noise ratio is a problem.[100] A good signal-to-noise ratio is preferable.

2.7 SUMMARY

The intention of this chapter was to get a feeling for the magnitude of the problems that we usually face during the analysis of the surface vibrational spectra, plus outline the methods and diagnostic techniques that allow us to resolve some of these problems. Of course, the way to look at it is to establish what caused certain spectral features and, on that basis, approach the problem using proper deconvolution methods. The reader should be continuously aware that understanding the origin of the spectral changes is the single most important item.

REFERENCES

1. L. H. Little, *J. Chem. Phys.*, 1961, **34**, 342.
2. L. H. Little, H. E. Klauser, and C. H. Amberg, *Can. J. Chem.*, 1961, **39**, 42.
3. D. R. J. Boyd, H. W. Thomson, and R. L. Williams, *Proc. Roy. Soc.*, 1952, **A212**, 42.
4. N. Sheppard and D. J. C. Yates, *Proc. Roy. Soc.*, 1956, **A238**, 69.
5. C. Cumming, *Can. J. Chem.*, 1955, **33**, 635.
6. G. Herzberg, *Infrared and Raman Spectra*, Van Nostrand, New York, 1945.
7. I. V. Demidenkova and L. D. Shcherba, *Izv. Akad. Nauk SSSR, Ser. Fiz.*, 1955, **22**, 1122.
8. C. G. Cannon, *Spectrochim. Acta*, 1958, **10**, 425.
9. F. P. Reding and D. F. Hornig, *J. Chem. Phys.*, 1951, **19**, 594.
10. G. C. Pimentel, M. O. Bulanin, and M. Van Thiel, *J. Chem. Phys.*, 1962, **36**, 500.
11. N. W. Cant and L. H. Little, *Can. J. Chem.*, 1964, **42**, 802.
12. D. J. C. Yates and P. J. Lucchesi, *J. Chem. Phys.*, 1961, **35**, 243.
13. M. W. Urban and E. M. Salazar-Rojas, *J. Appl. Polym. Sci.*, 1990, **28**, 1593.
14. J. Pritchard, in *Vibrations in Adsorbed Layers*, Conference Records Series KFA, H. Ibach and S. Lehwald, eds., 1978.
15. N. V. Richardson and A. M. Bradshaw, *Surf. Sci.*, 1979, **88**, 255.
16. O. S. Heavens, *Thin Film Physics*, Methuen, London, 1970.
17. M. Schefler, *Surf. Sci.*, 1979, **81**, 562.
18. A. A. Lucas and G. D. Mahan, *Surf. Sci.*, 1978, **68**, 121.
19. G. D. Mahan and A. A. Lucas, *J. Chem. Phys.*, 1978, **68**, 1344.
20. S. Efrima and H. Metiu, *Surf. Sci.*, 1980, **92**, 433.
21. S. Efrima and H. Metiu, *Surf. Sci.*, 1981, **108**, 329.
22. G. Blyholder, *J. Phys. Chem.*, 1964, **68**, 2773.
23. G. Blyholder, *J. Phys. Chem.*, 1975, **79**, 756.
24. E. J. Baerends and P. Ros, *Intr. J. Quantum Chem.*, 1978, **12**, 169.
25. D. E. Ellis, E. J. Baerends, H. Adachi, and F. W. Averill, *Surf. Sci.*, 1977, **64**, 649.
26. P. S. Bagus and B. O. Roos, *J. Chem. Phys.*, 1981, **75**, 5961.
27. J. C. Bertolini, G. Dalmai-Imelik, and J. Rousseau, *Surf. Sci.*, 1977, **68**, 539.
28. H. Ibach and G. Somorjai, *Appl. Surf. Sci.*, 1979, **3**, 293.
29. M. Primet, J. M. Basset, M. V. Mathieu, and M. Prettre, *J. Catal.*, 1973, **29**, 213.
30. E. L. Garfunkel, J. E. Cromwell, and G. A. Somorjai, *J. Phys. Chem.*, 1982, **86**, 310.
31. S. Efrima and H. Metiu, *Surf. Sci.*, 1981, **109**, 109.
32. S. Efrima, *Surf. Sci.*, 1982, **114**, L29.

33. A. Ortega, F. M. Hoffmann, and A. M. Bradshaw, *Surf. Sci.*, 1982, **119**, 79.

34. R. P. Eischens and W. A. Pliskin, *Adv. Catal.*, 1958, **10**, 1.

35. R. M. Hammaker, S. A. Francis, and R. P. Eischens, *Spectrochim. Acta*, 1965, **21**, 1295.

36. M. Scheffler, *Surf. Sci.*, 1979, **81**, 562.

37. G. D. Mahan and A. A. Lucas, *J. Chem. Phys.*, 1978, **68**, 1344.

38. M. Moskovits and J. E. Hulse, *Surf. Sci.*, 1978, **78**, 397.

39. P. Hollins and J. Pritchard, *Surf. Sci.*, 1979, **89**, 489.

40. H. Conrad, G. Ertl, J. Koch, and E. E. Latta, *Surf. Sci.*, 1974, **43**, 462.

41. R. M. Hammaker, S. A. Francis, and R. P. Eischens, *Spectrochim. Acta*, 1965, **21**, 1295.

42. M. Moskovits and J. E. Hulse, *Surf. Sci.*, 1978, **78**, 397.

43. R. M. Hammaker, S. A. Francis, and R. P. Eischens, *Spectrochim. Acta*, 1965, **21**, 1295.

44. H. Pfnür, D. Menzel, F. M. Hoffmann, A. Ortega, and A. M. Bradshaw, *Surf. Sci.*, 1980, **93**, 431.

45. M. Scheffler, *Surf. Sci.*, 1979, **81**, 562.

46. G. D. Mahan and A. A. Lucas, *J. Chem. Phys.*, 1978, **68**, 1344.

47. S. Efrima and H. Meitu, *Surf. Sci.*, 1981, **108**, 329.

48. S. Efrima and H. Meitu, *Surf. Sci.*, 1981, **109**, 109.

49. B. N. J. Persson and R. Ryberg, *Phys. Rev.*, 1981, **B24**, 6954.

50. M. Moskovits and J. E. Hulse, *Surf. Sci.*, 1978, **78**, 397.

51. G. Blyholder, *J. Phys. Chem.*, 1964, **68**, 2773.

52. G. Blyholder, *J. Phys. Chem.*, 1975, **79**, 756.

53. S. Efrima and H. Metiu, *Surf. Sci.*, 1981, **109**, 109.

54. S. Efrima, *Surf. Sci.*, 1982, **114**, L29.

55. R. Hoffmann, *Acc. Chem. Res.*, 1977, **4**.

56. A. M. Bradshaw and F. M. Hoffmann, *Surf. Sci.*, 1978, **72**, 513.

57. J. C. Tracy and P. W. Palmberg, *J. Chem. Phys.*, 1969, **51**, 4852.

58. A. Ortega, A. Garbout, F. M. Hoffmann, W. Stenzel, R. Unwin, K. Horn, and A. M. Bradshaw, *Proc. IVC-8, ICSS-4*, and *ECOSS-3*, Cannes, 1980.

59. R. M. Hammaker, S. A. Francis, and R. P. Eischens, *Spectrochim. Acta*, 1965, **21**, 1295.

60. R. Shigeishi and D. A. King, *Surf. Sci.*, 1976, **58**, 379.

61. K. Horn and J. Pritchard, *J. Physique*, 1977, **38**, C4–C164.

62. R. Shigeishi and D. A. King, *Surf. Sci.*, 1976, **58**, 379.

63. A. Crossley and D. A. King, *Surf. Sci.*, 1977, **68**, 528.

64. A. Crossley and D. A. King, *Surf. Sci.*, 1980, **95**, 131.

65. P. Hollins and J. Pritchard, *Surf. Sci.*, 1979, **89**, 489.

66. B. N. Person and R. Ryberg, *Phys. Rev.*, 1981, **B24**, 6954.

67. G. D. Mahan and A. A. Lucas, *J. Chem. Phys.*, 1978, **68**, 1344.

68. M. Scheffler, *Surf. Sci.*, 1979, **81**, 562.

69. P. Hollins and J. Pritchard, *Surf. Sci.*, 1979, **89**, 489.

70. P. Hollins and J. Pritchard, *Chem. Phys. Lett.*, 1980, **75**, 378.

71. P. Hollins and J. Pritchard, *Surf. Sci.*, 1979, **89**, 489.

72. R. L. Park and H. H. Madden, *Surf. Sci.*, 1968, **11**, 188.

73. R. J. Bhem, K. Christmann, and G. Ertl, *J. Chem. Phys.*, 1980, **73**, 2984.

74. J. C. Tracy and P. W. Palmberg, *J. Chem. Phys.*, 1969, **51**, 4852.

75. H. Pfnür, D. Menzel, F. M. Hoffmann, A. Ortega, and A. M. Bradshaw, *Surf. Sci.*, 1980, **93**, 431.

76. T. E. Madey and D. Manzel, *Jpn. J. Appl. Phys.*, Suppl. 2, Part 2, 1974, 229.

77. M. Scheffler, *Surf. Sci.*, 1979, **81**, 562.

78. P. Hollins, *Surf. Sci.*, 1981, **107**, 75.

79. L. A. Gribov, *Intensity Theory for Infrared Spectra of Polyatomic Molecules*, Consultants Bureau, New York, 1964.

80. H. S. Gold, *Anal. Chem.*, 1976, **48**, 1540.

81. K. Okata and H. Ishida, *Appl. Spectrosc.*, 1988, **42**, 952.

82. Y. J. Chabal and A. J. Sievers, *Phys. Rev. Lett.*, 1980, **44**, 944.

83. H. Metiu and W. E. Palke, *J. Chem. Phys.*, 1978, **69**, 2574.

84. B. N. J. Persson, *J. Phys.*, 1978, **C11**, 4251.

85. J. W. Gadzuk, *Phys. Rev.*, 1979, **B20**, 515.

86. A. A. Lucas and G. D. Mahan, in *Vibrations in Adsorbed Layers*, H. Ibach and S. Lehward, Eds., KFA, Jülich, 1978.

87. M. Moskovits and J. E. Hulse, *Surf. Sci.*, 1978, **78**, 397.

88. B. N. J. Persson and R. Ryberg, *Phys. Rev.*, 1981, **B24**, 6954.

89. E. D. Lipp, *Appl. Spectrosc.*, 1986, **40**(7), 1009.

90. T. B. Hirschfeld, *Anal. Chem.*, 1976, **48**, 721.

91. E. R. Malinowski and D. G. Mowery, *Factor Analysis in Chemistry*, Wiley, New York, 1980.

92. J. L. Koenig and M. J. M. Tovar, *Appl. Spectrosc.*, 1981, **35**, 543.

93. S. B. Lin and J. L. Koenig, *J. Polym. Sci., Polym. Phys. Ed.*, 1982, **20**, 2277.

94. J. Liu and J. L. Koenig, *Anal. Chem.*, 1987, **59**, 2609.

95. C.-N. Ho, G. D. Chistian, and E. R. Davidson, *Anal. Chem.*, 1978, **50**, 1108.

96. C.-N. Ho, G. D. Chistian, and E. R. Davidson, *Anal. Chem.*, 1980, **52**, 1071.

97. G. Horlick, *Anal. Chem.*, 1973, **45**, 319.

98. D. E. Honigs, G. M. Hieftje, and T. Hirschfeld, *Appl. Spectrosc.*, 1982, **36**, 223.

99. S. Davies, K. J. Packer, A. Baruya, and A. I. Grant, in *Maximum Entropy in Action*, B. Buck and V. A. Macaulay, Eds., Oxford Sci. Publ., 1991.

100. L. K. DeNoyer and J. G. Dodd, *Amer. Lab.*, March 1990.

CHAPTER 3

EXPERIMENTAL VIBRATIONAL
SURFACE TECHNIQUES

3.1 INTERACTIONS OF LIGHT WITH MATTER

The primary intent of this chapter is to present an introduction to various sampling techniques of IR and Raman spectroscopy. It is, however, by no means intended to be all-inclusive; after all, whole books and monographs have been written on highly specialized techniques and theories. The main purpose of this chapter is to provide the basis for a novice spectroscopist to be able to "move around" and provoke questions related to each particular technique described. Furthermore, it should be realized that only proper use of these methods will lead the reader to successful utilization of a particular sampling approach. However, before we can address the issues related to surface-related techniques, it is necessary to realize why we have the opportunity to use vibrational spectroscopy in the surface and interfacial studies. Sensitivity as well as the signal-to-noise ratio were the key issues in the introduction of Fourier transform techniques.

In classifying surface techniques, let us consider the process of light interaction with a matter. Using a simplistic view, such as that depicted in Fig. 3.1, it is quickly recognized that depending on the experimental configuration, various rays may be detected and the total amount of incident energy can be represented as a sum of reflected, elastically and inelastically scattered, and absorbed rays as a function of wavelength. For example, when measuring inelastically scattered photons, which is the principle of Raman scattering, the source and the detector are on the same side of a sample although other configurations are possible. In contrast, transmission IR spectroscopy is an absorption process, and the detector is located on the opposite side of the detector because I_T/I_0 is measured. This however, is not the case for most surface IR measurements where surface selectivity is required, imposing limitations on the source–sample–detector geometry.

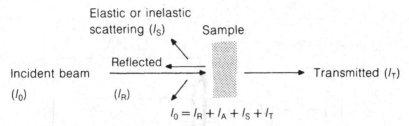

Figure 3.1   Interaction of light with materials. Positioning and sensitivity of source–detector configuration will decide which portion of light will be detected after light impinges on a sample.

3.2   INFRARED SPECTROSCOPY: DISPERSIVE VERSUS TRANSFORM SPECTROSCOPY

Infrared spectroscopy has long been recognized as a powerful method for the characterization of materials. Since it is based on the absorption of radiation in the IR frequency range due to vibrations of molecules and atoms within the vicinity of the molecular segments, it provides valuable structural and quantitative information. Prior to FT-IR (Fourier transform–infrared spectroscopy), IR spectroscopy was carried out using a dispersive instrument utilizing, initially, a moving in front of the detector slit. Later on, the stationary slits with moving prisms or gratings to dispersed IR radiation were employed instead. We will return to the concept of the movable slit system because it formulates the basis for Hadamard spectroscopy. In the case of the movable prism or grating system, a scanning mechanism is calibrated for the wavenumbers and interfaced to a recorder, the spectrum, that is, the energy transmitted through a sample as a function of a wavenumber, was obtained. All these methods, however, have limitations because slits would cut about 95% of initial radiation. Therefore, both sensitivity and resolution are poor.

Broadly defined, the difference between dispersive spectrometry and interferometry is a difference of domains. While both techniques can use the same sources, the spectra are acquired by different means. A dispersive spectrometer utilizes a grating or a prism to disperse radiation from the source into spectral elements. A detector then measures the energy of each spectral element. This is schematically illustrated in Fig. 3.2A. Because a dispersive spectrometer measures the energy from each scattered element along with the frequency domain calibrated in the units of turning velocity of a dispesive element, vibrational and rotational bands are observed directly from the spectrum. An interferometric detection, on the other hand, involves splitting radiation from the source into two optical paths. The radiation from each path is reflected by mirrors and returned along the same path to recombine constructively or destructively, depending on the phase difference of the two optical paths or the distance from the beamsplitter.

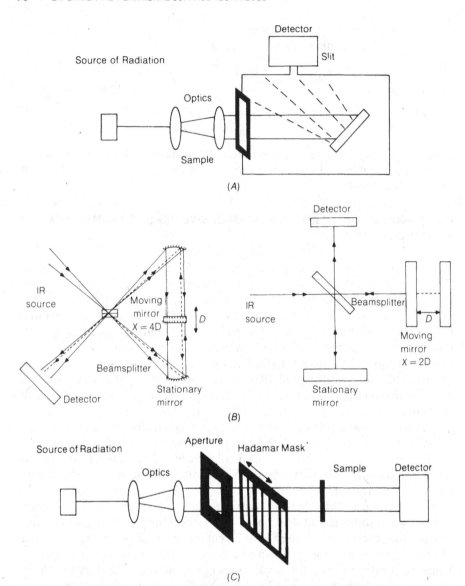

Figure 3.2   (*A*) A schematic representation of dispersive IR instrument; (*B*) an illustration of the two commonly used Fourier transform configurations—Michelson and Ganzel interferometers; (*C*) a schematic representation of Hadamar transform interferometer.

The resulting output is an interferogram, that is the amount of energies as a function of optical path difference. In contrast to dispersive spectroscopy, the spectrum is presented this time in the energy domain, usually expressed in reciprocal centimeters. In an effort to obtain spectrum, the digitized interferogram is obtained using Michelson or Ganzel interferometers shown in Fig. 3.2*B*.

The difference between these two is the optical paths of the light. Because of convenience, the Michelson interferometer is the most commonly used interferometer. This interferometer has two mutually perpendicular mirrors, one stationary and one movable. Bisecting the two vertical planes of both mirrors is a beamsplitter, a divider of source beam into two equal beams travelling along both vertical planes of the interferometer. When incident beams hit a beamsplitter, 50% of the light is reflected and 50% is transmitted throughout the beamsplitter. These two beams travel along the two vertical paths and are reflected back to the beamsplitter, where they recombine and are redirected to the detector. Now, depending on the distance of the movable mirror from the beamsplitter, the paths of both light beams may be the same or may be different. When the movable and stationary mirrors are at the same distance from the beamsplitter, the path lights will be identical resulting in a coherent addition on the beamsplitter, producing a maximum flux at the detector. This maximum flux is often referred to as the "center burst." On the other hand, when the movable mirror is displayed from an equal distance, the optical paths of both beams will change, and this difference will cause each wavelength of source radiation to destructively interact with itself at the beamsplitter. What the detector is going to "see" is the sum of the fluxes for each individual wavelength. For a monochromatic source of frequency v, the interferogram is a cosine function with frequency v, whereas for a polychromatic source it is a sum of many individual cosine interference patterns, which can be represented by the following expression:

$$I(x) = \sum_x A(v)[1 + \cos(2\Pi vx)]dv \qquad (3.1)$$

Solving this integral involves a determination of $I(x)$ at $x = 0$ and $x = \infty$, and comparing the difference, which is the actual interferogram expressed as

$$F(x) = I(x) - I(\infty) = \sum_x A(v)\cos 2\Pi vx\, dx \qquad (3.2)$$

Now taking Fourier transform, one can obtain $A(v)$, that is, an actual spectrum. As one may note, this approach involves a conversion from time to frequency (or energy) domains.

In any process of interaction of electromagnetic radiation with matter, it is important to realize not only the advantages of each situation but also the limitations. Although FT-IR provides an excellent opportunity for the study of surfaces, there are potential traps where band distortions or intensity changes might have been interpreted as chemical or physical changes within the studied system. In reality they might have been simply artifacts attributed to various reflection or scattering processes.

Other types of transformation utilized in spectroscopic measurements are essentially the concept of movable slip in front of the detector such as that shown in Figure 3.2C. This concept is based on the same principle as the lighted letters that move in a street light commercial or signs advertising casinos in Las Vegas.

The switching on and off of the light bulbs that form a lighted sign make the letter or message appear to move in a certain direction. If we visualize a slit or a mask moving in front of the detector, the detector will detect different wavelengths in a time of movement, similar to our eyes detecting a moving lighted sign. The signal is then Hadamard-transformed and converted to a spectrum.

3.3   EXPERIMENTAL INFRARED SPECTROSCOPY

The remaining portions of this chapter are meant to provide a critical look at the instrumental principles utilized in vibrational spectroscopy of surfaces. Although at this point FT-IR spectoscopy dominates the spectroscopy world, in principle, all techniques discussed below can apply to the dispersive as well as Hadamard transform modes.

3.3.1   Transmission Spectroscopy

This is the most commonly used IR technique to obtain IR spectra of powders in which a sample is dispersed in the IR-transparent medium. In a typical experiment, a small amount of sample is ground with IR-transparent powder (KBr, CsI, NaCl, etc.), which is pressed together to form a pellet. Although this sampling

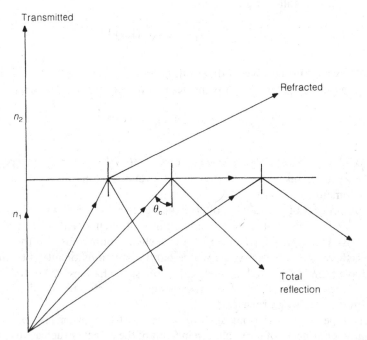

Figure 3.3   The effect of refractive index changes on the light path as a function of incidence angle.

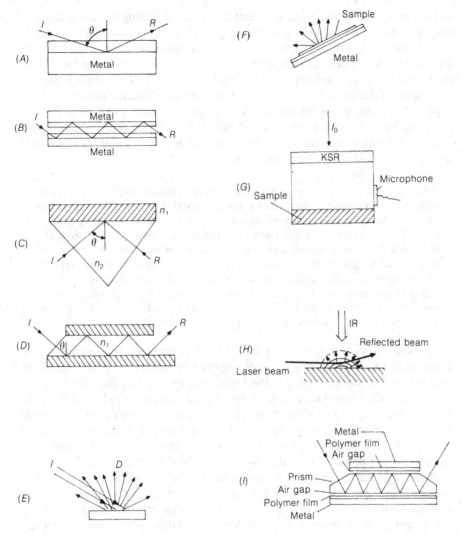

Figure 3.4 Surface IR techniques: (*A*) single reflection RA (reflection–adsorption) setup—light penetrates the sample and is reflected by the metal mirror θ should be between 75 and 98.5°); (*B*) multiple reflection RA setup—light penetrates the sample and is reflected by a metal mirror, (*C*) single reflection IRS (internal reflection spectroscopy) setup; light passes through the IRS element and is totally reflected at ($\theta > \theta_c$ ($n_1 \sin \theta_c = n_2 \sin 90°$); (*D*), multiple-reflection setup (attenuated total reflectance); (*E*), diffuse reflectance setup; diffusively scattered light is scattered and collected by mirrors and redirected to the IR detector; (*F*) emission spectroscopy setup—the sample is heated and the emitted light is analyzed by the IR detector; (*G*) photoacoustic setup—the incident, modulated light produces pressure fluctuations detected by a sensitive microphone; (*H*) photothermal beam deflection setup—the incident, modulated light produces refractive index fluctuations that cause the changes of the laser beam deflections; (*I*) surface electromagnetic wave (SEW) spectroscopy setup; (*J*) configuration used in ellipsometric spectroscopy.

technique provides an easy and sufficient method of obtaining spectra, difficulties may arise as a result of water absorption or undesired reactions with the salts.

If the sample is a thin, free-standing film, the transmission technique provides an excellent means of obtaining spectra. The film is simply placed in the sample compartment of the instrument, and the spectrum is run. This technique is particulary useful in the analysis of polymeric films. Obviously transmission spectroscopy will not allow one to obtain spectra from surfaces, although one could naively think that because light passes through two surfaces, it should be possible to do so. Of course, a contribution from the bulk is several orders of magnitude stronger, therefore shading the surface modes.

Interestingly enough, those who are trying to use spectroscopy as the last resort tool often run into various problems. They certainly hope that this one IR spectrum may answer all the questions they have been asking in the last 5 years. Because this is not the case, the very first item they learn while running a transmission spectrum of a polymeric film is the appearence of a "wavy baseline" in the spectrum, or so-called interference fringes. While in the majority of cases interference fringes obscure further analysis, it also gives a novice spectroscopist a lifetime opportunity to precisely determine film thickness that is given by $b = 5N/(v_1 - v_2)$, where N is the number of complete cycles between wavenumbers v_1 and v_2. How to eliminate interference fringes: roughen the surface or change film thickness.

It should also be remembered that FT-IR is particularly sensitive to the absorbance value, which should absolutely be kept within the Beer's regime or even below. Therefore, excessively thick films may be troublesome. This is related to the fact that a commonly utilized apodization process is achieved by a triangular (sinc^2) apodization function. This function was found to exhibit linear behavior of up to 0.8 absorbance unit; above that value it becomes nonlinear.[1] This is why the concentration of the species of interest for any quantitative analysis should be such that it does not exceed these values.

When light passes through two media having different refractive indices, and the media are in contact with each other, the path of the light is distorted. The magnitude of this distortion will depend on the angle of incidence at which the light enters the medium and the optical densities of both media. Figure 3.3 depicts the physical processes that affect light passing through two materials in optical contact with different refractive indices. Light is transmitted at a 90° angle of incidence (Φ), and partially reflected at $\Phi < \Phi_c$, or totally reflected at $\Phi > \Phi_c$. When the angle of incidence Φ is greater than Φ_c, the light is totally reflected. This is the basis for internal reflection spectroscopy. Figure 3.4 illustrates the majority of other IR spectroscopic methods, among which internal and external reflection, diffuse reflectance, and photoacoustics are the most commonly utilized in surface analysis.

3.3.2   Internal Reflection Spectroscopy

Internal reflection spectroscopy (IRS) is a technique that measures the optical spectrum of a sample that is in contact with an optically denser and transparent medium. This is shown in Fig. 3.4C, D. Originally developed for a single reflection (Fig. 3.4C) by Fahrenfort,[2,3] the technique was later modified experimentally by increasing a number of reflections, giving rise to the attenuated total reflectance spectroscopy, useful in various applications. A modified version of the internal reflection technique utilizing multiple reflections was originated to study surfaces and thin films.[4]

3.3.2.1 *Attenuated Total Reflectance*
The first applications of internal reflection spectroscopy were demonstrated by Harrick[5] and Fahrenfort.[6] The experimental setup of the internal reflection technique is shown in Fig. 3.4D. The IR light passes through the optically denser crystal and reflects at the surface of the sample. According to Maxwell's theory, the propagating light passing through an optically thin, nonabsorbing medium forms a standing wave, perpendicular to the total reflecting surface. If the sample absorbs a fraction of this radiation, the propagating wave interacts with the sample and becomes attenuated, giving rise to attenuated total reflectance. The reflectance R of the attenuated wave can be expressed as

$$R = 1 - kd_e \tag{3.3}$$

where d_e is the effective layer thickness and k is absorptivity of the layer. The energy loss in the refractive wave is referred to as *attenuated total reflectance* (ATR). In order to increase the sensitivity, multiple reflections are employed. With N-fold reflections, the total reflection R^N is expressed as

$$R^N = (1 - kd_e)^N \tag{3.4}$$

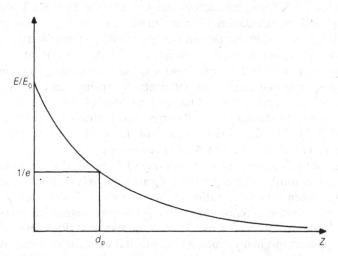

Figure 3.5 A magnitude of the electric field of evanescent wave as a function of its depth of penetration (z). The d_p is defined as a distance required for the amplitude of electric field to fall to e^{-1} value.

It should be mentioned that the effective layer thickness represents the thickness necessary in transmittance measurements to obtain the same absorbance as that from a single reflection at the phase boundary of a medium.

The depth of penetration of the IR radiation into the sample depends on the angle of incidence, the frequency of incident light, and the refractive index difference between ATR element and the sample. The depth of penetration d_p is given by

$$d_p = \frac{\lambda_0}{2\Pi n_1 (\sin^2 \theta - n_{21}^2)^{1/2}} \tag{3.5}$$

where λ_0 is the wavelength of light in the internal reflection crystal, θ is the angle of incidence, and n_{21} is the refractive index ratio of the sample and crystal. Figure 3.5 illustrates a decay of the evanescent wave with the depth of penetration.

3.3.2.2 Problems with Attenuated Total Reflectance   In an ATR experiment, the measured reflectivity spectrum is a complex function of absorption and refractive index spectra. Because these quantities are effected by experimental conditions, in order to compare the results obtained under various conditions, ATR spectra should be converted to the optical constants. If quantitative analysis is desired, the optical constants can be converted to the Beer–Lambert absorbance spectrum, providing a means for quantitative analysis. The problem is, however, how to accurately determine optical constants from ATR measurements. Because many early approaches exhibited certain limitations, Kramers–Kronig transforms (KKT) appeared to be the most promising. Although several algorithms for calculating optical constants from an ATR spectrum were proposed, the results could be obtained depending on what quantities were related by transformation, how KKT was carried out numerically, and how KKT was incorporated in the ATR calculations. Recent studies[7] undeniably demonstrate that the approach proposed by Bertie and Eysel[8] in 1985 and modified by the same authors[9] in 1992 is not suitable for applications with strong ATR intensities and, at best, can be considered for weak-band analysis. In contrast, Dignam et al.[10] proposed a method that can be used efficiently for strong bands and fails when weak bands are analyzed. On the basis of analysis of the possible deviation sources in both algorithms, a new ATR correction algorithm was proposed and is shown in Fig. 3.6. This algorithm is applicable for weak and strong bands and appears to serve well quantitative ATR spectroscopy.

Another often forgotten problem with the ATR surface depth profiling is the fact that the relationships for depth profiling are valid only for homogeneous surfaces, that is, when the concentration of the surface species does not change. To overcome this problem, reflection theory for stepwise stratified surfaces was sucessfully used to determine concentration changes at various depths.[11]

As was pointed out many times, a fundamental difference between ATR and transmission spectra is a position of bands and their intensities. The frequency shifts along with the sensitivity difference between high- and low-wavenumber

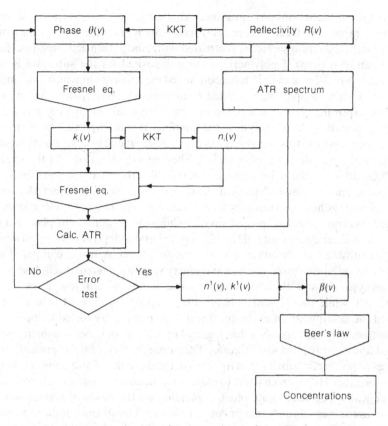

Figure 3.6   Urban–Huang algorithm for converting surface ATR spectra to absorbance. (Adopted from ref. 7.)

regions, and spectral distortions near critical angle conditions, are the primary concerns in ATR quantitative analysis. Further details concerning ATR limitations and theory can be found in the literature.[12]

3.3.2.3 *Selected Applications*

The depth profile measurements are achieved by varying the angle of incidence and/or using ATR crystals with different optical densities. ATR FT-IR has been utilized[13] in studies of morphology and structrue of polyurethanes at various depths. Subtraction of two spectra obtained at two different depths yielded a spectrum of the material that is representative of that region of the sample. This study showed that there are two distinctly different morphological forms of polyurethane: one on the surface and the other in the bulk of the polyurethane films. By changing the angle of incidence, Jacobsen[14] demonstrated a similar approach for obtaining intermediate-layer spectra. Although there are examples of the surface depth profiling in earlier literature,[15,16] only a few reports dealt with the problem of adhesion between two phases.

If the interface contains substances that are brittle or soluble in water, a relatively rapid interfacial failure will occur. Such interfacial properties often develop if the substrate surface is pretreated with zinc galvanized steel. Interfacial failure can also occur if polymeric coating deposited on the substrate is zinc-enriched.[17] Makishima et al.[18] have conducted experiments in which zinc and tin galvanized steel samples were coated with long-oil alkyd paints followed by outdoor exposure. The interfacial products were analyzed by peeling off the coating followed by ATR analysis of the interface. It was found that the formation of zinc soaps and oxidation products at the coating–metal interface is responsible for the substantial decrease of adhesion. They also determined that the effect of Ca_2PbO_4 added to the paint formulation, significantly enhanced adhesion.

O'Brien and Hartman[19] utilized ATR spectroscopy to interpret the mechanism of paint adhesion to wood by using a cellulose/epoxy resin model system. An infrared spectrum of cured epoxy resin on cellulose was obtained by placing a cast epoxy resin sheet on one side of the IRE crystal and a dry, pressed, reconstituted cellulose substrate on the other side. The recorded spectrum was compared with that of cast cellulose/epoxy mixture and epoxy resin-impregnated cellulose paper. The analysis showed that the characteristic bands of cellulose were heavily diminished, while the 775-cm^{-1} band due to epoxy was absent. These studies allowed the determination of chemical reactions responsible for adhesion.

Continuously increasing technological interest in polymeric substrates has created another problem to overcome. The apparent lack of polar groups on the surfaces of polymeric substrates is reflected in the difficulty of attaching coatings to such substrates. Hence, in an effort to improve adhesion, the surfaces are physically or chemically modified. Quite often, depending on the mode of application, the surfaces are subjected to plasma or corona discharge treatments in the presence of various gases. The choice of treatment is dictated by the chemical composition of the substrate or suitability of a particular treatment. One recently studied system is gas–plasma modifications of silicone elastomers.[20,21] A combination of spectroscopic measurements and dynamic mechanical analysis provides a suitable approach to establish that on the plasma surface treatments, only ammonia–plasma leads to an increase of the log E' modulus values. Argon and carbon dioxide treatments produce oxidative products that decrease the storage modulus the above glass transition temperature (T_g). These results indicate that in order for silicone elastomers to enhance their adhesion to other materials and be able to utilize them in artificial organs and biocompatible applications, surface treatments may be essential. Furthermore, these studies have also shown that in the previous assessments,[22] the surface crosslink density increases as a result of gas/plasma treatments are not necessarily the case.

In an effort to improve adhesion between metal and polymer surfaces it is common practice to modify the surface of a polymer by either acid etching or corona discharge treatments. In addition, it is necessary to remove traces of impurities on a metal surface in order to achieve desirable interfacial bonding. For example, surface contamination that results from the processing of aluminum sheets is a severe problem. Moreover, the presence of water on the surface of aluminum often leads to the conversion of amorphous Al_2O_3 to hydrous oxides such as AlOOH (boehmite) or $Al(OH)_3$ (bayerite).[23-25] To prevent these processes, sodium heptag-luconate and phosphonic acids are often utilized. While chemical treatments are utilized to improve bonding between the Al surface and polymers, physical treatments may also enhance this process. Such procedures as roughening of Al surfaces are employed along with various etching treatments [sulfuric acid or sodium dichromate, phosphoric (PAA), or chromic acid anodization][26] in an effort to enhance mechanical interlocking with the overlaid polymer. PAA treatments additionally lead to the formation of an insoluble $AlPO_4$ monolayer, which diminishes the rate of conversion of Al_2O_3 to $Al(OH)_3$.[4] Recently, Golander and Sultan[27] spectroscopically investigated how surface modification of polyethylene and aluminum affects interlaminar adhesion. Both ESCA (electron spectroscopy for chemical analysis) and multiple internal reflection FT-IR spectroscopy measurements indicated the presence of hydroxyl, ester, and carbonyl groups on the $KMnO_4/H_2SO_4$-treated polyethylene surfaces. The presence of sulfonate and sulfate groups in localized areas of crevices was detected along with the surface vinyl species. Although the majority of these species were detected earlier by ESCA,[28] the authors attributed the band at $1710 \, cm^{-1}$ to the COOH species. This assignment was confirmed by the presence of C—O stretching vibrations in the $1260-1000\text{-}cm^{-1}$ region as well as the C—H bending modes in the $1180-1085 \, cm^{-1}$. While these assignments may indeed suggest the appearance of the COOH surface species, it is possible that, in the presence of water-soluble cations, the COOH will dissociate to the COO^-/cation pairs, giving rise to the symmetric and asymmetric C—O stretching vibrations that would absorb in the regions mentioned above.[29] Regardless of data interpretation and the mechanisms responsible for bonding, the adhesion between polyethylene (PE) and aluminum can be improved by surface modifications of either surface as was demonstrated by the T-peel adhesion test. Apparently, the oxidation of the PE surface as well as incorporation of divalent cations before lamination enhance adhesion. Modifications of the aluminum surface with a double bond containing titanate leads to a threefold increase of adhesion, and on mechanical testing, failure occurs in poly- ethylene.

One of the important considerations in the ATR measurements is a contact between the sample and the ATR crystal. Although the most common practice is to use a rectangular crystal, more recent novel applications with circular crystals have been demonstrated. A circular cell, schematically depicted in Fig. 3.7A, and originally developed for study of aqueous solutions,[30-32] can be successfully used for the studies of films and fibers.[33] Figure 3.7B illustrates two possible alignments of fibers with respect to the ATR crystal. These are parallel and perpendicular. The spectra recorded with these two configurations provide information about the orientation of surface species in respect to the fiber axis. It was found that that the intensities of the C—N stretching and C—H out-of-plane normal vibrations change as the orientation changes. This is because the fibers are known to be highly crystalline, and as light passes through the crystal and penetrates the surface of the fiber, it becomes polarized differently in parallel and perpendicular directions; in other words the interaction of the electric vector of electromagnetic radiation is different for parallel and perpendicular alignments.

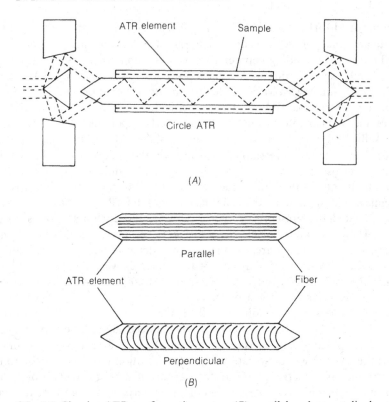

Figure 3.7 (*A*) Circular ATR configuration setup; (*B*) parallel and perpendicular alignment of fibers with respect to a crystal axis.

This difference is reflected by the intensity changes in the IR spectra. From such spectra it is possible to calculate the optical constants of the fibers for both parallel and perpendicular alignments with the use of the methods previously described in the literature.[34–36]

Further applications of the technique are currently demonstrated to establish the film–air and film–substrate interfaces in latex.[37] While an intentional addition of small molecules to polymeric systems was recognized many years ago as means of improving such properties as processibility, the question as to how their behavior may affect such properties as adhesion or durability has not been addressed. For example, plasticizers are often introduced to a polymer matrix in an effort to lower the temperature of processibility by effectively lowering overall T_g of the system. These monomeric species may have, however, a significant impact on adhesion to plastic substrates if they migrate to a film–air interface. While the use of plasticizers may be avoided, in the case of latex films, the presence of surfactants during the latex preparation is an absolute necessity governed by the formation of surfactant micelles responsible for the latex particle formation. It is, therefore, appropriate to address the question as to how the presence of unavoidable surfactants may influence the latex film properties, in particular, adhesion. This issue is especially important for surfactant molecules that may exhibit lack of compatibility with the polymer latex network after latex coalescence. Under these circumstances, one may expect phase separation, diffusion, or enhanced mobility, as these small molecules are free to move with the network. One can speculate that if the surfactant molecules are capable of migrating in some preferential direction across the film, the direction of its propagation will affect film properties such as adhesion and durability. This is because the majority of surfactants are water-soluble, and if their concentration at the film–substrate interface is increased, they may be washed away, consequently leading to adhesion failure. One of the most striking observations made using ATR FT-IR spectroscopy was the fact that it is possible to monitor surfactant exudation to either interface. The direction of exudation was found to be dependent on surface tension of the substrate and external forces imposed on the film. Surface tension of a substrate may affect distribution of sodium dioctyl–sulfosuccinate (SDOSS) surfactant molecules across the ethyl acrylate/methacrylic acid (EA/MAA) latex.[38–40] This topic is discussed further in Chapter 6.

3.3.3 External Reflection Techniques

When IR light impinges on a metal surface, an electric field is generated near the surface. The intensity of the electric field is determined by the angle of incidence of the radiation. A standing wave is formed near the surface at a normal incidence angle because the waves are shifted in phase by 180°. As a result, there is almost no interaction between the sample and the radiation, and no IR spectrum is produced. A similar situation appears for all angles of incidence when perpendicular polarized light is applied (s or perpendicular component). Metallic surfaces impose strict dipole moment selection rules demanding that the incident light should have a component parallel to the plane of incidence (p or parallel component). Under these circumstances, the phase shift for the component polarized parallel to the surface changes. When the phase shifts are different than 180°, an electric field is generated at the surface of the metal with the dominating component normal to the surface.

Thin polymer films adsorbed onto the oxidized surfaces of metals are important in areas such as corrosion inhibition, adhesion, and bonding or lubrication. Because of this, there has been a great deal of interest in developing spectroscopic techniques for characterizing such sampling situations. The technique called *reflection–absorption* exhibits characteristics that make it extremely useful for studying the chemical composition and molecular structures of thin films on the surfaces of metals. Because vibrational spectroscopy is sensitive to bonds rather than atoms, it can be used to determine functional groups present on the surface. Because IR radiation is a gentle source of excitation and the technique has no requirement of ultrahigh vacuum, surfaces with even weakly bonded species may be studied. While the basic principles for single and multireflections are illustrated in Fig. 3.4, parts *D* and *E*, respectively, in an effort to determine the

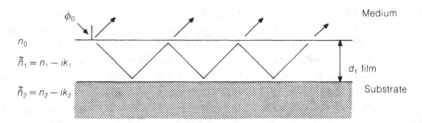

Figure 3.8   A model for reflection–absorption (RA) experiment.

conditions necessary for obtaining infrared spectra of thin polymer films on metal surfaces, it is useful to consider the model shown in Fig. 3.8.

The metal substrate and the polymer film have complex refractive indices, $n_2 = n_2 - ik_2$, $n_1 = n_1 - ik_1$, respectively. The surrounding medium, which is usually air, has a real refractive index of n_0. When parallel polarized light is reflected from the film deposited on a substrate back into the surroundings, the amplitude of the total electric field is given by

$$r = \frac{r_1 + r_2 e^{-2i\delta}}{1 + r_1 r_2 e^{-2i\delta}} \qquad (3.6)$$

where

$$r = \frac{n_0 \cos \phi_1 - \tilde{n}_1 \cos \phi}{n_0 \cos \phi_1 + \tilde{n}_1 \cos \phi_0} \qquad (3.7)$$

$$r_2 = \frac{\tilde{n}_1 \cos \phi_2 - \tilde{n}_2 \cos \phi_1}{\tilde{n}_1 \cos \phi_2 + \tilde{n}_2 \cos \phi_1} \qquad (3.8)$$

and

$$\delta = \frac{2\Pi n_1 d_1 \cos\phi_1}{\lambda} \qquad (3.9)$$

The quantities ϕ_1 and ϕ_2 can be determined from ϕ_0, n_0, $\tilde{n}_1$ and $\tilde{n}_2$ using Snell's law:

$$n_0 \sin \phi_0 = \tilde{n}_1 \sin \phi_1 \qquad (3.10)$$

$$\tilde{n}_1 \sin \phi_1 = \tilde{n}_2 \sin \phi_2 \qquad (3.11)$$

Because $\tilde{n}_1$ and $\tilde{n}_2$ have complex components, ϕ_1 and ϕ_2 are also complex and do not represent angles of refraction. The intensity of the reflected beam from the surface, or reflectivity, is given by $R = rr^*$, where r^* is the complex conjugate of r.

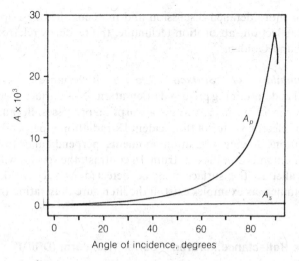

Figure 3.9 Dependence of the absorption factors for the parallel (A_p) and perpendicular (A_s) polarization at the wavenumber of maximum adlayer (adsorbed layer).

The intensity of the absorption band is obtained from the following ratio:

$$A = \frac{R - R_0}{R_0} \tag{3.12}$$

where R_0 is the reflectivity for $k_1 = 0$ and R is the reflectivity for k_1 not equal to zero.

In the case where radiation is polarized in a perpendicular incidence plane, exact analysis holds except for the expressions for r_1 and r_2, which are different:

$$r_1 = \frac{n_0 \cos \phi_0 - \tilde{n}_1 \cos \phi_1}{n_0 \cos\phi_0 + \tilde{n}_1 \cos \phi_1} \tag{3.13}$$

$$r_2 = \frac{\tilde{n}_1 \cos \phi_1 - \tilde{n}_2 \cos\phi_2}{\tilde{n}_1 \cos\phi_1 + \tilde{n}_2 \cos\phi_2} \tag{3.14}$$

Using these expressions and solving Eq. 3.12 for A gives the A values very small, indicating that for perpendicular polarization, for all angles of incidence we do not expect to observe any bands. In contrast, however, the band intensities for parallel-polarized incident radiation exhibit different characteristics. Although at low angles of incidence (below 50°) the band intensities are also negligible above 50° it rises sharply reaching a maximum of around 88°.[41-43] Thus, almost grazing angle and parallel polarization of the incident beam are desirable for RA experiments. Figure 3.9 illustrates the intensity of an electric field at the metal surface for the parallel- and perpendicular-polarized light as a function of angle of incidence. For more detailed discussion and mathematical descriptions of the theory of the reflection–absorption technique, the reader is referred to Swalen and Robolt[44] and Golden.[45]

3.3.3.1 *Orientation of Surface Species* Reflectance–absorption can

be quite useful in determining preferred orientation of the surface species and also for determining mechanisms governing adsorption processes. Because in the RA experiment the electric vector of the incident IR radiation is perpendicular to the surface, vibrations having transition moments perpendicular to the surface should appear enhanced in the spectrum. In contrast, the species with transition moments parallel to the surface may be detected as very weak or remain undetected. Numerous examples exist in the literature illustrating this phenomenon.

3.3.4 Diffuse Reflectance Infrared–Fourier Transform (DRIFT) Spectroscopy

The basic principle behind diffuse reflectance spectroscopy is that light impinging on a solid or powdered surface is diffusively scattered in all directions.[46] With a proper optical setup this scattered light is collected and redirected to the detector. Figure 3.4E shows a schematic diagram of the diffuse reflectance experiment. A general theory describing this process for powdered samples was developed by Kubelka and Munk.[47,48] The theory relates the sample concentration to the scattered radiation intensity and therefore is often called the *Beer–Lambert law of diffuse reflectance*. The Kubelka–Munk function is expressed by the following equation:

$$f(R) = \frac{(1 - R_\infty)^2}{2 R_\infty} = \frac{k}{s} \tag{3.15}$$

where R_∞ is the absolute reflectance of the layer, s is the scattering coefficient, and k is the molar absorption coefficient. The theory predicts that, for every band in the spectrum, a linear relationship should be obtained between the molar absorption coefficient and the maximum value of $f(R)$, if s is held constant. Because the value of the scattering coefficient s, depends on the particle size, quantitative analysis should be performed using only the same particle size. Subsequently, Eq. (3.15) can be rewritten as

$$f(R) = \frac{(1 - R_\infty)^2}{2 R_\infty} = \frac{c}{k'} \tag{3.16}$$

where c is the concentration and k' depends on the particle size and the molar absorptivity of the sample through relationship $k' = s/2.303 \, \varepsilon$. At higher concentrations the Kubelka–Munk (KM) relationship becomes nonlinear. This is attributed to large changes in the refractive index in the spectral regions of the intense absorption bands. This causes an increase of the specular reflectance, giving rise to Reststrahlen effects.[49–52] A number of useful papers have been published on the theory of diffuse reflectance spectroscopy.

Diffuse reflectance spectroscopy has been used for many years to measure UV–VIS and near-IR (NIR) spectra of powders. The mid-IR region, however, never developed because the same optical systems used in the UV–VIS and NIR were insufficient. Since 1978, however, when Fuller and Griffiths[53] first reported diffuse reflectance measurements utilizing FT-IR spectroscopy, many applications of this technique have been demonstrated. These applications include studies of glass fibers,[54,55] analysis of coals,[56] carbonaceous electrode materials,[57] pigment photodeposition on paints,[58] and modified polyesters.[59] In this review, however, we will focus only on those applications of DRIFT that are relevant to adhesion.

High sensitivity, surface selectivity, and the nondestructive nature of this procedure have made the DRIFT experiments a valuable characterization tool. In adhesion science, in addition to the above-mentioned advantages, it is equally important to be able to perform surface depth profiling. In this manner, one can gain molecular level information about the species forming at the interface. McKenzie et al.[60,61] have demonstrated this ability of DRIFT. It was achieved by using the so-called KBr overlayer technique. By placing the KBr powder over the glass fibers treated with silane coupling agents, it is possible to render the scattering of light from the glass substrate isotropic while, simultaneously, detecting the intensities of the bands due to adhesion-enhancing coupling agents. This reduction of substrate intensity is attributed, in part, to a greater reflection from the near-surface layers, which results in lesser penetration into the substrate. Although there has been a considerable amount of data published on the KBr overlayer technique, the nature of interaction of absorbed water in polyaramid (Kevlar) fibers is of great interest in adhesion of composite materials. By using the KBr overlayer method, Chatzi et al.[62] have found various water species absorbed in Kevlar fibers. Increasing amounts of the KBr overlayer leads to higher selectivity, which was demonstrated by the appearance of two well-resolved peaks at 3640 and 3560 cm^{-1}. Without the overlayer, both peaks appear as shoulders of the broad N—H stretching band.

It is apparent that this technique has only recently been utilized in the studies of surface and interfacial species that are responsible for adhesion enhancement.[63] It has certain drawbacks, such as maintaining an even distribution of KBr powder over the substrate and the adverse effect of the varied grain size on the intensities of IR bands. This technique can be especially useful in adhesion science since it provides a convenient means of obtaining IR surface spectra. When all conditions are kept constant, which is always a challenge for an experimentalist, the linearity of the Kubelka–Munk function with respect to concentration can be achieved, although in a relatively narrow range.

3.3.5 Photoacoustic (PA) FT-IR Spectroscopy

3.3.5.1 *Principles of Photoacoustic Detection*   Photoacoustic spectroscopy utilizes the detection of an acoustic signal emitted from a sample due to absorption

of modulated radiation. A sample is placed in an acoustically isolated chamber to which a sensitive microphone is attached. On absorption of modulated light, heat is generated within the sample. Its release leads to temperature fluctuations at the sample surface. This is depicted in Fig. 3.4G. The frequency of temperature fluctuations is in phase with the modulation frequency. These temperature fluctuations at the sample surface cause pressure changes in a surrounding gas, which, in turn, generate acoustic waves in the sample chamber. The pressure changes of the gas are detected by a sensitive microphone, and the obtained electrical signal is Fourier-transformed.

Figure 3.10 schematically depicts the generation of a photoacoustic signal. The modulated infrared radiation with the intensity I_0 enters the sample with refractive index n and absorption coefficient β. The intensity of the IR radiation diminishes exponentially as it penetrates the sample, giving rise to the intensity at depth x:

$$I(x) = I_0(1 - n) \exp(-\beta x) \tag{3.17}$$

The amount of light absorbed within the thickness x is equal to

$$E(x) = \beta I(x) = \beta I_0(1 - n) \exp(-\beta x) \tag{3.18}$$

The depth of optical penetration is defined as optical absorption length L_β and is

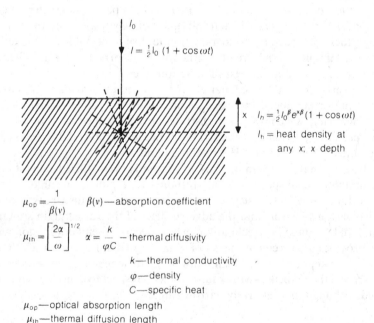

Figure 3.10   Schematic representation of photoacoustic detection (B) and generation of photoacoustic FT-IR signal.

inversely proportional to β:

$$L_\beta = \frac{1}{\beta}$$

(3.19)

In other words, L_β is the distance from the surface at which the initial IR intensity I_0 attenuates to $(1/e)I_0$. The absorbed energy is released in a form of heat that is transferred to the sample surface. The efficiency of the heat transfer is determined by the thermal diffusion coefficient of the sample a_s and the modulation frequency of the incident radiation ω:

$$a_s = \left[\frac{\omega}{2\alpha}\right]^{1/2}$$

(3.20)

where α is the thermal diffusivity ($\alpha = k/\rho C$; thermal diffusion conductivity/(density $\times$ specific heat). The thermal diffusion length μ_{th} is related to the thermal diffusion coefficient a_s as

$$\mu_{th} = \frac{1}{a_s}\left[\frac{2\alpha}{\omega}\right]^{1/2}$$

(3.21)

The amount of heat periodically transferred to the surface through the sample is then equal to

$$H(x) = E(x)\exp(-a_s x) = \frac{\beta I_o(1-n)}{\exp[(\beta + a_s)x]}$$

(3.22)

Since the photoacoustic effect is a two-stage process, the intensity of the photoacoustic signal depends on the sample optical and thermal properties. Rosencwaig and Gersho[64] have classified the samples as optically transparent or opaque and thermally thin or thick depending on the relationship between the absorption coefficient and thermal diffusion length. A detailed discussion of the effect of materials' properties and their effect on PA FT-IR intensities can be found in the literature.[65]

3.3.5.2 Selected Applications

The capability of surface depth profiling is one of the most appealing features of PA FT-IR since, with the appropriate modulation frequencies of the Michelson interferometer, interfaces have been studied. In an effort to demonstrate the capability of PA FT-IR spectroscopy to the surface depth profiling, Urban and Koenig[66] have performed studies on a double-layer PVF_2-on-PET. Figure 3.11 shows photoacoustic FT-IR spectra of a 6-μm overlayer of PVF_2 on PET obtained with various modulation frequencies. When the modulation frequency is fast, the spectrum contains primarily information from the surface (top PVF_2 layer). The decreasing modulation frequency causes an increase of the thermal diffusion length, resulting in a PA spectrum that

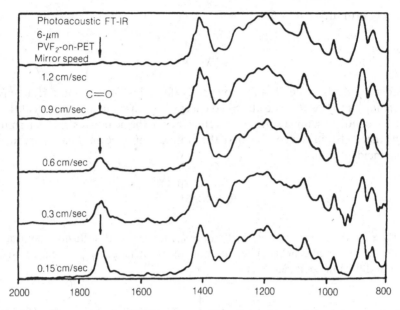

Figure 3.11 Photoacoustic FT-IR spectra recorded with various mirror velocities (sample: 6-μm PVF$_2$-on-PET). (Reproduced with permission from ref. 66, copyright 1986, Society for Applied Spectroscopy.)

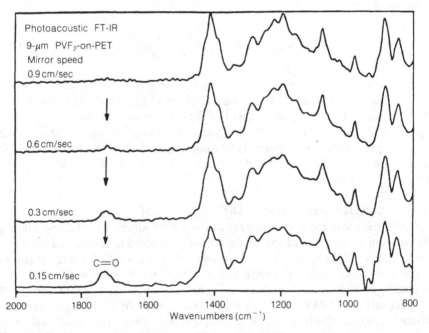

Figure 3.12 Photoacoustic FT-IR spectra recorded with various mirror velocities (sample: 9-μm PVF$_2$-on-PET). (Reproduced with permission from ref. 66, copyright 1986, Society for Applied Spectroscopy.)

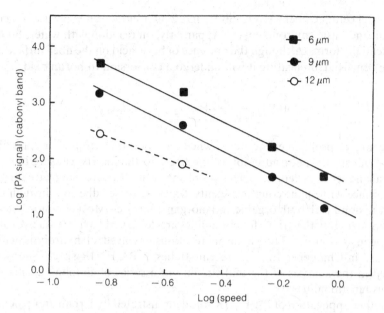

Figure 3.13 Log of C=O intensity plotted as a function of log ω. (Reproduced with permission from ref. 66, copyright 1986, Society for Applied Spectroscopy.)

consists of information obtained from the bulk. Figure 3.12 illustrates similar measurements with the 9-μm PVF$_2$-on-PET and indicates the same features: the intensity of the carbonyl band at 1738 cm^{-1} decreases as the mirror speed increases. A log–log plot of the integrated intensity of the carbonyl band as a function of mirror velocity (or modulation frequency; $\omega = 4\Pi V v$) is shown in Fig. 3.13. In both cases, the 6- and 9-μm thick PVF$_2$ film, a straight line is obtained. The value of the slope represents the power of the modulation frequency in the photoacoustic equation and corresponds to $-\frac{3}{2}$ equation for thermally thick and optically thin samples.

In composite materials, fibers are often treated to increase adhesion to a substrate. Yang et al.[67] have examined the depth distribution of polymer adhesives on the surfaces of cotton, poly(ethylene terephthalate) (PET), and glass fibers. The depth profiling studies of cotton fibers treated with a polyurethane agent indicated a greater concentration of the agent on the surface and a substantial decrease in its concentration toward the center of the fiber. Similarly, a concentration gradient was detected between the skin and the core of PET fibers. Since the PA technique is advantageous for surface studies, numerous articles have been published concerning the changes of coal surfaces,[68] surface treatments of silica surfaces[69] along with their photochemical reactions, corrosion of iron surfaces, adsorption on silica surfaces, and the Bronsted acid centers on silica–alumina substrate.

In composite technology, coupling between fibers and a polymer matrix is known to play a crucial role in achieving good interfacial properties. As a consequence, PA FT-IR has been utilized to monitor the extent of degradation of the Nextel ceramic fiber (Al_2O_3, SiO_2, and B_2O_3) surface on exposure to acidic, neutral, and basic environments.[70] Apparently, on reaction with water, $B(OH)_3$ preferentially forms. Although the presence of boric acid on the fiber surface may not be beneficial, on heating it will undergo a conversion to boron oxide:

$$B(OH)_3 \xrightarrow[H_2O]{heat} HBO_2 \xrightarrow[H_2O]{heat} B_2O_3$$

A common problem encountered when incorporating ceramic or glass fibers into composites is poor adhesion of the fibers to the matrix, causing low, dry, flexural and tensile strength.[71] This problem is usually overcome by treating the fiber surface with silane coupling agents, which increase adhesion through their ability to bond with both organic and inorganic surfaces. Most commonly, silane coupling agents with various functionalities are deposited on fibers via hydrolysis in aqueous solutions. The chemical reactions involved in hydrolysis will be discussed in Chapter 6. In more recent studies,[72] PA FT-IR spectroscopy was utilized in the analysis of thermal stability of a series of coupling agents with various functionalities.

Another application of PA FT-IR was demonstrated by Urban and Koenig[73] who showed that using highly polarizable insert gas (xenon) in the photoacoustic cell it is possible to determine orientation of the surface species. Several applications of the technique used to monitor orientation of the silane coupling agents were demonstrated[74] along with the studies of silane surface chemistry,[75] accessibility for bonding in polyaramid fibers,[76] and orientation of Langmuir–Blodgett films on poly(tetrafluoroethylene).[77] Because xenon intensifies the transition moment parallel to the surface, the strong intensity enhancement of the aromatic C—H stretching bands in the 3000-cm^{-1} region can be obtained and is attributed to the fact that aromatic rings are parallel to the surface. Hence, the accessibility of the hydrogen bonded C=O⋯H—N groups for bonding is restricted. This effect explains the fairly poor adhesion of Kevlar fibers to composite matrices.

Temperature studies can also be conducted using the temperature photoacoustic FT-IR cell[78,79] that was developed for that purpose. The advantages of photoacoustic FT-IR spectroscopy over other techniques arise from the fact that the sample is in its native state. Hence, it is a suitable tool to monitor crosslinking reactions at various temperatures. Figure 3.14 illustrates a comparison between PA and transmission FT-IR sensitivity during the network formation of poly(dimethylsiloxane) (PDMS) and triethoxysilane (TES).[80,81] The spectral intensity changes are plotted as a function of viscosity in a double-log scale. This sensitivity difference between the two FT-IR modes arises from the fact that the photoacoustic technique is also sensitive to thermal property changes during the crosslinking reactions, and hence the intensities of photoacoustically detected infrared bands change. After all, it is a spectroscopic method that measures the amount of light converted to heat, and, when thermal properties change, the intensity of photoacoustically detected bands will change also.

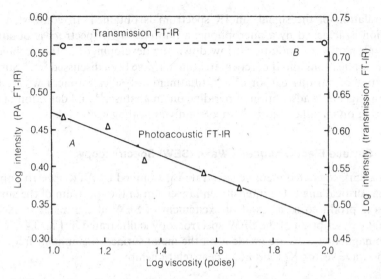

Figure 3.14 Log intensity ratio (2962/700 bands) plotted as function of log viscosity during the curing process of the PDMS/TES system: (*A*) photoacoustic detection; (*B*) transmission detection. (Reproduced with permission from ref. 81, copyright 1986, American Chemical Society.)

Typically, coatings for plastics have low adhesion to the substrate and interfacial failure may occur when stress is applied. Since a common problem encountered in the film technology and coatings industry is adhesion to substrates, rheophotoacoustic FT-IR, a novel technique, has been developed in our laboratory.[82,83] In essence, it is a stress–strain device built into the photoacoustic sample compartment that, when connected to an FT-IR spectrometer, allows one to monitor molecular changes on applying stress to the substrate. If the sample consists of two layers, the coating and the substrate, and stress is applied to the substrate only, the coating will follow elongation of the substrate if interfacial forces are high. On the other hand, failure of the interfacial bonding will result in the formation of voids at the interface, followed by separation of the coating from the substrate. Rheophotoacoustic FT-IR will be discussed further in Chapter 8.

3.3.6 Photothermal Deflection Spectroscopy

In principle, photothermal deflection spectroscopy is a derivative of the photoacoustic effect. This technique uses a "mirage" detector, where a deflection of a laser beam caused by refractive index changes in the surrounding gas is detected.[84] As depicted in Fig. 3.4*H*, a broad IR source is incident on the surface and, as a result of reabsorption, heat is generated at the sample surface, which causes fluctuation in the refractive index of the surrounding gas. When the laser beam passes through this zone, the beam is deflected at the modulation frequency. These small deflections are detected by a position-sensing detector. On demodulation of the signal, an IR spectrum is obtained. In contrast, if the detection is achieved by a microphone, a photoacoustic spectrum is obtained. Although hardware limitations slow down the development of this technique, several applications and theoretical treatments have been discussed.[85,86] Surface properties of various carbons,[87-91] treatments of silica surfaces with silane coupling agents,[92] adsorption of pyridine on an anatase,[93] and detection of rust appearing on metallic objects[94] are certainly appealing examples.

3.3.7   Surface Electromagnetic Wave (SEW) Spectroscopy

When a surface-coated metallic substrate is examined by ATR spectroscopy at the near-critical angle, the intensity enhancement of the spectrum of the surface species is produced as a result of excitement of SEW at the metal surface. A schematic description of the SEW spectroscopy is illustrated in Fig. 3.4I. When electromagnetic waves are incident on the metal surface, the angular frequency and the wave vector K_α follows a linear relationship:

$$K_\alpha = \frac{\omega}{c}$$

(3.23)

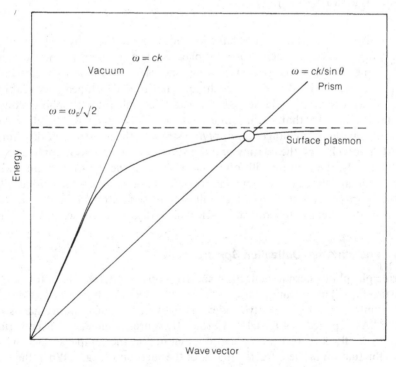

Figure 3.15 Dispersion curves for IR radiation from a vacuum, optical crystal and a surface of plasmon. The SEW can be produced at the open circle.

The surface plasmons of the metal exhibit the nonlinear relationship described by

$$K_\alpha = \frac{\omega}{c} R_e \left[\frac{\varepsilon(\omega)}{\varepsilon(\omega) + 1} \right]^{1/2} = R_e(K_\alpha) + i I_m(K_\alpha) \qquad (3.24)$$

where R_e and I_m are real and imaginary parts, $\varepsilon(\omega)$ is the dielectric constant of metal, and K_α is the one-dimensional component of the wave vector of SEW in the x direction. When a prism is utilized as an incident medium, the dispersion relationship becomes

$$K_\alpha = \frac{\omega}{c} n_p \sin \theta \qquad (3.25)$$

where n_p is the refractive index and θ is the angle of incidence. Solving Eqs. 3.25 and 3.26, we can calculate the conditions for exciting SEW. These conditions are schematically presented in Fig. 3.15.

The sensitivity of SEW spectroscopy was effectively presented by Hartstein et al.[95] when a very thin layer of silver was evaporated on a 4-nitrobenzoic acid layer. The resulting spectrum was drastically enhanced, as shown in Figure 3.16. Such observations were impossible without the use of a silver overlayer and the intensity enhancement is a function of the metal thickness. Tracing D of Fig. 3.16 shows that the intensities reached a maximum when the silver thickness was 6 nm. Other reports of increased sensitivity in the presence of silver layer on p- or m-nitrobenzoic acid have also been reported in the literature.[96]

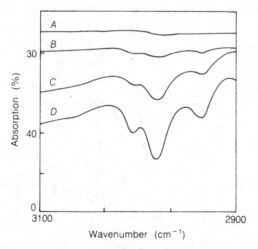

Figure 3.16 A monomolecular layer of p-nitrobenzoic acid on an ATR prism. Spectra A, B, C, and D were recorded with the different-thickness silver overlayers: 0, 1.6, 3.2, and 6.0 nm, respectively. (Reproduced with permission from ref. 95, copyright 1980, American Institute of Physics.)

3.3.8 Infrared Ellipsometry

Ellipsometry is a well known technique in the UV–VIS region for determining the thickness of very thin films deposited on a metallic substrate. Of course, in addition to the capability of monitoring a thickness, optical parameters such as absorption and dispersion characteristics of the coating can be determined. Because the theory of ellipsometry has been extensively documented and reviewed,[97–99] only the basic concept of this technique that relates to the applications in thin-film technology and adhesion will be briefly outlined.

A schematic setup for the static ellipsometric measurements is depicted in Fig. 3.17. The measured quantities are the differential phase retardation Δ and the differential amplitude ϕ. They are related to parallel and perpendicular polarizations R_s and R_p by following equations:

$$\frac{r_p}{r_s} = \tan \phi \exp(i\Delta) \tag{3.26}$$

$$\frac{R_p}{R_s} = \frac{r_p r_p^*}{r_s r_s^*} \tag{3.27}$$

where r_p and r_s are Frensel reflection coefficients and r_p^* and r_s^* are their respective complex conjugates. Both Δ and ϕ are determined by measuring the reflection spectra at various azimuth positions of the polarizers and compensator by the use of Eqs. 3.26 and 3.27. In dynamic ellipsometry, the spectrum is obtained at a fundamental frequency of the photoelastic modulator with a circular polariz-

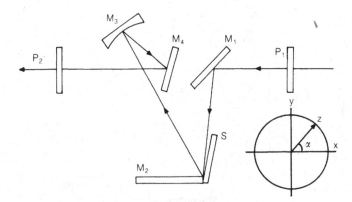

Figure 3.17 Schematic diagram of the reflection ellipsometry setup: P1 and P2—linear polarizers; M1, M2, and M4—flat mirrors; M4—spherical mirror; S—sample; The azimuth angles of the x axis is horizontal; the y axis is vertical; the z vector represents the passing axis of a polarizer.

ation and the spectrum obtained at fundamental frequency taken twice with linear polarization. These quantities are used to calculate the Δ and ϕ.

One of the most appealing features of ellipsometry is the capability to determine the refractive index and the absorption index spectra of a surface species.[100] In addition, anisotropic optical properties of surface oriented species can also be determined.[101] This technique, coupled with the use of dispersive IR spectrometers, has also been utilized successfully to determine the thickness of ion-implanted layers on silicon wafers,[102] along with optical constants of metals in the IR region.[103] FT-IR spectrometers have also been utilized for ellipsometric studies, in both static[104,105] and dynamic[106] modes of detection. Although this technique has not gained wide popularity, it is quite apparent that the sensitivity of monitoring thin films and their adhesion to metallic substrates is extremely high.

3.3.9 Emission Spectroscopy

When an IR source is removed from the IR spectrometer and replaced with a sample to be studied, an IR detector will detect IR rays emitted by the sample. The magnitude of the emitted signal depends on the temperature of the sample according to Stefan's law, which states that the emissivity is proportional to the fourth power of the temperature difference between the sample and the detector. Since the amount of emitted light depends on the temperature difference, the emission spectrum is usually measured from a hot sample to a cooler detector, although the reverse can also apply. In practice, the emissivity is obtained by ratioing the spectrum of the measured emission to a blackbody source.[107] A variety of blackbody references have been used,[108,109] but unfortunately, none of them produce ideal blackbody radiation.

Following Kirchhoff's law,[110] which states that at a given temperature, the absorbance of the sample is equal its emissivity, one can measure emissivity and translate it to absorbance.[111] The measured emission intensity $S(v, T)$ can be expressed as the sum of several components:

$$S(v, T) = R(v)[E(v, T) H(v, T) + B(v) + I(v) P(v)] \qquad (3.28)$$

where v is the infrared frequency, T the temperature, $R(v)$ the instrument response function, $E(v, T)$ the emittance of the sample, $H(v, T)$ the Planck function, $B(v)$ the background radiation, $I(v)$ the background radiation reflected from the sample, and $P(v)$ the reflectance from the sample. In the case of blackbody reference, because $E = 1$ and $P = 0$, the intensities obtained from the reference material are expressed by the following equations:

$$S_1(v, T_1) = R(v)[H(v, T_1) + B(v)] \qquad (3.29)$$

$$S_3(v, T_2) = R(v)[H(v, T_2) + B(v)] \qquad (3.30)$$

The following expressions will be valid for the sample:

$$S_2 = R(v)[H(v,T_1)E(v,T_1) + B(v) + I(v)P(v)]$$ (3.31)

$$S_4(v, T_2) = R(v)[H(v, T_2)E(v, T_2) + B(v) + I(v)P(v)]$$ (3.32)

The emittance E can be calculated as the ratio

$$E = \frac{S_4 - S_2}{S_3 - S_1}$$ (3.33)

The area of particular interest and application for emission spectroscopy is thin films on metal substrates. Because supporting metals give a relatively weak emission of only 4% of the blackbody at the corresponding temperature, several studies have been conducted on emission of lubricants and surface impurities from metal surfaces.[112–116]

Thin films deposited on a metal substrate can also be analyzed with the highest efficiency if the sample is oriented at the grazing angle with respect to the light path. This is schematically depicted in Figure 3.18. Greenler[117] has shown that the angular intensity distribution of light emitted by molecules adsorbed on a metal surface is highest when the viewing angle is between 70 and 80°. Figure 3.19 illustrates this relationship. When reflection and refraction at the film–air

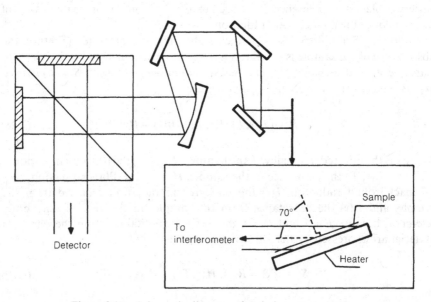

Figure 3.18   Schematic diagram of emission spectroscopy.

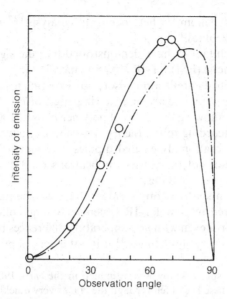

Figure 3.19 Angular dependence of emissivity per area for a Cu_2O film of 160-nm thickness on a gold mirror at 630 cm^{-1}. The dotted line represents theoretical prediction (without self-absorption). (Reproduce with permission from ref. 117).

interface are ignored, the amplitude of the emitted light A is expressed by

$$A = (A_1^2 + A_2^2 + 2A_1 A_2 \cos \delta)^{1/2} \qquad (3.34)$$

where δ is the phase angle between the direct and reflected radiations. The δ angle may also be defined as the change of phase angle caused by the reflection at the metal surface and path difference:

$$\delta = \delta_m + {}^1\delta_p \qquad (3.35)$$

where $\delta_p = 4\pi(n - \sin^2\phi)/(\lambda \cos\phi)$ and n is the refractive index of the film and λ is the wavelength of light in a vacuum.

The quantitative nature of the emission IR spectroscopy has been utilized in various applications. A linear relationship has been found for Cu_2O deposited on gold[118] and for poly(vinyl acetate) on an aluminium plate.[119] Greenler's predictions have been confirmed by Blanke et al.[120] for stearate films deposited on gold, and by Kember et al.[121] for aluminum oxide on aluminum.

It is quite obvious that the fourth-power dependence favors the use of a cooler or hotter detector in order to maintain the energy flux between them. In practice, however, cooler detectors allow one to detect the spectra even of the sample at room temperature.[122,123] Other applications of emission FT-IR spectroscopy include time-resolved catalytic effect of Pt on Pd surfaces,[124] oxide films on copper[125] and molybdenum,[126] heterogeneous catalysis,[127] and chemisorption of Pt supported on zeolites.[128]

Nagasawa and Ishitani[129] have demonstrated that the signal-to-noise (S/N) ratio of electromagnetic (EM) FT-IR thin polymer film layer spectra is comparable with that obtained in reflection absorption experiments. This was achieved by the collection of sample spectra at a viewing angle of 70°. This improvement enhances sensitivity allowing detection of polymer films of $100 = \overset{\circ}{A}$ thickness on metallic surfaces. According to the theory, emission intensity is proportional to the population of vibrationally excited dipoles determined by the Boltzmann distribution law. Many of the earlier applications of EM spectroscopy have previously been reviewed by Bates.[130]

Although emission spectroscopy seems to have an adequate sensitivity and low noise level offered by new FT-IR systems allowing collection of emission spectra with minimal or even with no temperature differences between the sample and the detector in a very short time,[131] it is not a very popular technique.

Although emission spectroscopy was estimated in the early 1980s as one of the future techniques, these hopes fell through the cracks very quickly. One lesson we should learn from the history of emission that the predictions of scientists are not always perfect.

3.4   RAMAN SPECTROSCOPY

After discussing selection rules of Raman spectroscopy, it is appropriate to design an experiment allowing for detection of Raman spectrum. When the monochromatic beam of photons passes through a nonabsorbing, optically homogeneous medium, it is attenuated by scattering process. While the majority of scattering is elastic, involving no change in the energy of the photons and being known as *Rayleigh scattering*, a few photons, about one in a million, are scattered inelastically in the effect known as *Raman scattering*. Because only a small fraction of initial radiation is of interest, there is an ongoing effort for sensitivity improvements and enhancement of spectral intensities. In the following sections, we will progressively introduce the reader to various advantages and problems associated with modes of Raman spectrum detection.

3.4.1   Conventional Raman Spectroscopy

Although the Raman effect was discovered at the turn of the twentieth century, it was not utilized as an analytical tool for many years because of severe limitations in monochromatic sources of excitation. With the advent of lasers (light-attenuated and stimulated emission of radiation), the situation has changed. Figure 3.20 illustrates all necessary components of a conventional Raman system. The three primary components are the laser, usually in a visible region serving as an excitation source; the monochromator, which is a dispersive

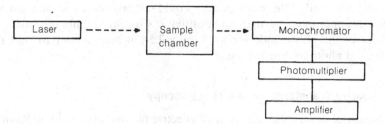

Figure 3.20   Components involved in Raman scattering experiment and their arrangements.

element; and a detecter, which in conventional detection is a photomultiplier tube interfaced with photon-counting electronics. Interestingly, these are lasers that added a lot of mystery and exoticism to Raman spectroscopy, although they are commonly used in many ways by an ordinary person. These three components constitute the Raman setup, and all of them are responsible for the overall spectral quality. While monochromaticity and intensity decide on the laser quality, double or triple monochromators are used to minimize intense elastically scattered components. The latter is due to the fact that an ordinary Raman signal is only 10^{-9} to 10^{-6} of the Rayleigh (elastic) scattering. In conventional Raman spectroscopy, the sensitivity of the photomultiplier tube is such that it allows detection of single photons.

The basic requirement for studying Raman scattering are (1) a monochromatic light source, usually a laser operating in the visible region; (2) dispersive element, usually a double of triple monochromator to disperse the components of the scattered light and reduce stray light; and (3) a photomultiplier and photon-counting detection system. The schematic diagram depicted in Fig. 3.20 illustrates the position of all components along with the three most commonly used geometries. It is important to realize that the quality of each element will influence the overall output, and therefore it should be chosen carefully. For example, a monochromatic laser is characterized by its wavelength λ, output power, and polarization.

Because of the inherent slowness of the traditional Raman spectrometer, an approach of multichannel Raman scanning was introduced. In essence, the entire system is the same as depicted in Fig. 3.20 except that instead of a photomultiplier tube, an optical multichannel detector is used. This device in principle remains a TV set screen. The detector consists of many photosensitive elements, each sensitive to a different wavelength. Instead of scanning the scattered light through a slit system, the dispersed from the sample light is focused on the optical multichannel detector, which, on enhancement by a photodiode array, monitors all wavelengths simultaneously. Such an arrangement provides certain advantages not attainable in an ordinary Raman scattering.[132] These include: a multiple spectral collection (one spectrum can be recorded every 100 msec), and there is no need for scanning, which makes the instrument work as a spectrograph. Of course, all of these advantages would not provide any enhancement if the multichannel detector had no sensitivity. Current technology offers the sensitivity of the photodiode for a 25-μm $\times$ 2.5-μm area similar to that of a cooled, good photomultiplier tube.

3.4.2 Fourier Transform Raman Spectroscopy

A key feature of an FT–Raman system is effective filtering of the intense Rayleigh component of scattered radiation; the intensity of this component can be 10^{10} stronger than that of Stokes component of the Raman spectrum. Because of this, a filter with an optical density of 10 at the laser frequency and a sharp transition from excitation to transmittance at the wavelength a bit shifted from the laser line are required. This can be accomplished by adding to a typical FT-IR instrument a long-wavelength near IR laser, for example, a Nd/YAG(neodymium/yttrium aluminum garnet) system lasing at 1.064 μm (9395 cm^{-1}) and a set of spatial filters. Figure 3.21 illustrates a schematic diagram of FT–Raman setup.

The essential feature of FT–Raman spectroscopy is elimination of fluorescence. This is commonly achieved by using Nd:YAG laser source excitation, giving rise to Raman scattering in near infrared region from 5398 to 9398 cm^{-1}, hence leaving fluorescing region of the spectrum in UV–VIS region. The FT–Raman setup depicted in Fig. 3.21 consists of the following components: a laser source and a FT interferometer equipped with the near-IR optics and detector for the near-IR region. The elimination of Rayleigh scattering is accomplished with the use of dielectric interference filters,[133] designed to pass all wavelengths longer than the laser wavelengths. It appears that the elimination of

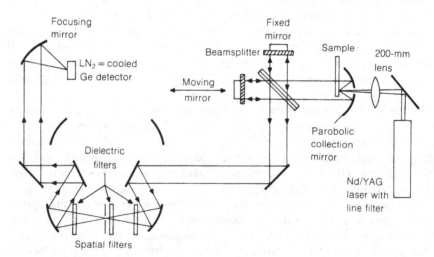

Figure 3.21 Schematic diagram of FT-Raman experiment. (Reproduced with permission from B. Chase, *J. Am. Chem. Sec.*, 1986, **108**, 7485.)

fluorescence is a major advantage of FT–Raman, and, for the most part, this is the only advantage.

3.4.3 Surface-Enhanced Raman Spectroscopy

Surface-enhanced Raman scattering was first reported in 1974 by Fleischmann et al.[134] Since that time, numerous theoretical[135-139] and experimental[140-144] studies have been undertaken to explain and explore the effect. Although there are many ways to schematically depict the generation of Raman enhancement, one model most commonly used visualizes a molecule attached to a metallic surface that has a distribution of charge. This is shown in Fig. 3.22. Since the electron charge density has its image on the metal surface, the impinging laser light will generate strong polarizability tensor changes during the vibration, thereby generating the enhanced Raman spectrum.

The area of particular interest that initiated SERS development concerned the surfaces of electrodes of silver and various other metals. In an electrochemical environment, SERS is highly sensitive to the potential applied to the electrode, and the enhancement factor can exceed 10^6. The currently accepted theory of the mechanism of SERS involves both long-range (electromagnetic) and short-range (chemical) effects. Surface roughness also plays a significant role in the development of the long-range contribution by distorting the translational symmetry along the metal surface. As a result, kinematic constraints on photon absorption or emission by plasmons are relaxed. A primary contributor to the long-range effects is the interaction of the molecular dipole with large induced fields on the metal surface. Charge-transfer excitation has been suggested as the result of short-range interactions. This effect is also thought to influence a redistribution of the electron density between metal conducting band levels and molecular ground or excited states.

Electrochemical studies utilizing SERS are of particular interest because one can determine the degree of adsorption by the electroactive species,[145] the

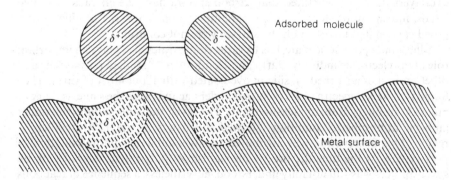

Figure 3.22 A schematic diagram of charge distribution on metal surfaces in SERS.

orientation of adsorbed species, and adjustments in the molecular energy levels of solution species in relationship to the Fermi level of the metal electrode.

Although SERS has enhanced and renewed interest in Raman spectroscopy, most research has continued to focus on the adsorption on metal surfaces, such as silver, gold, palladium, platinum, and nickel, which results in large enhancement effects. The first enhanced Raman effect of pyridine adsorbed on metallic rhodium prepared both as an aqueous colloid and as a supported metal was observed by Kissinger et al.[145] The observed spectral changes led the authors to suggest that pyridine is tightly bound to the rhodium colloid since these bands would not disappear by introducing an acidic environment. This, in turn, provided some evidence that the chemical models of the surface-enhanced Raman effect cannot be neglected. Another study in 1984 confirmed this finding.[146]

Since the observation of the Raman intensity enhancement of pyridine adsorbed electrochemically on rough silver surfaces,[147] many studies have been conducted ranging from small molecules, including benzoic acid,[148] terephthalic acid,[149] pyridinecarboxaldehyde,[150] p-nitrobenzoic acid,[151-153] and p-aminobenzoic acid[154-156] to macromolecules including poly(p-nitrostyrene),[157] polystyrene,[158] poly(4-vinylpyridine),[159] and polyacrylamide.[160] The output of these studies has shown that the SERS enhancement is related either to a long-range electromagnetic enhancement of electric field at the surface of metals or to a short-range mechanism involving charge transfer between adsorbed molecules and the metal surface. One interesting observation encountered while investigating polymer/metal interfaces was the appearance of two bands near 1350 and 1590 cm^{-1}.[161] These bands have been observed in numerous compounds under various experimental conditions often overshadowing the spectra of interest. Otto[162] attributed these bands to a carbonate adsorbed on the surface. Incidentally, Tuinstra and Koenig[163] were the first investigators to obtain Raman spectra of graphite single crystals. These spectra also contained a single band at 1575 cm^{-1}, whereas the spectra of all other graphitic materials contained an additional band near 1355 cm^{-1}. Annealing experiments showed that the 1355-cm^{-1} band was due to small graphite crystallites and sensitive to the changes of the crystallite size. Other SERS studies[164] of carbon and benzoic acid with silver overlayers have demonstrated that carbon spectra have the characteristic two bands, indicating the formation of graphitic carbon. Thus, the originally proposed carbonates adsorbed on the surface were not confirmed.

Adhesion of polymeric materials to various substrates also plays an important role in the electronic industry. Parry and Dendram is[165] studied polystyrene and other molecules adsorbed on silicon overcoated with about 40 Å of silver. They found that SERS spectra of PS recorded under vacuum were the same as that obtained in ordinary Raman measurements. However, the same sample run under ambient atmospheric conditions showed the bands in the 1200–1600-cm^{-1} region. This observation was attributed to silver-catalyzed photooxidation of polystyrene. At the time the precise mechanism of polystyrene degradation on silver surfaces was still not fully understood; nevertheless, a sequence of reactions leading to polystyrene degradation can be envision as illustrated in Fig. 3.23.

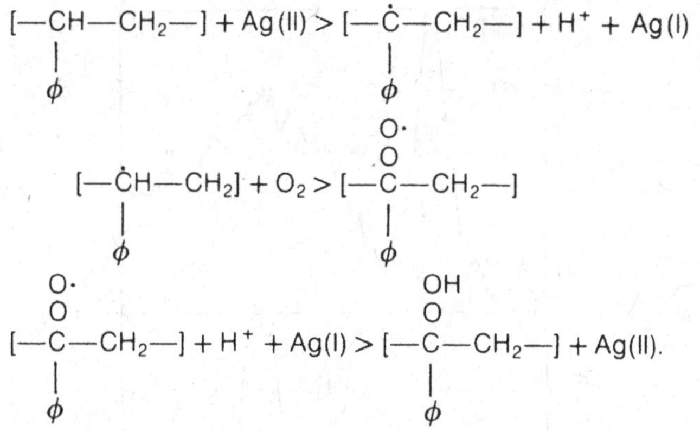

$$[-CH-CH_2-] + Ag(II) > [-\overset{\bullet}{C}-CH_2-] + H^+ + Ag(I)$$
$$\underset{\phi}{|} \qquad\qquad \underset{\phi}{|}$$

$$[-\overset{\bullet}{C}H-CH_2] + O_2 > [-\overset{\overset{\textstyle O\cdot}{\textstyle \overset{|}{O}}}{\underset{\phi}{\overset{|}{C}}}-CH_2-]$$

$$[-\overset{\overset{\textstyle O\cdot}{\textstyle \overset{|}{O}}}{\underset{\phi}{\overset{|}{C}}}-CH_2-] + H^+ + Ag(I) > [-\overset{\overset{\textstyle OH}{\textstyle \overset{|}{O}}}{\underset{\phi}{\overset{|}{C}}}-CH_2-] + Ag(II).$$

Figure 3.23 The proposed mechanism of degradation of polystyrene on a metal surface.

Although SERS has been recognized as a powerful tool in interfacial chemistry, only in recent years have a few applications of SERS appeared that have taken advantage of the unique sensitivity and selectivity to adsorbed molecules on surfaces. Brandt[166] has utilized this technique to obtain the first molecular spectra of 1,1'-diethyl-2,2'-cyanine molecule (DEC^+) by partial development of photographic film containing AgBr microcrystals to generate conditions favorable to SERS and fluorescence quenching. Such an approach allowed a successful elimination of the gelatin background spectrum. From these studies it was concluded that the essential factor responsible for obtaining good signal-to-noise (S/N) ratio in Raman spectra is the presence of silver metal, which quenches fluorescence. As a result, no Raman activity was detected from gelatin, and the enhancement was observed from the adsorbed dye. Besides these factors, further understanding of the adsorption processes have resulted. Specifically, DEC^+ does not irreversibly adsorb onto silver metal, and both the structure and surface chemistry of active surface sites are responsible for the enhancement of Raman intensities.

In an effort to establish the structure–property relationship, it is necessary to combine macroscopic properties with the chemical and structural information on a molecular level. The wettability of a solid is one of many examples that may be greatly affected by molecules adsorbed on a solid surface and ultimately may influence adhesion. Amphiphilic molecules, possessing a polar head group bonded to a long, nonpolar tail, are commonly used modifiers of wettability and the solid–liquid interfacial tension. However, one of the most important aspects of their modifying action is their structure and conformational changes occurring during liquid–solid contact and interface formation since these changes may significantly affect the wetting action. SERS was utilized to monitor the changes occurring during wetting of silver island surfaces covered with 1-hexadecane thiol ($C_{16}H_{33}SH$).[167] Figure 3.24 illustrates the region of SERS spectra that is sensitive to the conformational changes of the amphiphile molecule and represents the

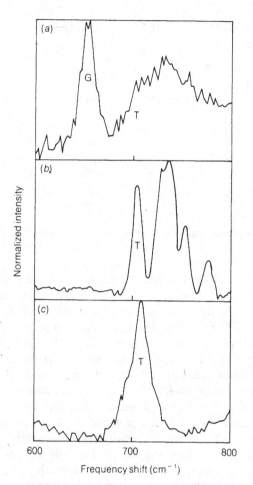

Figure 3.24   S—C stretching region in Raman spectrum of $C_{16}H_{33}SH$. (Reproduced with permission from ref. 167, copyright 1983, American Institute of Physics.)

S—C stretching region. It should be noted that, going from the liquid-phase spectrum to the spectrum of the 1-hexadecane thiol adsorbed on the surface, the band at $700\ cm^{-1}$ dominates the latter (Fig. 3.24C). Since the C—S stretching mode is acutely sensitive to the molecular conformations near the sulfur head group,[168] the authors concluded that the dominating conformation near the surface is that of trans conformation. In addition, this band is also sensitive to the surface coverage and shifts to lower energies with decreasing surface coverage.

In contrast to the S—C band, which provides conformational information near the surface, the intensity of the C—C stretching normal mode will convey conformational details from the more distant carbon tail. Figure 3.25 illustrates spectral details in the $1000–1150$-cm^{-1} region sensitive to these changes along with the conformational band assignment. On the basis of band intensity

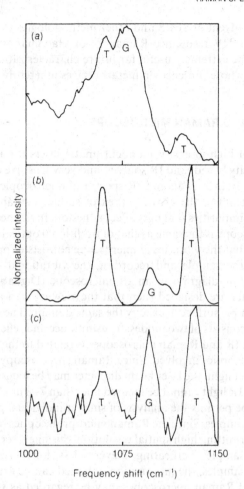

Figure 3.25   C—C stretching region in Raman spectrum of $C_{16}H_{33}SH$ (T = *trans*; G = *gauche*). (Reproduced by permission from ref. 167, copyright 1983, American Institute of Physics.)

changes, these data indicate that the predominant conformation of the hydrocarbon tail is trans, with the gauche bonds at the end of the trans sequence. In the same study, in order to further understand how adsorbed molecules may act on contact with overlaying fluids, the amphiphile molecules were precoated on silver surface, which was covered with various neat fluids, and the SERS spectra were recorded. It was found that the presence of the less polar than water (chloroform) fluids may significantly reduce the number of trans segments in the film and the process is reversible. Although a hydrocarbon tail of the amphiphile molecule was affected by the presence of nonpolar solvents, these studies also showed that the C—S bond remained unchanged, indicating that near-surface groups exhibit unchanging trans conformation.

Thus, with the advent of SERS and other methodologies to improve Raman sensitivity, such as UV resonance Raman[169] or Hadamar transform, Raman spectroscopy can be extremely useful for future characterization of surface and interfacial species whose presence on metal surfaces inherently affects adhesion.

3.5 INFRARED AND RAMAN MICROSCOPY

The availability of high-intensity monochromatic lasers for Raman measurements, high-intensity broadband IR sources, and very sensitive detectors make it possible to obtain both Raman and IR spectra of microscopic samples. This is particularly important in the growing area of surface contamination and the analysis of surface impurities that may affect adhesion. In essence, the principle of vibrational microscopy is the same as that of traditional optical microscopy. The only difference is that the vibrational microscope consists of optics appropriate only for viewing the sample and recording the vibrational spectrum. As an example, a schematic diagram of an IR microscope (Digilab) is presented in Fig. 3.26. The optics are designed such that the viewing area and the IR beam impinge on the sample surface in exactly the same domain. The universal version of the microscope consists of two modes: transmittance and reflection. One major difference between IR and Raman microscopes is related to the light wavelength restrictions and the optics involved. Since Raman spectroscopy often uses either visible or ultraviolet light, the laser beam diameter may be focused as low as 1 μm diameter; whereas IR light cannot be focused finer than 7–8 μm. Thus, the use of a Raman microprobe permits the analysis of smaller individual particles or zones in heterogeneous samples. Since the Raman microprobe optics focuses a laser to a small diameter, permitting high spatial resolution, this fine laser focus may create problems due to sample overheating. Nevertheless, for thermally and photochemically stable samples, the information obtained can be extremely valuable. Thus, both IR and Raman microscopes may be regarded as surface molecular analysis techniques that complement and, in many sampling situations, substitute the established technique of scanning electron microscopy. Heating sample stages, originally developed for optical microscopy, have been adapted to the vibrational microscopes.[170] Although both Raman and IR microscopies provide useful analytical tools, their main use becomes essential in troubleshooting or quality-control situations. Hence, there is limited literature data describing specific research in this area.[171–175]

Recent advances in the field of FT-IR spectroscopy have expended the capabilities of IR spectroscopy as an analytical tool. An extension of FT-IR spectroscopy for microscopic applications has occurred by allowing samples to be viewed with the optical microscope. The optical microscope contains optics for viewing the sample and the IR optics for recording the spectra. Both optical systems are designed to give coincident light viewing of a small area such that the IR spectrum is recorded for the exact region that is viewed.

Because of the potential benefits of performing Raman spectroscopy with FT

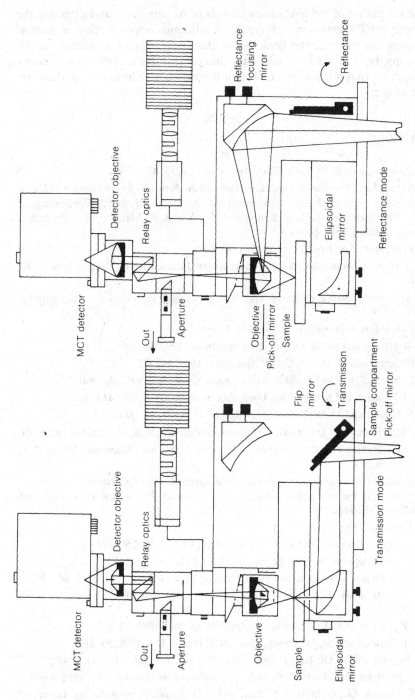

Figure 3.26 A schematic diagram of FT-IR microscope. (Reproduced by permission from Digilab Users Manual.)

near-IR excitation of Nd/YAG laser instead of Ar–ion laser, and detecting the latter with a FT spectrometer, there is a continuous ongoing effort in Raman microscopy. In essence, the principle is the same as that presented for IR microscopy, but the spacial resolution enhancement due to diffraction is better and it appears that it is possible to obtain Raman spectra from a 1-μm-diameter spot or larger.

REFERENCES

1. R. J. Anderson and P. R. Griffiths, *Anal. Chem.*, 1975, **47**, 2339.
2. J. Fahrenfort, *Attenuated Total Reflection—New Principle for Production of Useful Infrared Spectra of Organic Compounds, Molecular Spectroscopy*, (Proceedings of IVth International Meeting, Bologna, 1959), Vol. 2, A. Mangini, Ed., Pergamon Press, London, 1962, p. 701.
3. J. Fahrenfort, *Spectrochim. Acta*, 1961, **17**, 698.
4. N. J. Harrick, *Internal Reflection Spectroscopy*, Interscience Publishers, New York, 1967.
5. N. J. Harrick, *Internal Reflection Spectroscopy*, 2nd ed., Harrick Sci. Corp., Ossining, N.Y., 1979.
6. J. Fahrenfort, *Spectrochem. Acta*, 1961, **17**, 698.
7. J. B. Huang and M. W. Urban, *Appl. Spectrosc.*, 1992, **46**, 1666.
8. J. E. Bertie and H. H. Eysel, *Appl. Spectrosc.*, 1985, **39**, 392.
9. J. E. Bertie, S. L. Zhang, and R. Manji, *Appl. Spectrosc.*, 1992, **46**, 1660.
10. M. J. Dignam and S. Mamiche-Afara, *Spectrochim. Acta*, 1988, **44A**, 1435.
11. J. B. Huang and M. W. Urban, *Appl. Spectrosc.*, 1992, **46**, 1666.
12. M. W. Urban, *ATR Spectroscopy; Theory and Practice*, ACS, Washington DC, 1994.
13. M. M. Coleman and P. C. Painter, *J. Macromol. Sci., Rev. Macromol. Mater. Res.*, 1977, **C16**, 197.
14. R. J. Jacobsen, *Fourier Transform Infrared Spectroscopy: Applications to Chemical Systems*, J. R. Ferraro and L. J. Basile, Eds., Academic Press, New York, 1979, and references cited therein.
15. T. Hirshfeld, *Appl. Spectrosc.*, 1977, **31**, 289.
16. A. Saucy, S. J. Simko, and R. W. Linton, *Anal. Chem.*, 1985, **57**, 871.
17. K. A. Van Oeteran, *Werkst. Korros.*, 1964, **10**, 427.
18. H. Makishima, T. Toyoda, and N. Nakamula, *Shikizai Kyokaishi*, 1971, **44**, 156.
19. R. N. O'Brien and K. Hartman, *Am. Chem. Soc., Div. Org. Plast. Chem. Pap.*, 1968, **28**(2), 236.
20. M. W. Urban and M. T. Stewart, *J. Appl. Polym. Sci.*, 1990, **39**, 265.
21. M. T. Stewart and M. W. Urban, *Proc. ACS, PMSE Div.*, 1988, **59**, 334.
22. N. Inagaki and K. Oh-Ishi, *J. Polym. Sci., Polym. Chem. Ed.*, 1985, **23**, 1445.
23. S. S. Pesetskii, N. L. Egorenkov, and S. V. Shcherbakov, *Kolloidn. Zk.*, 1981, **43**, 992.
24. J. S. Ahearn, G. D. Davis, T. S. Sun, and J. D. Venables, in *Adhesion Aspects of Polymeric Coatings*, K. L. Mittal, Ed., Plenum Press, New York, 1983, pp.281–299.

25. O. D. Hennemann and W. Brockmann, *J. Adhes.*, 1981, **12**, 297–315.

26. G. D. Davis, T. S. Sun, J. S. Ahearn, and J. D. Venables, *J. Mater. Sci.*, 1982, **17**, 1807.

27. C. G. Golander and B. A. Sultan, *J. Adhes. Sci. Techn.*, 1988, **2**(2), 125.

28. J. C. Erikson, C. G. Golander, A. Baszkin, and L. Ter-Minassin Saraga, *J. Colloid Interface Sci.*, 1983, **100**, 381.

29. N. B. Colthup L. H. Daly, and S. E. Wiberley, *Introduction to Infrared and Raman Spectroscopy*, Academic Press, London, 1975.

30. E. G. Barlick and R. G. Messerschmidt, *Amr. Lab.*, Nov. 1984.

31. M. W. Urban, J. L. Koenig, S. B. Shih, and J. Allaway, *Appl. Spectrosc.*, 1987, **41**, 590.

32. M. W. Urban and J. L. Koenig, *Appl. Spectrosc.*, 1987, **41**, 1028.

33. A. M. Tiefenthaler and M. W. Urban, *Appl. Spectrosc.*, 1988, **42**, 163.

34. J. P. Hawranek, P. Neelakantan, R. P. Young, and R. N. Jones, *Spectrochim. Acta*, 1976, **32A**, 75.

35. J. P. Hawranek, P. Neelakantan, R. P. Young, and R. N. Jones, *Spectrochem. Acta*, 1976, **32A**, 85.

36. J. E. Bertie and H. H. Eysel, *Appl. Spectrosc.*, 1985, **39**, 392.

37. M. W. Urban, and K. W. Evanson, *Polym. Commun.*, 1990, **31**, 279.

38. K. W. Evanson and M. W. Urban, *J. Appl. Polym. Sci.*, 1991, **42**, 2287.

39. K. W. Evanson , T. A. Thorenstenson, and M. W. Urban, *J. Appl. Polym. Sci.*, 1991, **42**, 2297.

40. K. W. Evanson and M. W. Urban, *J. Appl. Polym. Sci.*, 1991, **42**, 2309.

41. S. A. Francis and A. H. Ellison, *J. Opt. Soc. Amr.*, 1959, **49**, 130.

42. G. W. Poling, *J. Electrochem. Soc.*, 1970, **117**, 520.

43. R. G. Greenler, *J. Chem. Phys.*, 1966, **44**, 310.

44. J. D. Swalen and J. F. Rabolt, *Fourier Transform Infrared Spectroscopy: Applications to Chemical Systems*, J. R. Ferraro and L. J. Basile, Eds., Academic Press, New York, 1979, p. 283.

45. W. G. Golden, p. 315, ref. 41 (above).

46. G. Kortum, *Reflectance Spectroscopy*, Springer-Verlag, New York, 1969.

47. P. Kubelka and F. Munk, *Z. Techn. Phys.*, 1931, **13**, 593.

48. P. Kubelka and F. Munk, *J. Opt. Soc. Amr.*, 1948, **38**, 448.

49. R. K. Vincent and G. R. Hunt, *Appl. Opt.*, 1968, **7**, 53.

50. P. R. Griffiths and F. Fuller, *Advanced Infrared and Raman Spectroscopy*, R. J. H. Clark and R. E. Hester, Eds., Heyden, London, 1981, Vol. 9., Chapter 2.

51. M. P. Fuller and P. R. Griffiths, *Am. Lab.*, 1978, **10**, 69.

52. W. M. Grim, J. A. Graham, and W. G. Fateley, *Transform Times*, 1983, **1**(Nov.), 1.

53. M. P. Fuller and P. R. Griffiths, *Anal. Chem.*, 1978, **50**, 1906.

54. R. T. Graf, H. Ishida, and J. L. Koenig, *Anal. Chem.*, 1984, **56**, 773.

55. M. W. Urban and J. L. Koenig, *Proc. SPIE, Int. Conf. FT-IR Spectrosc.*, Ottawa, Canada, 1985.

56. M. P. Fuller, *Coal*, 1982, **61**, 529.

57. R. M. Ianniello, H. J. Wiek, and A. M. Yacynych, *Anal. Chem.*, 1983, **55**, 2067.

58. D. B. Chase, R. L. Amey, and W. G. Holtje, *Appl. Spectrosc.*, 1982, **36**, 155.

59. J. A. Davies and A. Sood, *Amr. Lab.*, 1984, **16**, 122.

60. M. T. McKenzie, S. R. Culler, and J. L. Koenig, *Appl. Spectrosc.*, 1984, **38**, 786.

61. S. R. Culler, M. T. McKenzie, L. J. Fina, H. Ishida, and J. L. Koenig, *Appl. Spectrosc.*, 1984, **38**, 791.

62. E. G. Chatzi, H. Ishida, and J. L. Koenig, *Appl. Spectrosc.*, 1986, **40**, 695.

63. K. C. Cole, A. Pilon, D. Noel, J. -J. Hechler, A. Chouliotis, and K. C. Overbury, *Appl. Spectrosc.*, 1988, **42**, 761.

64. A. Rosencwaig and A. Gersho, *Science*, 1975, **190**, 556.

65. M. W. Urban, S. R. Gaboury, W. T. McDonald, and A. M. Tiefenthaler, *Chemistry Series*, C. Craver and T. Provder, Eds., 42271, 1990, Amer. Chem. Soc., Washington DC, Chap. 17.

66. M. W. Urban and J. L. Koenig, *Appl. Spectrosc.*, 1986, **40** (7), 994.

67. C. Q. Yang, T. J. Ellis, R. R. Breese, and W. G. Fateley, *Polym. Mat. Sci. Eng.*, 1985, **53**, 169.

68. B. S. H. Royce, Y. C. Teng, and J. B. Enns, *Ultrason. Symp. Proc.*, 1980, 652

69. R. J. Bell, *Introductory Fourier Transform Spectroscopy*, Academic Press, New York, 1972.

70. A. M. Tiefenthaler and M. W. Urban, *Composites*, 1989, **20**, 585.

71. B. Arkles and W. Peterson, *Proc. 35th Ann. Conf. Reinf. Plast./Comp. Inst. Soc. Plast. Ind.*, 1980, **20-A**, **1-20-A**, 2.

72. A. M. Tiefenthaler and M. W. Urban, *Composites*, 1989, **20**(2), 145.

73. M. W. Urban and J. L. Koenig, *Appl. Spectrosc.*, 1985, **39**, 1051.

74. M. W. Urban and J. L. Koenig, *Appl. Spectrosc.*, 1986, **40**, 513.

75. E. G. Chatzi, M. W. Urban, and J. L. Koenig, *Markomol. Chem., Mcromol. Symp.*, 1986, **5**, 99.

76. E. G. Chatzi, M. W. Urban, H. Ishida, and J. L. Koenig, *Polymer*, 1986, **27**, 1850.

77. E. G. Chatzi, M. W. Urban, H. Ishida, J. L. Koenig, A. Laschewski, and H. Ringsdorf, *Langmuir*, 1988, **4**, 846.

78. S. G. Gaboury and M. W. Urban, *Proc. ACS, PMSE Div.*, 1989, **60**, 875.

79. M. W. Urban and J. L. Koenig, *Anal. Chem.*, 1988, **60**, 2408.

80. S. R. Gaboury and M. W. Urban, *Polym. Preprints*, 1988, **29**, 356.

81. M. W. Urban and S. R. Gaboury, *Macromolecules*, 1989, **22**, 1486.

82. W. F. McDonald, H. Geottler, and M. W. Urban, *Appl. Spectrosc.*, 1989.

83. W. F. McDonald and M. W. Urban, *J. Adhes. Sci. Eng.*, 1990, **4**, 751.

84. A. C. Boccarra, D. Fournier, and J. Badoz, *Appl. Phys. Lett.*, 1980, **36**, 130.

85. A. C. Boccarra, D. Fournier, and W. Jackson, *Am. Opt. Lett.*, 1980, **5**. 377.

86. D. Fournier, A. C. Boccarra, N. M. Amer, and R. Gerlach, *Appl. Phys. Lett.*, 1980, **37**, 519.

87. C. Mortarra and M. J. D. Low, *Spectrosc. Lett.*, 1982, **15**, 689.

88. M. J. D. Low, C. Morterra, and A. G. Saverdia, *Spectrosc. Lett.*, 1982, **15**, 415.

89. M. J. D. Low and C. Morterra, *Carbon*, 1983, **21**, 275.

90. C. Morterra and M. J. D. Low, *Carbon*, 1983, **21**, 283.

91. M. J. D. Low, C. Morterra, and A. G. Severdia, *J. Mol. Catal.*, 1983, **20**, 311.

92. M. J. D. Low and G. A. Parodi, *Spectrosc. Lett.*, 1978, **11**, 581.

93. C. Morterra, M. J. D. Low, and A. G. Severdia, *Appl. Surf. Sci.*, 1985, **20**, 317.

94. M. J. D. Low, M. Lacroix, and C. Morterra, *Spectrosc. Lett.*, 1982, **15**, 57.

95. A. Hartstein, J. R. Kirtley, and J. C. Tsang, *Phys. Rev. Lett.*, 1980, **45**, 201.

96. A. Hatta, Y. Suzuki, and W. Suetaka, *Appl. Phys.*, 1984, **A35**, 135.

97. R. M. Azzam and N. M. Bashara, *Ellipsometry and Polarized Light*, North-Holland, New York, 1977.

98. M. Born and W. Wolf, *Principles of Optics*, 6th ed., Pergamon Press, Oxford, 1980.

99. P. S. Hauge, *Surf. Sci.*, 1980, **96**, 108.

100. R. W. Stobie, B. Rao, and M. J. Dignam, *Appl. Opt.*, 1975, **14**, 999.

101. J. A. Bardwell and M. J. Dignam, *Fourier Transform Infrared Characterization of polymers*, H. Ishida, Ed., Plenum press, New york, 1987.

102. R. R. Schaefer, *J. Phys. Colloq.*, 1983, **C10**, 87.

103. J. R. Adams, J. R. Ziedler, and N. M. Bashara, *Opt. Commun.*, 1975, **15**, 115.

104. A. Roseler and W. Molgedey, *Infrared Phys.*, 1984, **24**, 1.

105. R. T. Graf, J. L. Koenig, and H. Ishida, *Anal. Chem.*, 1986, **58**, 64.

106. R. T. Graf, J. L. Koenig, and H. Ishida, *Appl. Spectrosc.*, 1986, **40**, 498.

107. A. Handi, *Essentials of Modern Physics Applied to the Study of the Infrared*, Pergamon Press, Oxford, 1966.

108. E. Stager and R. Rasmus, *Appl. Spectrosc.*, 1974, **28**, 376.

109. P. R. Griffiths, *Appl. Spectrosc.*, 1972, **26**, 73.

110. J. Houghton and S. D. Smyth, *Infrared Physics*, Oxford University Press, Oxford, 1966.

111. T. R. Harrison, *Radiation Pyrometry*, Wiley, New York, 1960.

112. V. W. King and J. L. Laurer, *J. Lubr. Techn.*, 1981, **103**, 65.

113. J. L. Laurer and M. E. Peterkin, *J. Lubr. Tech.*, 1975, **97**, 145.

114. J. L. Laurer and M. E. Perkin, *Am. Lab.*, 1975, **7**, 27.

115. J. L. Laurer and M. E. Peterkin, *J. Lubr. Tech.*, 1978, **98**, 230.

116. J. L. Laurer, *J. Lubr. Techn.*, 1979, **101**, 67.

117. R. G. Greenler, *Surf. Sci.*, 1977, **69**, 647.

118. K. Makinouchi, K. Wagatsuma, and W. Suetaka, *Bunko Kenkyu*, 1980, **29**, 23.

119. T. Matsui, K. Tani, S. Ohashi, and S. Tanaka, *Bunko kenkyu*, 1982, **31**, 360.

120. J. F. Blanke, S. E. Vincent, and J. Overend, *Spectrochim. Acta*, 1976, Part A, **32A**, 162.

121. D. Kember, D. H. Chenery, N. Sheppard, and J. Fell, *Spectrochim. Acta*, 1979, Part A, **35A**, 455.

122. M. J. D. Low, *Appl. Spectrosc.*, 1968, **22**, 463.

123. R. J. Brown and B. G. Young, *Appl. Opt.*, 1975, **14**, 2927.

124. D. A. Mantell, S. B. Ryali, and G. L. Haller, *Chem. Phys. Lett.*, 1983, **102**, 37.

125. D. Kembler and N. Sheppard, *Appl. Spectrosc.*, 1975, **29**, 496.

126. L. M. Gratton, S. Paglia, F. Scattaglia, and M. Cavallini, *Appl. Spectrosc.*, 1978, **32**, 310.

127. P. C. M. Van Woerkom, P. Blok, H. J. van Veenendaal, and R. L. de Groot, *Appl. Opt.*, 1980, **19**, 2546.

128. M. Primet, P. Flouillous, and B. Imelik, *J. Catal.*, 1980, **61**, 553.

129. Y. Nagasawa and A. Ishitani, *Appl. Spectrosc.*, 1984, **38**, 168.

130. J. B. Bates, in *Fourier Transform Infrared Spectroscopy — Application to Chemical Systems*, J. R. Ferraro and L. J. Basile, Eds., 1978, Vol. 1, Chapter 3, New York, Academic Press, and references cited therein.

131. D. Compton, private communication.

132. A. Champion and W. H. Woodruff, Anal. Chem., 1987, 59, 1299A.

133. D. Bruce Chase, *J. Amr. Chem. Soc.*, 1986, **108**, 7485.

134. M. Fleischmann, P. J. Hendra, and A. J. McQuillan, *Chem. Phys. Lett.*, 1974, **26**, 163.

135. R. K. Chang and T. Furtak, *Surface Enhanced Raman Scattering*, Plenum Press, New York, 1982.

136. H. Metiu and P. Das, *Ann. Rev. Phys. Chem.*, 1984, **35**, 507.

137. T. E. Futrak and D. Roy, *Surf. Sci.*, 1985, **158**, 126.

138. J. I. Gersten, R. L. Birke, and J. R. Lombardi, *Phys. Rev. Lett.*, 1979, **43**, 147.

139. S. S. Jha, *Surf. Sci.*, 1985, **158**, 190.

140. H. Baltruschat and J. Heitbaum, *Surf. Sci.*, 1986, **166**, 113.

141. Q. Feng and T. M. Cotton, *J. Chem. Phys.*, 1986, **90**, 983.

142. H. Ishida and A. Ishitani, *Appl. Spectrosc.*, 1983, **37**, 450.

143. T. Watanabe, O. Kawanami, H. Katoh, and K. Honda, *Surf. Sci.*, 1985, **158**, 341.

144. T. Vo-Dinh, M. Y. K. Hiromoto, G. M. Begun, and R. L. Moody, *Anal. Chem.*, 1984, **56**, 1667.

145. P. T. Kissinger, C. R. Preddy, R. E. Shoup, and W. R. Heineman, "Fundamental Concepts of Analytical Electrochemistry", *in Laboratory Techniques in Analytical Chemistry*, P. T. Kissinger and W. R. Heineman, Eds., Marcel Dekker, New York, 1984, Chapter. 2.

146. W. L. Parker, R. M. Hexter, and A. R. Siedle, *Chem. Phys. Lett.*, 1984, **107**(1), 96.

147. D. J. Jeanmaire and R. P. Van Duyene, *J. Electroanal. Chem.*, 1977, **84**, 1.

148. C. Y. Chen, I. Davoli, G. Ritchie, and E. Burnstein, *Surf. Sci.*, 1980, **101**, 363.

149. M. Moskowits and J. S. Suh, *J. Phys. Chem.*, 1984, **88**, 1293.

150. J. C. Tsang, J. R. Kirtley, and J. A. Bradley, *Phys. Rev. Lett.*, 1979, **43**, 772.

151. R. Dornhaus, R. E. Benner, R. K. Cheng, and I. Chabay, *Surf. Sci.*, 1980, **101**, 367.

152. C. A. Murray, D. L. Allara, and M. Rhinewine, *Phys. Rev. Lett.*, 1981, **46**, 57.

153. P. G. Roth, R. S. Venkatachalam, and F. J. Boerio, *J. Chem. Phys.*, 1986, **85**, 1150.

154. J. S. Suh, D. P. DiLella, and M. Moskovits, *J. Phys. Chem.*, 1983, **87**, 1540.

155. J. C. Tsang, P. Avouris, and J. R. Kirtley, *J. Chem. Phys.*, 1983, **79**, 493.

156. R. S. Venkatachalam, F. J. Boerio, and P. G. Roth, *J. Raman Spectrosc.*, 1988, **19**, 281.

157. D. L. Allara, C. A. Murray and S. Bodoff, in *Physicochemical Aspects of Polymer Surfaces*, Vol.1, K. L. Mittel, Ed., Plenum Press, New York, 1983.

158. D. B. Parry and A. L. Dendramis, *Appl. Spectrosc.*, 1986, **40**, 656.

159. P. G. Roth and F. J. Boerio, *J. Polym. Sci.; Part B: Polym. Phys.*, 1987, **25**, 1923.

160. J. S. Suh and K. H. Michaelin, *J. Raman Spectrosc.*, 1987, **18**, 409.

161. R. S. Venkatachalam, F. J. Boerio, M. R. Carnevale, and P. G. Roth, *Appl. Spectrosc.*, 1988, **42**, 1207.

162. A. Otto, *Surf. Sci.*, 1978, **75**, L392.

163. F. Tuinstra and J. L. Koenig, *J. Chem. Phys.*, 1970, **53**, 1126.

164. J. C. Tsang, J. E. Demuth, P. N. Sanda, and R. J. Kirtley, *Chem. Phys. Lett.*, 1980, **76**, 54.

165. D. B. Parry and A. L. Dendramis, *Appl. Spectrosc.*, 1986, **40**, 656.

166. E. S. Brandt, *Appl. Spectrosc.*, 1988, **42**, 882.

167. C. J. Sandroff, Sandroff, S. Garoff, and K. P. Leung, *Chem. Phys. Lett.*, 1983, **96** (5), 547.

168. D. W. Scott and M. Z. El-Sabban, *J. Mol. Spectrosc.*, 1969, **30**, 317.

169. A. A. Sanford, *Anal. Chem.*, 1984, **56**, 720.

170. E. Payen, M. C. Dhamelincourt, P. Dhamelincourt, J. Grimblot, and J. P. Bonnello, *Appl. Spectrosc.*, 1982, **36**, 30.

171. M. A. Harcock, L. A. Lentz, B. L. Davie, and K. Krishnan, *Appl. Spectrosc.*, 1986, **40**, 210.

172. D. J. Skovanek, *J. Coat. Techn.*, 1989, **61**, 31.

173. J. A. Reffner, R. G. Messerschmidt, and J. P. Coates, in *Microbeam Analysis*, R. H. Geiss, Ed., 1987, p. 177..

174. M. E. Lacy, *Proc. 28th Ann. Tech. Meeting Inst. Environmental Science*, 1982.

CHAPTER 4

NORMAL VIBRATIONS OF SMALL MOLECULES ON SURFACES

4.1 FACTORS GOVERNING SURFACE VIBRATIONAL ACTIVITY; SURFACE VIBRATIONAL FREEDOM

As was shown in Chapter 1, diatomic molecules exhibit only one vibration along the chemical bond. Its frequency is given by

$$v = \frac{1}{2\Pi c}\sqrt{\frac{k}{\mu}} \tag{4.1}$$

where k is the force constant, the reduced mass, and c the velocity of light. For homonuclear X—X molecules with the D_{xh} point group symmetry, the vibration is not IR-active but is Raman-active. For the heteronuclear A—B molecules (C_{xv}), the A—B stretching vibration is both IR- and Raman-active. While molecules of the bulk have defined symmetry, the same molecule at the surface will exhibit different symmetry due to the loss of one or more sites. A typical case is schematically depicted in Fig. 4.1A, where the X_2 molecule with the initial gas-phase D_{xv} group symmetry, along with the surface site B, undergoes symmetry changes. A new B—X—X bonding has a symmetry of C_s.

For the surface species depicted above, only those symmetry elements that exhibit mirror planes perpendicular to the interface or rotational axes normal to the surface will be permitted. What that means from the groups theory point of view, is that only C_n and C_{nv} groups are allowed on the surface to which molecules are being attached as a result of adsorption. The situation, however, will be different if multilayer molecules are attached to the surface. The Langmuir–Blodgett films are a good example. One can consider such systems as two phases containing sub-

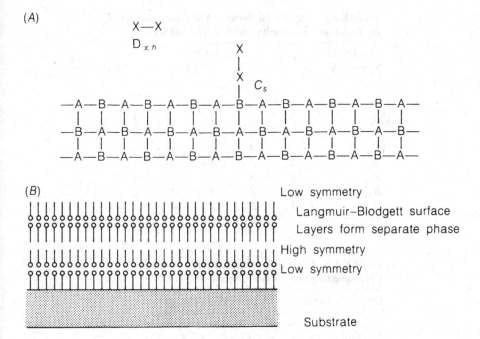

Figure 4.1   Symmetry changes between the surface and bulk environment for atom B.

strate and adsorber bonded by an usually low-symmetry interface. An example of the monolayer and multilayer molecules with different symmetries are shown in Fig. 4.1*B*.

The symmetry of the species at the interface will be affected by two factors: crystal structures of a substrate and the symmetry of the adsorbed molecules. For example, if one considers the gas-phase isolated molecule with the symmetry of D_{2h}, one can construct useful correlation tables illustrating how vibrational gas-phase modes change as a result of adsorption. Table 4.1 shows the correlations between the symmetry species indicated by the molecule having D_{2h} symmetry with other symmetry sites on a substrate. Of course, the character tables for different point groups and different symmetries will provide the answer to which mode is IR- or Raman-active or inactive.

Let us go back to a main trust and consider the two factors that define the surface symmetry and therefore vibrational selection rules of the adsorbed species. They are the crystal structures, or point groups, in the case of amorphous species of the substrates possessing localized symmetries, and the symmetry of the adsorbates. Table 4.2 summarizes various bulk crystal structures and their effect on the allowed symmetry elements at the surface giving various point groups. Having established theoretical aspects of the group theory considerations, we should keep in mind that the situations described above refer to cases with low surface

TABLE 4.1 Correlation between Vibrational Modes of the Gas-Phase Symmetry and Site Symmetries

Gas-Phase Symmetry D_{2h}	Site—Symmetries				
	C_{2v}	C_2	C_s	C_s^1	C_1
A_g	A_1	A	A'	A'	A
A_u	A_2	A	A''	A''	A
B_{1g}	A_2	A	A''	A''	A
B_{1u}	A_1	A	A'	A'	A
B_{2g}	B_1	B	A'	A''	A
B_{2u}	B_2	B	A''	A'	A
B_{3g}	B_2	B	A''	A'	A
B_{3u}	B_1	B	A'	A''	A

TABLE 4.2 Crystal Structures of Substrates and Corresponding Local Point Groups

Crystal Structure	Point Group
Square	$C_{4v}, C_{2v}, C_4, C_2, C_s, C_1$
Hexagonal	$C_{6v}, C_{3v}, C_{2v}, C_6, C_3, C_2, C_s, C_1$
Rectangular	C_{2v}, C_2, C_s, C_1
Oblique	C_2, C_1

coverage. Physically, it means that the adsorbed species do not interact with each other. As we saw earlier, an extensive surface coverage may lead to other effects. With this in mind, let us consider various surface situations leading to symmetry changes.

4.2 GAS ATOMS AND DIATOMIC MOLECULES ADSORBED ON METAL SUBSTRATES

4.2.1 Atoms Adsorbed on Metals

When atom X, with zero vibrational degrees of freedom, becomes attached to a metal surface with a cubic face, such as shown in Fig. 4.2A, additional vibrational modes will be introduced. Since such surface structure exhibits C_{4v} symmetry, the M—X bond stretching will have A_1 symmetry along with the parallel to the surface degenerate bending mode with the E symmetry.

Under different circumstances, gas phase atom X may form bridge structures shown in Fig. 4.2B, also with the C_{4v} symmetry. In the cases discussed above, since the species formed have the same point groups (the same symmetry elements), the same IR and Raman modes will be active. However, because the number of in-

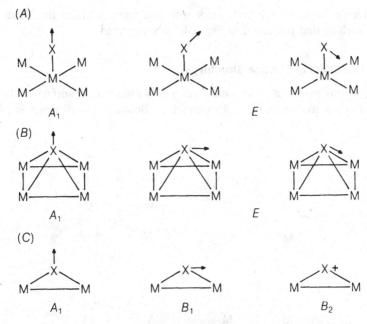

Figure 4.2 (*A*) Normal surface vibrations resulting from adsorption of an atom X on C_{4v} symmetry site in a normal to the surface position; (*B*) normal surface vibrations resulting from the adsorption of an atom X on C_{4v} symmetry site in a bridge position; (*C*) normal surface vibrations resulting from the adsorption of an atom X on C_{4v} symmetry site in a twofold bridge position.

volved atoms is different as compared to the previous example, the number of vibrations will also change.

When an atom X present in the gas phase forms a twofold bridge with a substrate, the symmetry of such species will be C_{2v} and the normal vibrations such as shown in Fig. 4.2C, will be active. It should be noted that as a result of the symmetry changes, instead of a doubly degenerated E mode, the selection rules for the C_{2v} point group predict two nondegenerated B_1 and B_2 modes.

4.2.2 Diatomic Molecules Attached to Metal Surfaces

When an XY molecule is being attracted to the surface and remains in a vertical position such as depicted in Fig. 4.3, the group symmetry will be C_{4v}. Again, the same vibrational modes as presented in the first two cases will be IR- and Raman-active, but the total number will be dictated by the number of atoms involved ($3n - 6$ rule; Chapter 1). This is illustrated in Fig. 4.3A. When the same molecule XY forms twofold bridge structures with the C_{2v} symmetry, the following modes ertinent to the XY species are expected and presented in Fig. 4.3B. Finally, if an

X_2 molecule forms bridge structures with the metal surface, the vibrational modes, such as that presented in Fig. 4.3C, are expected.

4.2.3 The Effect of Surface Structures

Figure 4.4 illustrates the arrangements of atoms found at the surface of the face-centered cubic structures (fcc) exposing the following crystallographic planes:

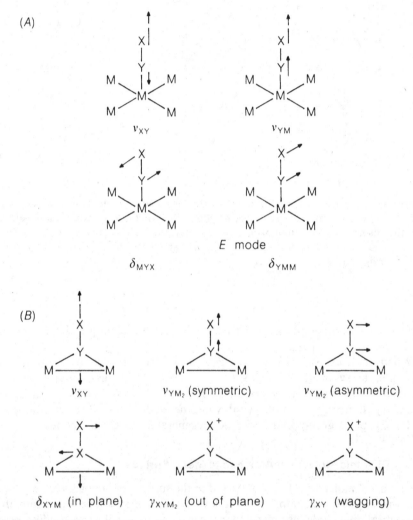

Figure 4.3 (*A*) Normal surface vibrations resulting from adsorption of XY molecule on C_{4v} symmetry site in a normal to the surface position; (*B*) normal surface vibrations resulting from the adsorption of XY molecule on C_{4v} symmetry site in a twofold bridge position; (*C*) normal surface vibrations resulting from the adsorption of XY molecule on C_{4v} symmetry site in a bridge position.

(C)

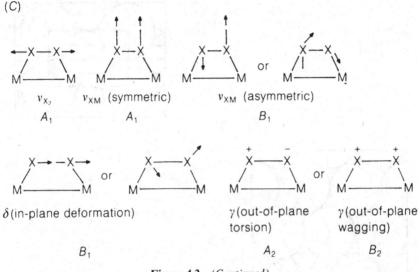

Figure 4.3 (*Continued*)

(100), (111), and (110). These planes belong to the square, hexagonal, and primitive rectangular two-dimensional lattice systems. The high symmetry potential adsorbate sites can be designated according to the nearest-neighbor metal atom coordination number in the site, indicating how many equivalent atoms can participate in surface interactions. For example, the (100) surface can adopt on-top, bridge and fourfold hollow structures, with the C_{4v}, C_{2v} and C_{4v} bare-site symmetries, respectively.

As was indicated in Chapter 2, many adsorption processes may lead to completely different spectroscopic results if the surface coverage is high and the adsorbate molecules are able to form ordered surface structures. While the first requirement can be achieved by the experimental designs, formation of the ordered surface structures will be determined by the nature of the crystal structures of the bulk. Therefore, the symmetry of the crystallographic unit cell is a necessary ingredient in the determination of the surface crystallographic unit cell, which, in turn, establishes the reciprocal surface lattice, often called the *surface Brillouin zone*. This knowledge is essential in determining the existence of elastic waves called *phonons*, propagating throughout the surface. The term *phonon* applies to all elastic waves propagating through the crystal, not necessarily the surface vibrations, and is characterized by a wave vector $k = 2\pi/\lambda$, where λ is the wavelength of a wave resulting from the vibrating atoms deviating from the equilibrium position. For simplicity purposes, let us consider direction x and characterize the surface Brillouin zone such as that depicted in Fig. 4.5A. Since $k = 0$ only when λ goes to infinity, k becomes π/a_x, where a_x is the dimension of the surface Brillouin zone representing the edge of the zone. Consider the motion of the adsorbate atom perpendicular to the metal atom chain in the direction A—M shown in Fig. 4.5B. Under these circumstances, each point of the Brillouin zone

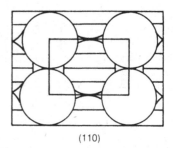

(110)

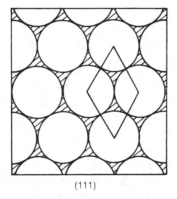

(111)

(100)

Figure 4.4 Surface arrangement of atoms present on the face-centered cubic (fcc) surface exposing (100), (111), and (110) planes.

has a single phonon arising from the A—M stretching motion. Atoms move in phase at the center of the zone at $k = 0$ or out of phase at the phase boundary when $k = \pi/2m$ (where m is the metal dimension).

In the case of two adsorbate molecules per unit cell, such as that presented in Fig. 4.5C, the adsorbate atom imposes the unit-cell dimensions as $3m$. One atom occupies a bridge site and one atom a top side. As a result, there are two A—M stretches in the unit cell that are perpendicular to the MM direction.

The case of two adsorbate atoms per unit cell is depicted in Fig. 4.5D. Each adsorbate atom contributes to the two symmetric and antisymmetric modes of vibrations. Interestingly, these two modes exhibit different symmetries relative to the mirror plane of the unit cell: they belong to symmetric and antisymmetric representations. It should be kept in mind, however, that in practice the observation of more than one frequency associated with each mode of vibration is referred to as *factor group* or *field splitting*.

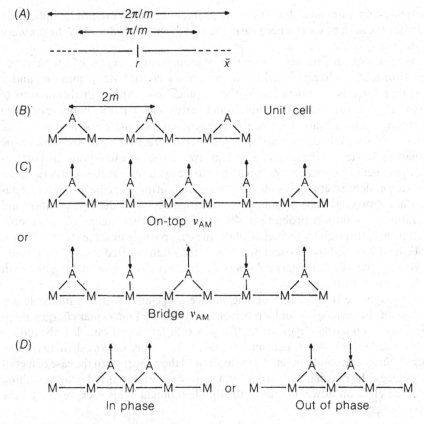

Figure 4.5 (*A*) Brillouin zone corresponding to metal lattice (dashed line); (*B*) one adsorbate atom per unit cell; (*C*) two nonequivalent adsorbate atoms per unit cell; (*D*) two equivalent adsorbate atoms per unit cell.

4.3 FORCE FIELD

As was pointed out in Chapter 1, an essential factor in the vibrational frequency and intensity calculations is the choice of the interatomic potential function or, using more commonly employed terminology, the force field. It should be also mentioned that it is practically impossible to determine from experiment the numerical values of all force constant $\mathbb{F}$ matrix elements in internal coordinates. In most cases, even including the constants obtained from the analysis of the vibration-rotation spectrum, experimental data available is smaller than the number of force constants. Thus it appears apparent that the basic concept is to search for a good model of vibrational potential. This is primarily attributed to the fact that when molecules have symmetry, a mathematically rigorous way to reduce the number of force constants would imply to remain in the space of internal coordinates. While such a task is not easily achievable for "ordinary molecules" in a nonadsorbed state, the problem becomes even more difficult to solve for the surface species because of the complexity introduced by the presence of surface.

As described in Chapter 1, a motion of atoms can be expressed by Newton's equation and is characterized by displacement coordinate q, mass m, and a restoring force is proportional to the displacement. Although the motion of atoms is prerequisite and, as the vibrational energy selection rules imply, exists even at absolute zero, vibrational spectroscopy requires that these nonideal harmonic oscillators are excited or decay as a result of interaction with electromagnetic radiation. In view of these considerations, two classes of excitations are particularly pertinent to surface vibrational spectroscopy. If the oscillator experiences a time-dependent interaction potential, the conservation principle provides a relationship between the energy gain of the vibrating oscillator during excitation and vibrational excitation probability distribution. Another means of vibrational excitation is through the influence of the moving particle in some sort of electrostatic field, coupled to the oscillator. The latter, often referred to as the *trajectory theory*, is, however, non-energy-conserving. Further discussion on the subject can be found in the literature.[1-3]

Although the total charge on the surface associated with the molecule adsorbed on the surface is usually represented as a sum of molecular charges along with induced polarizibilities within the substrate, let us first establish the monopole contributions. When a chemical bond to the surface is formed, there is some sort of charge transfer between the molecule and the substrate. In the case of metallic surfaces, the net charge distribution will be such that induced image resulting from the electron flow will induce the dipole moment. This is expressed by the following equation:

$$\mu_d = 2fe[s_0(v) + q(t)] \tag{4.2}$$

where $(f \times e)$ represents the fraction of electrons being transferred, s_0 is the equilibrium distance of the static charge from the image plane, and q is a small displacement from its equilibrium position. Figure 4.6 illustrates the molecular

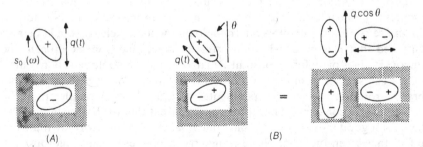

Figure 4.6 The distribution of molecular and induced image charges for (*A*) molecule having a permanent monopole charge and (*B*) molecule having a dynamic moment.

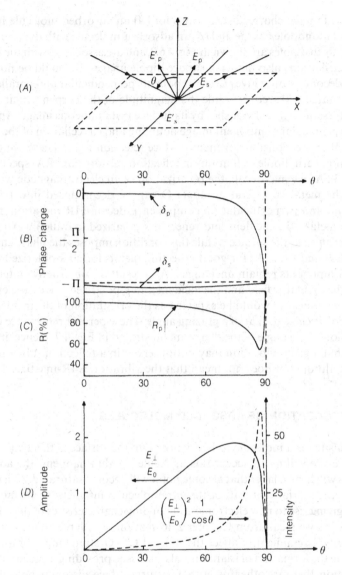

Figure 4.7   The IR reflectivity from metal surfaces: (A) the incident and reflected s- and p-polarized electric field vectors; (B) the angular dependence of phase change; (C) the reflectance of the p-polarized component; and (D) the normalized amplitude (left side) and intensity (right side) variation of the component of the p-polarized electric fied acting normal to the surface.

and induced image charge distributions for (*A*) an adsorbed molecule having a permanent monopole charge and (*B*) an adsorbed molecule with dynamic dipole moment.[4] Both dipoles are shown in Fig. 4.6*B*, and are associated with the intrinsic charge distributions also inducing surface dipole images. It should be noted that vectorial decomposition gives two components perpendicular and parallel to the substrate surface. Moreover, while the magnitude of the perpendicular component doubles, the parallel vanishes by its surface instantaneous image. Apparently, the perpendicular component image doubling and cancellation of the parallel component are responsible for many surface selection rules. These are the processes that govern dipole excitations in reflection–absorption (RA) spectroscopy (Chapter 3). As we may recall, the polarization of an electromagnetic wave incident on the metal surface at an angle Θ can be decomposed into parallel (*s* component) and perpendicular (*p* component). Because IR radiation does not penetrate metals, the incident and reflected *s*-polarized (parallel) wave components will vanish at the surface. While this condition imposes the phase change by 90° for ideal and near 90° for good reflecting metals for all *s*-polarized components, *p* components remain unchanged. As a result of constructive interactions between the oscillator field and the incident and *p*-polarized waves, the enhancement of the intensities would be expected as the orientation of the incident wave propagation is close to that of a grazing angle. These percent reflectance changes as a function of the angle of incidence are illustrated in Fig. 4.7. Hence, it is quite obvious that a given vibration may not observed in a reflection–absorption IR spectrum, although it does not mean that the vibration is IR-inactive.

4.4 PERTURBATION OF ADSORBED MOLECULES

A prerequisite for a molecule to be adsorbed on the surface is the availability of the surface site willing to accept this molecule. As that happens, the adsorbate molecules will form a layer that at some point will become saturated. As a result of overall or local surface saturation, the intramolecular interactions are most likely to occur, giving rise to new electronic and bonding characteristics of the adsorbate molecules. As we recall from Chapter 2, formation of ordered islands of carbon monoxide will significantly affect the normal C—O stretching vibrational energy. While the formation of islands is only one issue providing a preferential area of adsorption, there are other forces that perturb surface adsorbed molecules. The structure–property relationships between adsorbate and chemically active surfaces that, for a given crystal structure, are often characterized by electronegativity are extremely useful in predicting the ionicity of its bonds with various adatom species. The difficulty, however, in developing further understanding and uniform theory of adsorbate interactions arises from the fact that low symmetry of the adsorbate molecules often limits the calculations. Because there are numerous theories regarding these interactions, and unfortunately many of them represent quite controversial issues, the reader is directed to the source literature.[5] One principle that should be kept in mind is that although solid surfaces introduce a similar effect on the adsorbate molecules as condensation of gases to liquid phase, the asymmetric nature of the force field plays an additional role. Whereas solution environment provides a highly symmetric environment, the presence of surface on one side of molecule induces different interactions with surroundings, distorting the molecules so that certain previously forbidden vibrations will appear. In fact, solvent surroundings on a molecule may lead to a loss of symmetry and appearance of otherwise inactive bands when solvent–solute interactions are strong. In view of these considerations, several specific and experimentally proven cases will be discussed in the following sections.

4.5 HYDROGEN BONDING

Although IR spectroscopy has been utilized for many decades in monitoring hydrogen bonding in polymers and surfaces, only during the 1980s have IR data been employed to estimate the thermodynamic quantities associated with hydrogen bond formation. In essence, depending on chemical composition of a sample, hydrogen bonding often occurs with carbonyl groups that exhibit tremendous affinity to accepting species having deficiency in electrons. A typical example of these species are hydrogen atoms on hydroxyl or amine groups that, as a result of the electron affinity of $C{=}O$ groups, will form associations such as

$$
\begin{array}{ccc}
\cdot\text{H}-\text{O}- & & \cdot\text{H}-\text{N}- \\[4pt]
\underset{\displaystyle\overset{\displaystyle\|}{\underset{\diagup\ \diagdown}{\text{C}}}}{\text{O}} & \text{or} & \underset{\displaystyle\overset{\displaystyle\|}{\underset{\diagup\ \diagdown}{\text{C}}}}{\text{O}}
\end{array}
$$

Because the surfaces of all oxides are covered with hydroxyl groups or water, the latter being another species that tends to form strong hydrogen-bonded structures, all play an important role in adsorption processes. While the role of hydroxyl groups will be discussed later on in this chapter, let us focus on such aspects of hydrogen bonding as the ability of surfaces to attract other species.

The first observation even inexperienced spectroscopists will make just by looking at an IR spectrum of any species that contains or adsorbs water vapor, is that even its traces adsorbed on the surface will lead to tremendous intensity changes in the $3500{-}3000\text{-cm}^{-1}$ region. Now, depending on the surface coverage, the intensity will vary but may also be influenced by the existence of other species.

These other species may be N—H or $C{=}O$ groups. It should be kept in mind that even though spectroscopic data in the N—H stretching region of IR spectra for polyamides and polyurethanes were utilized to estimate thermodynamic parameters associated with hydrogen bonding, only in the last 10 years was it recognized that the absorption coefficient of the N—H groups may vary as a function of temperature.

4.6 CARBON MONOXIDE

When carbon monoxide becomes attached to the metal surface, it undergoes symmetry changes from C_v to either C_{4v}, C_{2v}, or C_s, depending on the bonding symmetry to the surface. For reference purposes, the reduction of local symmetry of CO when adsorbed on the (100) plane of face-centered cubic metal surface and the conversion of translational and rotational degrees of freedom into vibrational modes at the surface are depicted in Fig. 4.8.

Carbon monoxide forms coordination compounds with transition-metal atoms. Usually the molecule is bonded to the metal through the carbon atom. With 10 valence electrons, 4 from the carbon atom and 6 from the oxygen, a CO molecule in the isolated state possesses a σ bond and two π bonds. Two electrons with ionization energy of 43 eV occupy the 1 σ molecular orbital, formed from the $2s$-oxygen and $2s$- and $2p_x$-carbon orbitals. The electron density is greater near the oxygen nucleus. Molecular 1 π_y and π_z orbitals derived from the carbon and oxygen atomic p_y and p_z orbitals are each occupied by two electrons. The remaining four valence electrons occupy 2σ and 3σ molecular orbitals. The nonbonding 2σ orbitals possesses two electrons with ionization potential of 19.7 eV, made largely from the oxygen $2s$ and $2p_x$ orbitals. The remaining two are lone-pair electrons, occupying the 3σ orbital, composed of $2s$ and $2p_x$ orbitals of carbon with the lowest ionization potential of 14.0 eV. This lone-pair contributes significantly to the formation of metal carbonyl compounds. Figure 2.6 illustrated the bonding involved in the formation of metal complexes.

The 3σ molecular orbital, having the lone-pair electrons, overlaps with an empty metal orbital to form a σ bond, linking the carbon monoxide and metal. There is also a backdonation of electrons from a filled metal $d\pi$ orbital into a vacant π^* antibonding orbital, which usually results in lowering the vibrational frequency. Vibrational frequencies of the CO molecule and their isotopes in various states are listed in Table 4.3.

TABLE 4.3 IR Frequencies of CO and ^{13}CO Molecule in Gaseous and Surface States

Molecule	State	Observed Frequency (cm^{-1})	Ref.
$^{12}C^{16}O$	Gaseous	2143	
$^{13}C^{16}O$	Gaseous	2096	
Cu—C≡O	Solid	2128	15
Pt—C≡O	Solid	2070	"
Ni—C≡O	Solid	2033	"
Pd—C≡O	Solid	2053	"
Pd C≡O Pd	Solid	1908, 1916	"

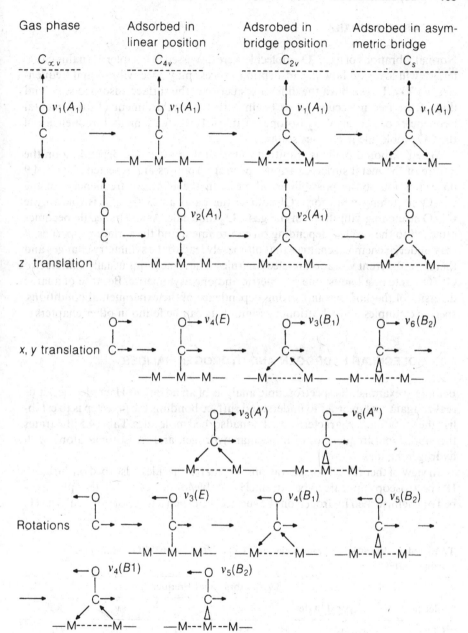

Figure 4.8 A local symmetry reduction of CO adsorption on the (100) fcc (face-centered cubic) metal surface along with conversion of rotational and translational degrees of freedom to surface vibrational modes.

4.7 CARBON DIOXIDE

Normal vibrations of free CO_2 molecule were discussed in Chapter 1. In this chapter, we will focus on how the adsorption process may affect vibrational frequencies of CO_2. To establish the differences between the surface adsorbed state and that of the free molecule, let us begin with the establishment of fundamental frequencies of various CO_2 isotopes. Table 4.4 lists fundamental frequencies of the CO_2 molecule in various states.

When carbon dioxide is admitted to the surface of metals, depending on the nature of the metal surface, various spectral responses are expected. Figure 4.9 illustrates four basic possibilities along with the tentative frequencies of the C—O stretching vibrations. Of course, the band at $2350 \, cm^{-1}$ is due to the C—O stretching vibrations in the gas CO_2 molecule. As the molecule becomes attached to the surface, depending on environment and the surface properties, it may undergo chemical changes that ultimately lead to the symmetry changes and therefore, different vibrational selection rules. The ionic forms usually exhibit two C—O stretching bands: one symmetric and one asymmetric. Because of a large diversity of the surfaces and strong dependence of the experimental conditions, specific examples and vibrational assignments can be found in other chapters.

4.8 MOLECULAR HYDROGEN AND HYDROGEN HALIDES

Before we examine the spectroscopic analysis of an adsorbed H_2 molecule, let us realize again that in order to understand surface bonding, the first step is to establish the vibrational characteristics of nonadsorbed molecules. Table 4.5 illustrates the vibrational frequencies of molecular hydrogen and its isotopes along with hydrogen halides.

In view of theoretical considerations for small molecules adsorbed on surfaces, IR spectroscopy appears to be extremely suitable for detection of either physisorbed or chemisorbed hydrogen on the surface. For such a reaction to occur, the H_2

TABLE 4.4 Vibrational Frequencies of $^{12}CO_2$, $^{13}CO_2$ and $^{14}CO_2$ Molecules in Various States

Molecule	Physical State	Vibrational Frequency (cm^{-1})			Ref.
		v_1	v_2	v_3	
$^{12}CO_2$	Gaseous	1337	667	2349	16
	Solid ($-190°C$)	—	660		17, 18
			653	2344	
	Aqueous solution	—	—	2342	19
$^{13}CO_2$	Gaseous	—	649	2284	20
	Solid ($-190°C$)		637	2280	
$^{14}CO_2$	Gaseous	—	632	2226	

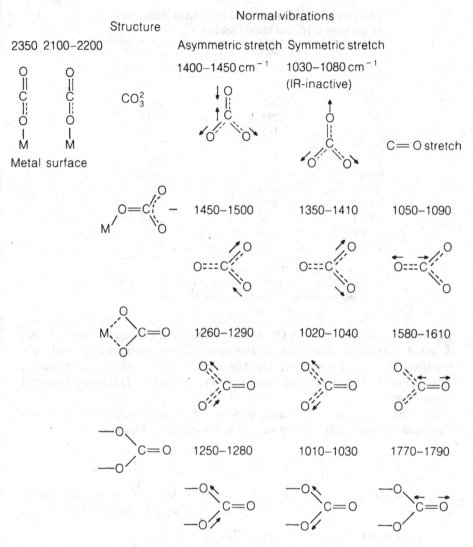

Figure 4.9 Infrared-active modes for selected structures of CO_2 on metal surfaces.

molecule must first dissociate, and the two produced atoms will attach to a metal surface. One of the first experiments demonstrating the capability of infrared spectroscopy has been shown in the early 1960's.[6] Molecular hydrogen has been adsorbed onto the γ-alumina supported platinum, giving rise to two infrared bands at 2105 and 2055 cm^{-1} at high surface coverages. As the surface coverage was diminished, the only band present in the IR spectrum was at 2055 cm^{-1}. This observation was attributed to the strength of the bonding of apparently two species among which the more weakly bonded ones were responsible for the higher energy band. In order to confirm the above band assignment, the same

TABLE 4.5 Experimental and Calculated Vibrational Frequencies of H_2 and Halide Species

Molecule[a]	Observed[b] Frequency (cm^{-1})	Theoretical[b] Frequency (cm^{-1})
H_2	4161	4395
HD	3632	3817
D_2	2993	3118
$H^{19}F$	3961	4138
DF	—	2998
TF	2443	2508
$H^{35}Cl$	2886	2990
$H^{37}Cl$	2884	—
$D^{35}Cl$	2091	2145
$D^{37}Cl$	2088	2142
$T^{35}Cl$	1739	1776
$T^{37}Cl$	1735	1772

[a] D—deuterium (^2H); T—tritium (^3H).
[b] From K. Nakomoto, *Infrared Spectra of Inorganic and Coordination Compounds*, Wiley-Interscience, New York, 1986.

experiment was repeated using D_2. Again, two bands at 1512 and 1480 cm^{-1} were detected at a high surface coverage, and as the surface coverage decreased, only the 1480-cm^{-1} band was detectable. Accordingly, the bands detected in the H_2 and D_2 experiments were attributed to the Pt—H or Pt—D stretching normal vibrations.

Four basic processes can occur during the chemisorption process of dissociated hydrogen molecules on the surface of nondefective MgO:

$$
\begin{array}{ccc}
 & H & H \\
 & | & | \\
1. \quad H_2 + MgO_s = & -Mg & -O-
\end{array}
\qquad
\begin{array}{l}
\text{(where } -Mg-O- \text{ represents chemi-} \\
\text{sorption on adjacent lattice sites)}
\end{array}
$$

2. $H_2 + MgO_s = OH_s^{2-} + MgH_s^{2+}$ (homolytic splitting)

3. $H_2 + MgO_s = OH_s^- + MgH_s^+$ (heterolytic splitting)

4. $H_2 + 2O_s^{2-} = 2OH_s^- + 2e_c$ (ionized splitting) (where e_c represents a conduction band electron).

4.9 NITROGEN

For the most part, chemisorption of nitrogen on metal surfaces is accomplished at low temperatures. Table 4.6 lists vibrational frequencies of the nitrogen molecule chemisorbed on metal surfaces.

TABLE 4.6   Observed Vibrational Frequencies for $^{14}N_2$, $^{14}N^{15}N$, and $^{15}N_2$ Isotopes Adsorbed on Ni Surfaces

Metal/adsorbate	Vibrational Frequency (cm^{-1})	Temperature	Ref.
Ni/$^{14}N_2$	2202	$-100\,°C$	21
Ni/$^{15}N_2$	2128	$-100\,°C$	
Ni/$^{14}N_2$, ^{14}N-^{15}N, $^{15}N_2$	2195, 2160, 2123	$-100\,°C$	
Ni/$^{14}N_2$, H$_2$	2202/2231–2255	$-100\,°C$	

It should be noted that these early experiments indicated the process of nitrogen chemisorption. From the preceding experimental results it was concluded that the presence of $Ni—N{\equiv}N^+$ species is responsible for such vibrational features. Theoretical calculations showed that for this species, the $N{\equiv}N$ stretching frequency should appear at $2207\,cm^{-1}$. The addition of H_2 premixed with N_2, gives rise to two bands and, more importantly, the decrease of the bands due to N_2 with the increasing concentration of molecular hydrogen.

4.10 WATER AND HYDROXYL GROUPS

The hydroxyl ion $[OH]^-$ is characterized by a sharp band in the $3760–3500\,cm^{-1}$ region. In general, the hydroxyl OH stretching band is sharper and absorbs at a higher frequency when associated with metal than in water molecules. On the other hand, water exhibits a much broader bands in the ranges of $3500\,cm^{-1}$ (OH stretching) and $1600\,cm^{-1}$ (OH bending). Table 4.7 shows the OH vibrational frequencies of various compounds, including metal hydroxides. It is also important to realize that there is a relationship between the electronegativity of the metal and the OH stretching frequency. West and Baney[7] have studied this phenomenon for a series of the $(C_6H_5)_3M—OH$ compounds.

Lithium hydride (LiH) and its isotopic counterpart lithium deuteride (LiD) are often used in ceramics. Since these materials are sensitive to environmental gases, in particular H_2O vapor, and may undergo numerous reactions that lead to the process known as corrosion. Some of these reactions are listed below:

1. $LiH + H_2O \rightleftharpoons LiOH + H_2$

2. $LiOH + LiH \rightleftharpoons Li_2O + H_2$

3. $2\,LiOH \rightleftharpoons Li_2O + H_2O$

4. $LiOH + H_2O \rightleftharpoons LiOH \cdot H_2O$

5. $2\,LiH + CO_2 \rightleftharpoons Li_2CO_3 + H_2O$

6. $2\,LiH \rightleftharpoons 2Li + H_2$

TABLE 4.7 Normal Vibrational Modes of Water- and Hydroxyl-Containing Species

Compound		Vibrational Frequencies (cm^{-1})			Ref.
	(gas)	3657	1595	3756	22
$H_2^{16}O$	(liquid)	3219	1627	3445	23
	(solid)	3400	1620	3220	24
$H_2^{18}O$	(gas)	3647	1586	3744	25
	(gas)	2727	1402	3707	22
$HD^{16}O$	(liquid)	2520	1455	3405	26
	(solid)	2416	1490	3275	24
$D_2^{16}O$	(gas)	2671	1178	2788	21
	(solid)	2520	1210	2432	24
$D_2^{18}O$	(gas)	2657	1169	2764	27
THO	(gas)	—	1324	3720	28
TDO	(gas)	—	—	2735	28
T_2O	(gas)	—	996	2370	28

	R		Raman	
OH (free radical)[a]	3596		—	29
OD (free radical)[a]	2680		—	29
LiOH	3647		3664	22–25, 30–33
$LiOH \cdot H_2O$	3574		3563	22–25, 30–33
NaOH	3637		3633	34
NaOD	2681		2681	
KOH	3600		—	27, 28, 35, 36
$Ca(OH)_2$	3644		3818	37, 38
$Mg(OH)_2$	3698		—	31, 32, 39, 40

[a] Matrix isolation data.
Adapted from K. Nakamoto, *Infrared and Raman Spectra of Inorganic and Coordination Compounds*, Wiley, New York, 1992.

When a solid LiH surface is exposed to water vapor, a LiOH corrosion layer is formed on the surface, which liberates H_2 gas.[8-10] Although the first three reactions are written as independent events, it should be remembered that they are interrelated; for example, reaction 3 in the reverse direction may compete with reaction 1. In addition to the surface changes associated with exposure to moisture, atmospheric CO_2 may also participate in the reaction with LiH, leading to the formation of surface layers of lithium carbonate (reaction 5) on lithium hydride. Another aspect is thermal stability of LiH. Apparently, LiH is thermally unstable and will decompose above 300°C. With the enhancement of oxygen solubility at temperatures approaching 600°C, the carbon from Li_2CO_3 is reduced to CH_4, resulting in surface enrichment in Li.[11,12] Diffuse reflectance spectrum of solid LiD, with a prior history of a H_2O exposure, indicates that the stretching frequencies of the hydroxyl species for LiOH are detected at 3675 cm^{-1}, whereas the LiOD counterpart exhibits the O—D stretching at 2705 cm^{-1}. On the other hand, the band at 3566 cm^{-1} and a broadband at approximately 3000 cm^{-1} are attributed to the Li—O—H and water stretching modes in $LiOH \cdot H_2O$, respectively.

The reaction allowing the replacement of hydroxyl groups from silica surfaces is refluxing silica with thionyl chloride.[13]

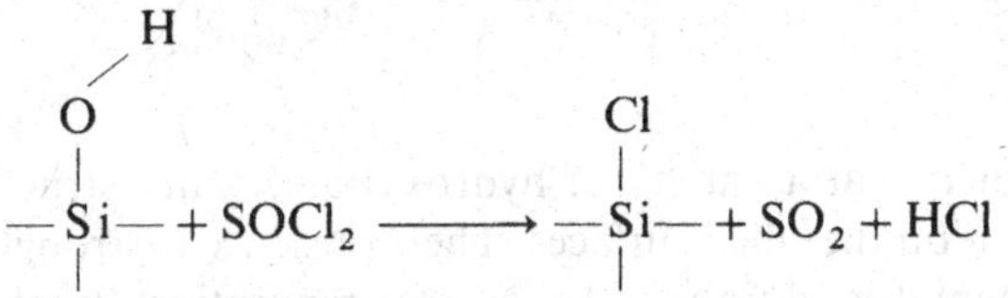

$$\begin{array}{c} \overset{\displaystyle H}{\diagup} \\ O \\ | \\ -Si- \\ | \end{array} + SOCl_2 \longrightarrow \begin{array}{c} Cl \\ | \\ -Si- \\ | \end{array} + SO_2 + HCl$$

Such chlorinated silica can be conveniently converted to silica amine functionalities by treating the surface with NH_3 gas:

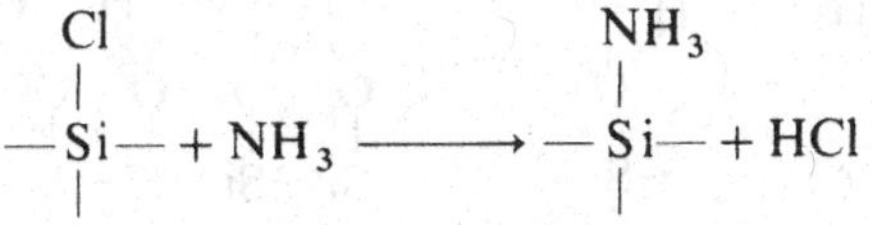

$$\begin{array}{c} Cl \\ | \\ -Si- \\ | \end{array} + NH_3 \longrightarrow \begin{array}{c} NH_3 \\ | \\ -Si- \\ | \end{array} + HCl$$

The occurrence of this reaction is detected by the presence of amine asymmetric and symmetric NH stretching bands at 3520 and 3445 cm^{-1}, respectively. Because HCl is produced in this reaction and can readily react with the excess of NH_3, surface ammonium chloride is formed, as demonstrated by the appearance of the bands at 3150, 3050, and 2805 cm^{-1}. Since these species are weakly bonded, vacuum heat treatment at 450°C eliminates these bands leaving only chemically bonded amine groups.

Quite often, a sharp IR-active band at 3770 cm^{-1} and a broadband at 3450 cm^{-1} are observed in the spectrum of silica surface. While the former is due to free hydroxyl groups, the latter is detected because of the presence of hydrogen-bonded hydroxyl groups. Interestingly, the broadband at 3450 cm^{-1} may be removed on heating the sample under vacuum at 800°C, but the 3770 cm^{-1} is still observed. These observations indicate that non-hydrogen-bonded species are far harder to be removed from the surface than are covalently bonded species. It was proposed that a direct relationship exists between the amount of water physically adsorbed on the silica surface and the surface hydroxyl group content. The two proposed dyhydroxylation reactions leading to removal of water and hydroxyl groups are depicted below:

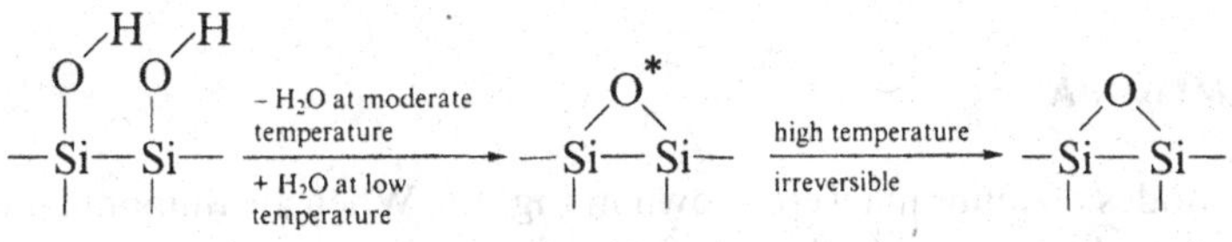

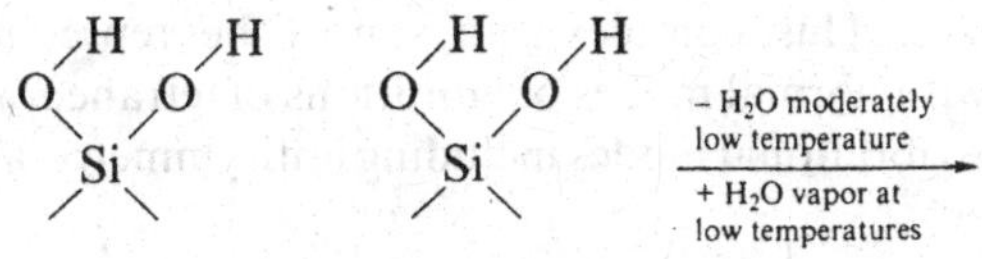

$$\begin{array}{c} \overset{H}{\diagup}\ \overset{H}{\diagup} \\ O\ \ \ O \\ | \ \ \ | \\ -Si-Si- \\ | \ \ \ | \end{array} \quad \underset{\text{+ H}_2\text{O at low temperature}}{\overset{\text{- H}_2\text{O at moderate temperature}}{\xrightarrow{\hspace{2cm}}}} \quad \begin{array}{c} O^* \\ \diagup\ \diagdown \\ -Si-Si- \\ | \ \ \ | \end{array} \quad \underset{\text{irreversible}}{\overset{\text{high temperature}}{\xrightarrow{\hspace{2cm}}}} \quad \begin{array}{c} O \\ \diagup\ \diagdown \\ -Si-Si- \\ | \ \ \ | \end{array}$$

$$\begin{array}{c} \overset{H}{\diagup}\ \ \overset{H}{\diagup} \\ O\ \ \ \ O \\ \diagdown\ \diagup \\ Si \\ \diagup\ \diagdown \end{array} \quad \begin{array}{c} \overset{H}{\diagup}\ \ \overset{H}{\diagup} \\ O\ \ \ \ O \\ \diagdown\ \diagup \\ Si \\ \diagup\ \diagdown \end{array} \quad \underset{\text{+ H}_2\text{O vapor at low temperatures}}{\overset{\text{- H}_2\text{O moderately low temperature}}{\xrightarrow{\hspace{2cm}}}}$$

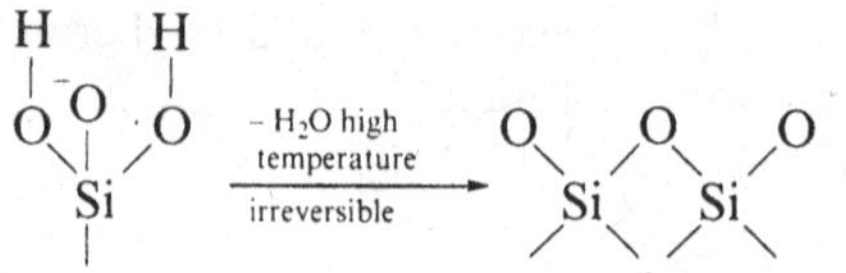

It should be noted that a variety of hydroxyl and water structures have been proposed to exist on the silica surfaces. Their presence is strongly influenced by specific experimental conditions and a sample preparation. Besides the structures depicted above, other structural features have been claimed as well.

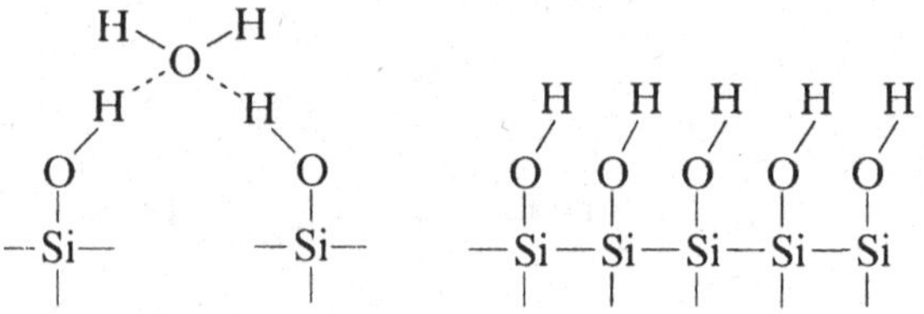

On elimination of $2H_2O$, structure **B** may result in structure **C**.

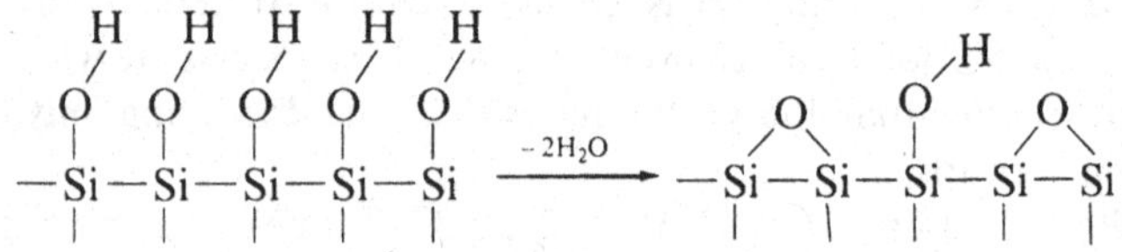

Because these reactions and surface structures are formed depend on many variables among which thermal history of the system, particle size, and experimental conditions are the dominating ones, the vibrational frequencies of the chemical bonds being formed as a result of the surface reactions may change. Although vibrational energies of the OH stretching modes on various substrates alone and in the presence of other adsorbates have been reported in the past,[14] their exact energies are not listed here because there are too many system-dependent factors involved. In general, the OH stretching bands appear in the $3775-3650 \, cm^{-1}$ region.

4.11 AMMONIA

Normal modes of ammonia were shown in Fig. 1.5. When an ammonia molecule becomes attached to the surface, the selection rules change again. One may consider the formation of relatively simple complex with the surface remaining typical amine complexes. This complex can simply be represented as the $M-NH_3$ association with normal modes of vibrations of tetrahedral $M-NH_3$ molecules. Although six vibrational modes including antisymmetric and symmetric NH_3 stretching, NH_3 degenerate and symmetric deformations, NH_3 rocking and M—N stretching (the latter will tell us if indeed there is a chemical bonding to the surface) are expected, surface reactions may lead to some complexity of the spectra due to surface coverage. Each region, however, will exhibit characteristic frequencies ranging from 3440–3070 to 3340–3040 cm^{-1} for the stretching vibrations, 1580–1710 cm^{-1} to 1400–950 cm^{-1} for degenerate and symmetric deformations, and 620–820 cm^{-1} for rocking to the 500-cm^{-1} region for the M—N stretching vibrations.

Although this chapter covers only fundamental aspects of small but commonly appearing species on metal surfaces, providing the frequency ranges for the species, our intention was to provide an introduction to more advanced surface vibrational analysis. Therefore, in the later chapters, we will refer to the above results and will take advantage in the more complex surface situations.

4.12   HYDROCARBONS ON SURFACES

When considering useful metal surface properties, an understanding of the structures and bonding is one of the primary concern because such practical aspects of coating metals with organic films, as adhesion or metal-catalyzed reactions are of great importance. Hence, the identification of the structures and bonding of chemisorbed hydrocarbons to metallic surfaces using vibrational spectroscopy and various aspects of it will be discussed in this chapter. However, before the analysis of the hydrocarbon vibrational spectra is given, it is necessary to define substrates. In most studies concerning metal surfaces, the surface preparation has a significant effect on the resulting bonding. Not only the surface purity, but also the atomic arrangements on the surface, may have an effect on vibrational spectra. As we recall from previous chapters, most studies have been performed on the face-centered cubic (fcc) metals as a convenient means for identifying the (111), (100), and (110) specifically cut surfaces. Such as those depicted in Fig. 4.4. Because of the highly reflective nature of metal surfaces, high-resolution electron energy-loss spectroscopy (EELS),[41] reflection–absorption (RA) IR spectroscopy,[42] and surface-enhanced Raman spectroscopy (SERS)[43] were most often utilized for that purpose.

Let us first formulate a theoretical basis for using the knowledge from the previous chapters and then consider a methyl molecule bonded to a metal surface. When a CH_3 molecule becomes attached to a single metal atom, vibrational selection rules predict the normal modes, which are depicted in Fig. 4.10. This case, however, should be considered as an ideal one that oversimplifies the issue because it assumes that only one terminal methyl group is attached to the central metal atom. When more CH_3 groups are attached, the symmetry and electronic changes within the system will obviously change the selection rules. This is why Table 4.8 provides energy ranges that would be expected for CH_3 and its dueterated counterpart. It should be kept in mind that the existence of intermolecular interactions may

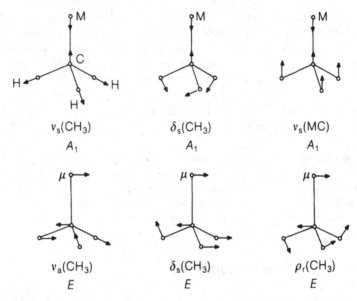

Figure 4.10 Normal vibrational modes of CH_3 attached to a metal atom.

TABLE 4.8 Frequency Ranges for Terminal Methyl and Its Deuterated Counterpart

	Energy Range (cm^{-1})	
Mode	CH_3	CD_3
ν_a	3050–2810	2275–2035
ν_s	2950–2750	2175–2000
δ_d	1475–1300	1125–960
δ_s	1350–1100	1000–870
ρ_r	975–620	750–475

Source: Adopted from ref. 43.

influence the symmetry of the entire unit, and the band intensities for each mode may vary with the number of methyl groups bonded to a metal atom.

4.12.1 Alkenes

4.12.1.1 Ethylene The hydrogenation reaction of ethylene to ethane

$$C_2H_4 + H_2 \longrightarrow C_2H_6$$

in the presence of catalysts has been extensively investigated because understanding of mechanistic aspects of the reaction may provide further information about adsorption of ethylene on metals. For that reason, D_2 was used in the exchange process

$$C_2H_4 + D_2 \longrightarrow CH_2{=}CHD + HD$$

Both reactions appear to follow the same kinetics, and in the presence of a mixture of H_2 and D_2, occur simultaneously with an approximate activation energy of 10.7 kcal/mol.

Several mechanisms and kinetics of ethylene adsorption on metals have been proposed in the past. Among the most extensively studied mechanisms are associative or dissociative adsorptions. For example, adsorption is considered as an associative process when the reaction shown in Fig.4.11A occurs. As shown, the process is followed by the dissociation to C_2H_6.

The deuterium exchange reaction, on the other hand, was proposed to proceed through the mechanism illustrated in Fig. 4.11B. In the case of the dissociative process, the situation is quite different. Although the adsorption of ethylene has been studied on such metal surfaces as Ni(111),[44,45] Ni(100),[46-48] Ni(110),[49-51] Ru(001),[52] the first studies have been conducted on Pt(111).[53] While in these studies EELS has been used as a measurement method, not too long ago RA FT-IR spectroscopy was explored.[54,55]

There are essentially two structures that exist at low temperatures and have een reliably assigned. They are depicted in Fig. 4.12. In the presence of C=C double bond, and depending on the nature of the surface, the bonded[55] (**A**) or a π-bonded[56] species (**B**) exists usually at lower temperatures, whereas structure **C** was detected at room temperature. Table 4.9 summarizes those metal surfaces that exhibit **A**, **B**, or **C**-type structures, whereas Table 4.10 lists the vibrational regions characteristic of these structures. Structures **A** and **B** exhibit a nondissociative nature; hence they retain the C_2H_4 formula.

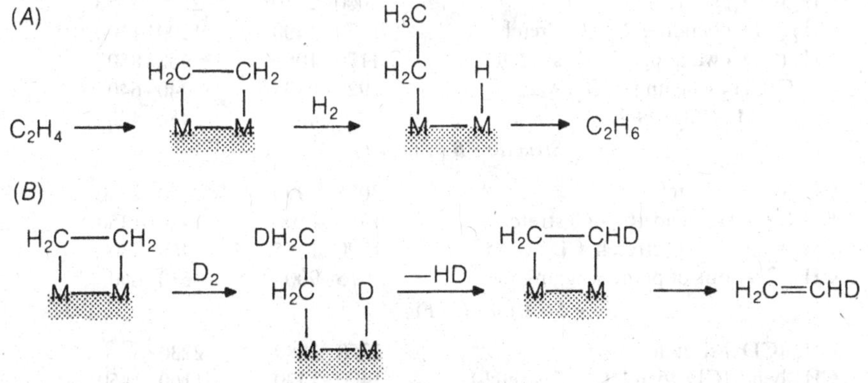

Figure 4.11 Process of adsorption of C_2H_4 followed by dissociation of C_2H_6; (A) in the presence of H_2; (B) in the presence of D_2.

TABLE 4.9 **Metal Surfaces Forming Structures A, B, and C, (Fig. 4.13) with Alkenes**

Structure Type[a]	Metal	Ref.
A	Ni(111)	72
	Ni(100)	90–93
	Ni(110)	49–51
	Ru(001)	45, 52
	Pt(111)	55, 87
	Pt(100)	88, 95
	Fe(110)	89, 96
	Cu(100)	56
B	Pd(111)	66, 67
	Pd(100)	53
	Pd(110)	90, 97
C	Pt(111)	91, 92, 98, 99
	Rh(111)	93, 94, 100, 101
	Pd(111)	66, 67
	Ir(111)	95, 102

[a] See Fig. 4.12.

TABLE 4.10 **Vibrational Frequency Ranges and Their Assignments for Structures A–C in Fig. 4.12**

Vibrational Mode	C_2H_4 (cm^{-1})	C_2D_4 (cm^{-1})
Structure A (Fig. 4.12)		
CH_2 (CD_2) stretch	3000–2910	2200–2150
CH_2 (CD_2) bending/C—C stretch	1470–1400	1200–1100
CH_2 (CD_2) wagging/C—C stretch	1170–1060	950–850
C—C/CH_2 wagging (CD_2) wag.	920–830	740–640
M—C (M—C) stretch	480–400	450–400
Structure B (Fig. 4.12)		
CH_2 (CD_2) stretch	3075–2990	2250–2230
C—C s/CH_2 bend/(C—C) stretch	1560–1500	1420–1350
CH_2 bend/C—C stretch/CD_2 bend	1290–1225	960–930
CH_2/CD_2 out-of plane wagging	915–900	680–660
Structure C (Fig. 4.12)		
CH_3/(CD_3) stretch	2950–2880	2230
CH_3 bend/(CD_3 bend; C—C stretch)	1400–1330	1160–1450
C—C stretch/(CC stretch/CD_2 bend)	1165–1080	1050–950
C—M_3 stretch/ C—M_3 stretch	460–410	430–400

Although in the early studies of chemisorption of ethylene on metal surfaces, structure **C** of Fig. 4.12 was suggested, it was not until later that CH_3CH–metal bonding was established.[56] For such a structure to exist, it is necessary that a hydrogen atom move from one end of the CC bond to the other during dissociation process. This is apparently the case for ethylidyne (Fig. 4.12C). Although the process of metal–CCH_3 formation involves severe symmetry changes, ethylidyne species can also be derived from such structures as di-σ or π-bonded ethylenes.[57] This is schematically depicted in Fig. 4.13. Table 4.11 provides a summary of band assignments for ethylene adsorbed on other metals. A comparison of C_2H_4 ad- sorbed on Rh(111) with the bands observed in model compound studies is given in Table 4.12.

In essence, the following three structures are expected for Rh(111):

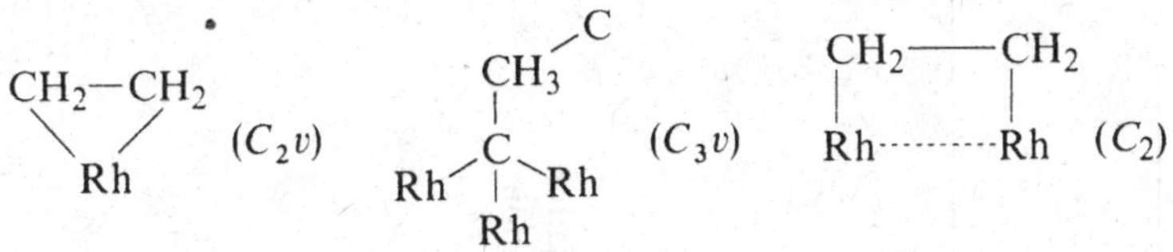

$$CH_2\!-\!CH_2 \quad (C_2v) \qquad \underset{Rh}{\overset{CH_3}{\underset{\displaystyle Rh\;|\;Rh}{\overset{|}{\underset{C}{\overset{C}{\diagdown}}}}}} \quad (C_3v) \qquad \underset{Rh\cdots\cdots Rh}{\overset{CH_2\!-\!-\!-\!CH_2}{|\qquad\;\;|}} \quad (C_2)$$

Spectral analysis and selection rules indicated that the C_2 symmetry is most likely present.

As one would expect, the C—C stretching mode will be affected by these configurations, and for C_{2v} geometry, C—C ring deformation is detected at 1271 cm^{-1};

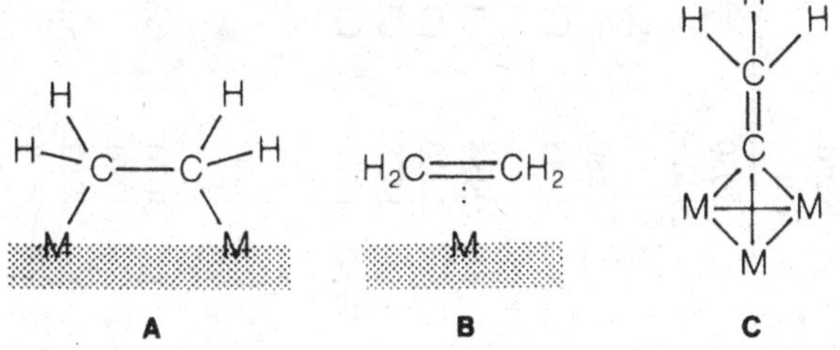

Figure 4.12 Low-temperative (**A** and **B**) and room-temperature (**C**) structures of C_2H_4 formed on metal surfaces.

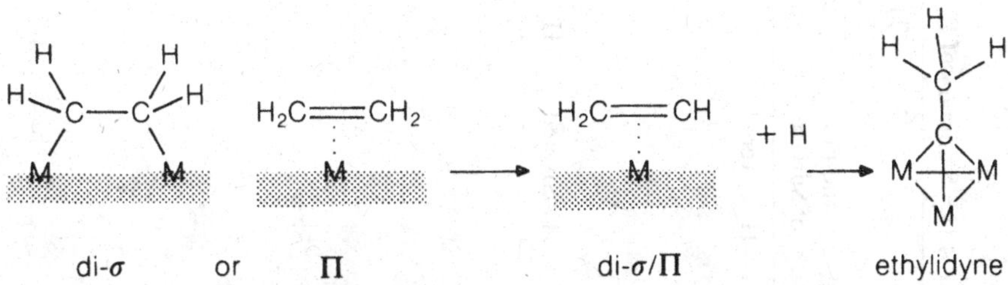

Figure 4.13 Formation of ethylidyne from di-σ and π-bonded C_2H_4 to metal surface.

TABLE 4.11 Summary of IR Band Assignments for Ethylene Chemisorbed on Silica-Supported Metals

Metal	Condition of Surface	Treatment Subsequent to C_2H_4 Adsorption	Adsorption bands (cm^{-1})	Assignment	Structure	Ref.
Ni	H₂-covered	—	3030	=CH stretch	Unassigned	A
			2940 sh		CH_2-CH_2 (bonded to M, M)	
			2880	Saturated CH stretch		
			1448	CH₂ deformation	plus HC—CH (bonded to M, M M M); CH₃ / CH₂ (bonded to M)	
Ni	H₂-covered	H₂ added	2955	CH₃ asymmetric stretch		A
			2920	CH₂ asymmetric stretch		
			1458	CH₃ asymmetric deformation		
			1378	CH₃ symmetric deformation		
Ni	Bare	—	3070	Olefinic CH stretch	Unassigned Dissociative adsorption to • give HC—CH (bonded to M, M M M) and C—C (bonded to M M M M)	A
			2970			
			2910	Saturated CH stretch		
			2860			

Metal			Wavenumber (cm⁻¹)	Assignment	Structure	Ref.
Ni	Bare	H₂ added	2955	CH₃ asymmetric stretch	$CH_2CH_2CH_2CH_3$ bonded to M	A
			2920	CH₂ asymmetric stretch		
			2860	CH₃, CH₂ symmetric stretch		
Pd	Bare	—	3030	=CH stretch	$HC{=}CH$ on $-M-M-$	B
			2970	CH₃ asymmetric stretch	$CH_3{-}CH{-}M$ or $CH_3{-}CH$ on $-M-M-$	
			2870	CH₃ symmetric stretch	and $HC{=}CH$ on $-M-M-$	B
Pd	Bare	H₂ added	2950	CH₃ asymmetric stretch	Equal amounts of $CH_2{-}CH_2$ on $-M-M-$ and $CH_3{-}CH_2{-}M$ or $CH_2CH_2CH_2CH_3{-}M$	B
			2920	CH₂ asymmetric stretch		
			2850	CH₃, CH₂ symmetric stretch		

REFERENCES:

A. W. A. Pliskin and R. P. Eischen, *J. Chem. Phys.* 1956, **24**, 482.
B. L. H. Little, N. Sheppard, and D. J. C. Yates, *Proc. Roy. Soc.* 1960, **A259**, 242.

TABLE 4.12 Summary of Raman Bands Observed for C_2H_4 and Model Compounds

C_2H_4 (Ref. A)	C_2D_4	$^{13}C_2H_4$	C_2H_4S (Ref. A)	gauche-1,2-$C_2H_4Br_2$ (Ref. B)	Zeiss' salt[b] (Ref. C)	C_2H_4 on Rh (111) (Ref. D)
1074	1091	1080	2990	3005	3019	3000
549.9	532.4	545	1445	2953	1623	2900
351	307.1	304.8	1120	1420	1342	Shoulder
			1040	1278	3108	1350
			625	1104	1236	1130
			3080	1019	3106	880
				898	810	450
			660	550	2990	
			2990	231	1444	
			1390	91	1007	
			949	3005	943	
			711	1420	949	
			3080	1245	331	
			823	1104	407	
			686	836	310	
				589	183	
				355	339	
					210	
					161	
					121	
					92	

^{a} RERERENCES:
A. W. L. Parker, A. R. Siedle, and R. M. Hexter, Langmuir, 1988, **4**, 999.
B. J. Davidson, M. Green, F. G. A. Stone, and A. Welch, J. Am. Chem. Soc., 1975, **97**, 7490.
C. J. Heaviside, P. J. Hendra, P. Tsai, and P. R. Cooney, J. Chem. Soc., Faraday Trans. 1, 1978, **74**, 2542.
D. S. Lehwald, H. Ibach, and E. J. Demuth, Surf. Sci., 1978, **78**, 577.
^{b} K [Pt μ_3(C_2H_4)]·H_2O.

for C_{3v}, C—C stretch is at 1163 cm^{-1}; and for C_2, symmetry C—C is around 1020 cm^{-1}.

4.12.1.2 Higher Alkenes   Although limited spectroscopic data are available for higher alkenes, essentially three structures have been reported for propene.[58-60] They are shown in Fig. 4.14. Similar to ethylene adsorption, propene forms di-σ species (structure **A**), which is stable at low temperatures (170 K). On heating to 270 K, it is converted to propylidyne (structure **B**). It should be kept in mind that the temperatures at which various structures exist depend on a metal substrate and the crystallographic planes to which the species are attached. For example, the temperatures listed above were reported for Pt(111), whereas on Rh(111) the conversion to propylidyne occurs at 200 K. When temperature is raised even further, propylidyne on Rh(111) decomposes[61] at 270 K to ethylidyne or a mixture of ethylidyne/methylidyne.[62] Apparently, the C—C bond near the metal surface cleaves. At higher temperatures, ethylidyne may be further decomposed to fragments, such as C_2H, giving two characteristic bands in the IR spectrum at 3025 and 825 cm^{-1}. In contrast to Rh(111) surfaces, the same heating process conducted on Pt(111) does not generate an intermediate, but the molecule decomposes directly.

Chemisorption of *cis*- and *trans*-1-butene on Pt(111) at low temperatures shows two distinguishable *cis* and *trans* isomers with the characteristic features of di-σ spectra.[58.59] At higher temperatures the isomers are not distinguishable. Apparently, a structure such as structure **C** in Fig. 4.14 exists and was assumed to be de- rived from dimethyl acetone. Isobutene has been also identified as giving a di-σ low-temperature spectra, but at higher-temperatures isopropylidyne was detected. Although various decomposition pathways for *n*-butene, *cis*- and *trans*-2-butene and 2-butyne have been proposed, the exact decomposition products still raise a lot of controversy.

4.12.2 Alkynes

4.12.2.1 Acetylene   The adsorption of acetylene at low temperatures also gives two distinguishable spectra attributed to the two types of structures (**A** and **B**) shown in Fig. 4.15. Although there is still a certain degree of uncertainty as to

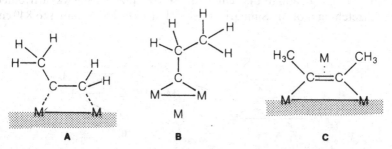

Figure 4.14 Structures resulting from propene chemisorption on metal surfaces.

TABLE 4.13   Vibrational Bands Observed for Structures A and B in Figure 4.15

Vibrational Band	C_2H_2 (cm^{-1})	C_2D_2 (cm^{-1})
Structure A (Fig. 5.15)		
C—H stretch/(C—D stretch)	2970–2880	2200–2140
C—C stretch/(C—C stretch)	1320–1100	1270–1060
C—H bend/(C—D bend)	990–810	700–630
C—H wag/(C—D wag)	780–630	550–450
Structure B (Fig. 5.15)		
C—H stretch/(C—D stretch)	3020–2930	2240–2190
C—C stretch/(C—C stretch)	1410–1260	1360–1210
C—H bend/(C—D bend)	990–870	730–620
C—H wag/(C—D wag)	770–670	570–500

Source: Adopted from refs. 63–65.

the nature of each structure, it is believed that structure **A** originates from a σ bonding of C=C to two metal atoms, but the π-bonding to the third one. Structure **B** is believed to involve four metal atoms, most likely close-packed surface face. Table 4.13 lists tentative band assignments associated with the presence of structures **A** and **B** depicted in Fig. 4.15.[63–65]

When deposited on a metal surface at low temperature, acetylene shows an interesting thermal evolution. For example, in the studies on Pd(111) surfaces,[66] the frequencies characteristic of structure **B** in Fig. 4.16 were detected at 150 K. As the temperature was raised to 250 K, di-σ/π bonded vinylidene species were detected. Furthermore, at 300 K, ethylidyne was observed followed by its decomposition at higher temperatures. From these and other studies,[67] the decomposition sequence of acetylene was proposed. Other structures and band assignments for acetylene adsorbed on selected metal surfaces are tabulated in Table 4.14.

Raman spectra of C_2H_2, C_2D_2, and $^{13}C_2H_2$ adsorbed on alumina-supported rhodium are presented in Fig. 4.16, and indicate that the spectrum of C_2H_2 contains two bands at 1476 and 1096 cm^{-1}, along with four bands at 3000, 2588, 2208 and 978 cm^{-1} (see ref. A in Table 4.15). Because the band at 1476 cm^{-1} shifts to 1408 cm^{-1} for C_2D_2, and to 1451 cm^{-1} for $^{13}C_2H_2$, this shifting was attributed to the CC stretching motion. Similarly, the band at 1096 cm^{-1} shifted to 849 cm^{-1}

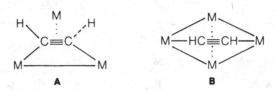

Figure 4.15 Structures resulting from adsorption of acetylene on metal surfaces.

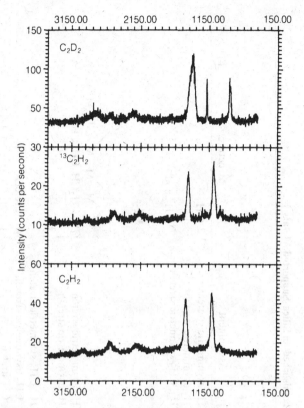

Figure 4.16 Roman spectra of C_2H_2, $^{13}C_2H_2$, and C_2D_2 adsorbed on alumina-supported rhodium. (Reproduced with permission from W. L. Parker, A. R. Siedle, and R. M. Hexter, *Langmuir*, 1988, **4**, 999. Copyrighted by American Chemical Society, Washington, DC.)

on deuteration and to 1081 cm^{-1} on ^{13}C substitution. Table 4.15 lists vibrational bands for adsorbed acetylenes and model compounds.

4.12.2.2 Higher Alkynes The studies conducted on methyl acetylene showed that this molecule is bonded to the surface through the $C\equiv C$ group and the entire molecule is parallel to the surface, giving rise to the reduced bond order.[62,65] In the case of dimethyl acetylene, similar structures were deduced.[68]

4.12.3 Saturated Hydrocarbons

Because the majority of adsorption studies of saturated hydrocarbons have been done of cyclic molecules, in particular cyclohexane, Table 4.16 compares vibrational frequencies for ethane and cyclohexane. Various studies of ethane adsorbed on Cu(110)[69,70] and Pt(111)[71] showed that ethane molecules are not

TABLE 4.14 Summary of IR Band Assignment for Acetylene Chemisorbed on Silica-Supported Metals

Metal	Condition of Surface	Treatment Subsequent to C_2H_2 Adsorption	Absorption Bands (cm^{-1})	Assignment	Structure	Ref.
Ni	H_2-covered or bare	—	2960	CH_3 asymmetric stretch	CH_3 and CH_2 and $-M-$ bridged structure	
			2920	CH_2 asymmetric stretch		
			2860	CH_3, CH_2 symmetric stretch		
Ni	H_2-covered or bare	H_2 added	2960	CH_3 asymmetric stretch	$CH_2CH_2CH_2CH_2CH_3$ on M M M M M	A, B
			2920	CH_2 asymmetric stretch		A
			2860	CH_3, CH_2 symmetric stretch		
Pd	Bare	—	2860	$=CH_2$ asymmetric stretch	Mixture of $H_2C=C$ and $HC=CH$ structures on M M	
			3090	$=CH_2$ asymmetric stretch		
			3030	$=CH$ stretch		

			$M-(CH=CH)_2-CH=CH_2$; C
			or
			C
			with small amount of
			$-M-$ or $M-(CH_2)_5-CH_3$ D

Pd	Bare	2960 weak shoulder	CH_3 asymmetric stretch
		2910	CH_2 asymmetric stretch
		2850	CH_2 symmetric stretch
	H$_2$ added		
Cu		2965–2959	
Ni		2925	as above
Pd		2865	

^{a} REFERENCES:

A. W. A. Pliskin and R. P. Eischens, *J. Chem.*, 1956, **24**, 482.

B. R. P. Eischens and W. A. Pliskin, *Advances in Catalysis*, Vol. X, 1958, Academic Press, New York, p. l.

C. L. A. Little, N. Sheppard, and D. J. C. Yates, *Proc. Roy. Soc.*, 1960, **A259**, 242.

D. C. P. Nash and R. P. De Sieno, *J. Phys. Chem.*, 1965, **69**, 2139.

153

TABLE 4.15 Comparison of Raman Band Assignments (From Ref. A, below) for C_2H_2 Adsorbed on 10% Rh and Various Model Compounds[a]

(Ref. B) $Os_3(CO)_9(CCH_2)$	(Ref. C) $Co_2(CO)_6\ C_2H_2$	(Ref. D) $(CH_3C)\ Co_3(CO)_9$	(Ref. A) C_2H_2 on 10% Rh	Assignment
2986	3116	2888	1476 (1451)	ν_{CH} (s)
1331	1402	1161		ν_{CC}
3047	3086	2930	1098 (1081)	δ_{CH} (s)
1470	894	1420		
1051	768	1356	981 (982)	δ_{CH} (as)
963		1004		
811				
2180	2359			ν_{CD} (s)
1282	1346	1161	1408	ν_{CC}
2140	2297	2192		ν_{CD} (as)
760	751	828	849	δ_{CD} (s)
		1031		
742	602	1002	717	δ_{CD} (as)

REFERENCES:

A. W. L. Parker, A. R. Siedle, and R. M. Hexter, *Langmuir*, 1988, **4**, 999.

B. J. Davidson, M. Green, F. G. A. Stone, and A. Welch, *J. Am. Chem. Soc.*, 1975, **97**, 7490.

C. H. Shirakawa, M. Sato, A. Hamano, S. Kawakawi, K. Soga, and S. Ikeda, *Macromolecules*, 1980, **13**, 457.

D. L. H. Dubois, D. G. Castner, and G. Somorjai, *J. Chem. Phys.*, 1980, **72**, 5234.

chemisorbed; instead, they are weakly bonded through the physisorption process, and the first layer of C—C is parallel to the surface. The detected vibrational bands are listed in Table 4.16 and agree with the selection rule predictions, which indicate that the symmetric C—H mode is weak in the IR spectrum because of distortions of the molecule due to interactions with the surface. In the unadsorbed state, such mode would not be seen in IR.

Vibrational frequencies of a cyclohexane molecule attached to Cu(111) surface[69,70] are listed in Table 4.16. The band assignments as well as temperature effect on the surface coverage dependence showed that for low surface coverages, the cyclohexane ring average position is parallel to the surface, giving rise to the modes not observed under ordinary circumstances. For example, studies on Ni(111) and Pt(111) at 140 K showed two bands attributed to the C—H stretching vibrations: one at 2900 cm^{-1}, in the normally expected for C—H region; and the second strong and broad centered at 2720 cm^{-1} for Ni(111) and 2590 cm^{-1} for Pt(111).[71] Although the 2900-cm^{-1} region is characteristic to the C—H stretching modes, the remarkably low frequencies of other bands were attributed to the "hydrogen-bonding type" of interactions of the CH bond with the metal surface. Apparently, in the case of cyclohexane, the presence of so-called soft modes (with lower then expected) is attributed to paraffinic hydrogen, which exhibits the highest electron density. Hence, it is an electron donor, whereas the electron-deficient metal becomes an acceptor.

A comparison of the "soft mode" vibrational frequencies for various substrates as well as the temperature experiments indicate, that on Pt(111), surface cyclohexane dissociates to benzene at 200 K.[71] In contrast, when cyclohexane is adsorbed on Ni(111), the molecule desorbs at 170 K. However, when Ni surface is in a stepped form, such as Ni[5(111) × (110)], the decomposition of cyclohexane to benzene is detected at 225 K.[72]

The issue of soft mode becomes even more complex in view of the adsorption studies on Rh(001) at 90 K of a series of cycloparaffins with a general formula C_nH_{2n}, where $n = 4, 5, 6,$ and 8.[73] While no separate lower C—H frequency was observed for cyclobutane, a significant lowering to 2610, 2516, and 2630 cm^{-1} for $n = 5, 6,$ and 8, respectively, was detected. The issue becomes even more complicated when, prior to saturated hydrocarbon adsorbtion, oxygen is preadsorbed on a metal surface.[73,74] In this case, no softmode bands were detected.

Although significant progress has been made in this area in the last 10 years, numerous issues still need further investigation and should be addressed before the

Table 4.16 Comparison of the C—H Stretching Regions for Ethane and Cyclohexane[a]

Ethane	2961 (as), 2861 (s), 1488 bending (2923, 2903; overtones)
Cyclohexane	2905 (s), 2895 (as), 2845 (eq), 2770 (s ax), 1444 (ben)

[a] eq—equatorial; ax—axial.

origin of spectroscopic features and associated surface species is resolved. It is thus obvious that novel approaches are needed, especially in surface preparation and cleaning. Apparently, there have been many published results that are difficult to reproduce. None of the investigations addressed the broad bands in the 2000–2800-cm^{-1} region as also being characteristic for ionic associations in salts.

4.12.4 Aromatic Hydrocarbons

4.12.4.1 Benzene While the surface structures that develop as a result of chemisorption of aliphatic hydrocarbons may lead to additional spectral complexities, in the case of the benzene ring it is generally agreed that at low and room temperatures benzene lies flat on a smooth metal surface. At first, it seems that the repulsive forces between π electrons and freely flowing electrons on a metal surface should prevent that from happening. However, the ability of both types of electrons to delocalize allows benzene to "stick" to a metal surface. As a result of the π-bonded benzene molecule, the gas-phase symmetry of D_{6h} is being reduced by eliminating a horizontal plane of symmetry to C_{6v}. As a consequence, the following D_{6h} modes are being converted to C_{6h} modes:

$$D_{6h} \qquad\qquad\qquad C_{6h}$$
$$A_{1g}(\nu_{C\text{-}H} \text{ and } \nu_{C\text{-}C}) \longrightarrow A_1$$
$$A_{2u}(\nu_{C\text{-}H}) \longrightarrow A_1$$

where $\nu_{C\text{-}H}$, $\nu_{C\text{-}C}$, and $\gamma_{C\text{-}H}$ correspond to ν_1, ν_2, and ν_4 modes in Herzberg's notation. Similarly to the aliphatic hydrocarbons, Table 4.17 lists the frequency ranges for benzene molecule adsorbed on the surface metals. It should be noted that the $\nu_{C\text{-}C}$ modes occur in the 920–800-cm^{-1} region, which is significantly lower than that observed for the nonadsorbed benzene at 992 cm^{-1}. The out-of-plane $\nu_{C\text{-}H}$ modes exhibit similar behavior and may vary depending on the nature of the substrate and crystal planes. Table 4.18 summarizes these differences.

For comparison, Table 4.19 summarizes selected vibrations of the gas-phase and Ag(111) surface-coordinated benzene molecules. The presence of highly delocalized π electrons on a benzene ring leads to a parallel or near-parallel orienta-

Table 4.17 Effect of Isotopic Substitution on C—H Stretching and Bending Frequencies in C_6H_6 and C_6D_6

Vibrational Band	C_6H_6 (cm^{-1})	C_6D_6 (cm^{-1})
$\nu_{C\text{-}H}/(\nu_{C\text{-}D})$	3040–2990	2300–2240
$\nu_{C\text{-}C}$	920–800	920–800
$\gamma_{C\text{-}H}/(\gamma_{C\text{-}D})$	910–700	660–510

TABLE 4.18 Effect of Metal Substrate on ν_{C-H} and ν_{C-H} Frequencies

Substrate	Frequency (cm^{-1})	
	ν_{C-H}	ν_{C-H}
Ag(111)	675	—
Ni(110)	700	510
Pd(100)	710	520
Pd(111)	720	525
Re(001)	740	535
Ni(111)	745	545
Ni(100)	750	540
Rh(111)	805	565
Pt(111)	835	605
Pt(110)	910	655

TABLE 4.19 Comparison of Benzene Frequencies in Gas Phase and in the Adsorbed State on Ag Surface

Mode	A_{1g}	E_{2g}	E_{1u}	A_{2g}	B_{2u}	A_{1g}	E_{1g}	A_{2u}
Gas phase[a]	3059	1596	1479	1346	1149	992	849	670
Ag(111)[a]	3030	1590	1480	1360	1155	1000	820	675

[a] All wavenumbers (cm^{-1}).

tion to the metal surface. It should be noted that there are some differences in vibrational frequencies between transition metal (Groups VIII and VIIA) and Ag or Cu (Group IB). While the latter exhibit only marginal changes in vibrational frequencies, the changes for transition-metal surfaces clearly indicate the alternating C—C bond lengths of the aromatic ring, and have been extended to closely resemble a single-bond character. In the case of Group IB, it has been speculated that a lack of the frequency shifts with respect to the gas-phase spectra is related to the position of benzene on the surface metal atom. Apparently, the benzene molecule may be centered over a single metal atom, forming a usual σ-donation/d-π^* backdonation type of bonding. It should be kept in mind that the site–symmetry to which the benzene molecule will adsorb will significantly affect the symmetries of fundamental vibrations in C_6H_6. This can be easily obtained by generating a correlation table that relates the molecular symmetries in the gas-ghase (D_{6h}) with the three most probable site symmetries, C_{3v}, C_{2v}, and C_s. This is illustrated in Table 4.20, which shows how gas-phase molecular vibrations can be correlated with the symmetry changes resulting from adsorption.

The C_6H_6 and C_6D_6 spectra on the following single-crystal metal faces are available: Ni(111),[76-79] Ni(100),[77] Ni(110),[80] Pt(111),[78,81] Pt(110),[82] Re(001),[83] Rh(111),[84,85] Pd(110),[86,87] Pd(100),[86,88] and Ag(111).[89]

TABLE 4.20 Symmetry–Site Effect on Symmetries of Fundamental Vibrational Modes in C_6H_6[a]

D_{6h}	C_{6v}	C_{3v} σ_v	C_{3v} σ_d	C_3	C_{2v} σ_v	C_{2v} σ_d	C_2	C_s σ_v	C_s σ_d	Modes Herzberg Numbering (129)	Modes Wilson Numbering
A_{1g}	A_1	A_1	A_1	A	A_1	A_1	A	A'	A'	ν_1, ν_2	ν_2, ν_1
A_{2g}	A_2	A_2	A_2	A	A_2	A_2	A	A''	A''	ν_3	ν_3
A_{2u}	A_1	A_1	A_1	A	A_1	A_1	A	A'	A'	ν_4	ν_{11}
B_{1u}	B_2^*	A_1	A_2	A	B_2^*	B_1^*	B^*	A'	A'	ν_5, ν_6	ν_{13}, ν_{12}
B_{2g}	B_1^*	A_1	A_2	A	B_2^*	B_1^*	B^*	A'	A'	ν_7, ν_8	ν_5, ν_4
B_{2u}	B_2^*	A_2	A_1	A	B_1^*	B_2^*	B^*	A''	A''	ν_9, ν_{10}	ν_{14}, ν_{15}
E_{1g}	E_1^*	E	E	E	$B_1^* + B_2^*$	$B_1^* + B_2^*$	$2B^*$	$A' + A''$	$A' + A''$	ν_{11}	ν_{10}
E_{1u}	E_1^*	E	E	E	$B_1^* + B_2^*$	$B_1^* + B_2^*$	$2B^*$	$A' + A''$	$A' + A''$	$\nu_{12}, \nu_{13}, \nu_{14}$	$\nu_{20}, \nu_{19}, \nu_{18}$
E_{2g}	E_2	E	E	E	$A_1 + A_2$	$A_1 + A_2$	$2A$	$A' + A''$	$A' + A''$	$\nu_{15}, \nu_{16}, \nu_{17}, \nu_{18}$	$\nu_7, \nu_8, \nu_9, \nu_6$
E_{2u}	E_2	E	E	E	$A_1 + A_2$	$A_1 + A_2$	$2A$	$A' + A''$	$A' + A''$	ν_{19}, ν_{20}	ν_{17}, ν_{16}

[a] NOTES:

1. The A_1, A, and A' modes are completely symmetric and dipolar-active on-specular.

2. The *starred symmetry symbols* are antisymmetric with respect to the twofold axis and therefore impact forbidden on-specular. They are also dipolar-forbidden. Other noncompletely symmetric modes may be impact-forbidden on-specular for certain relative orientations of symmetry planes and the plane of incidence of electrons (see text).

3. σ_v denotes symmetry planes perpendicular to the benzene ring that also include CH bonds; σ_d denotes such planes that bisect CC bonds.

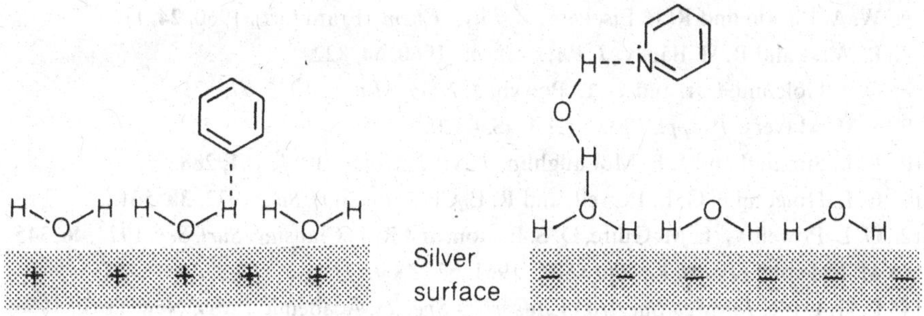

Figure 4.17 Orientation of pyridine on silver metal surfaces. (Adopted from ref. 43.)

4.12.4.2 Pyridine In-situ electrochemical Raman measurements on rough metal surfaces have been recognized as a source of enhancement of Raman vibrational modes and are commonly referred to as surface-enhanced Raman spectroscopy (SERS). The system that has been examined fairly extensively is that of pyridine/silver, which is for the most part a function of applied potential. Although there has been an ongoing controversy as to the occurrence and intensities of the various Raman bands observed in SERS, it is apparent that at approximately 1036, 1025 and 1008 cm^{-1} the three bands were detected near the Raman bands of pyridine in aqueous media. While the band at 1025 cm^{-1} is most likely attributed to the Lewis acid N-coordinated to a silver surface, the 1036 and 1008 cm^{-1} were assigned to pyridine physisorbed to the water layer adsorbed to silver. Interestingly, the latter bands are sensitive to the applied voltage, and both bands exhibit a shift that has been attributed to the orientation changes on the applied voltage. Figure 4.17 illustrates the possible adsorption processes on a silver electrode on anodic and cathodic potentials.[43]

The interactions between silver surfaces and pyridine molecules has received special attention because of almost 10^6 enhancement of the pyridine bands in a SERS spectrum. Numerous theories have been proposed to explain the effects,[103–105] but the difficulty arose in explaining distinctly different enhancements for atomically rough surfaces, argon-etched surfaces, anodized electrodes, colloidal particles in aqueous media, and argon matrices at low temperatures.[106]

REFERENCES

1. E. E. Nikitin, *Theory of Elementary Atomic and Molecular Processes in Gases*, Claredon Press, Oxford, 1974.
2. A. W. Kleyn, J. Los, and E. A. Gislason, *Phys. Rep.*, 1982, **90**, 1–71.
3. A. A. Lucas and M. Sunjic, *Prog. Surf. Sci.*, 1972, **2**, Part 2, 243.
4. J. W. Gadzuk, in *Vibrational Spectroscopy of Molecules on Surfaces*, J. T. Yates, Jr., and, T. E. Madey, Eds., Plenum Press, London, 1987.
5. P. J. Feibelman, *Ann. Rev. Phys. Chem.*, 1989, **40**, 261, and references cited therein.

6. W. A. Pliskin and R. P. Eischens, *Z. Phys. Chem.* (Frankfurt), 1960, **24**, 11.

7. R. West and R. H. Baney, *J. Phys. Chem.*, 1960, **64**, 822.

8. C. E. Holcombe, Jr. and G. L. Powell, *J. Nucl. Mater.*, 1973, **47**, 121.

9. S. M. Mayers, *J. Appl. Phys.*, 1974, **45**, 4320.

10. R. L. Simandl and J. F. McLaughlin, *J. Nucl. Mater.*, 1981, **101**, 288.

11. C. E. Holcombe, G. L. Powell, and R. E. Clausing, *Surf. Sci.*, 1972, **30**, 561.

12. G. L. Powell, G. E. McGuire, D. S. Easton, and R. E. Clausing, *Surf. Sci.*, 1974, **46**, 345.

13. M. Foleman, *Trans. Faraday Soc.*, 1961, **57**, 2000.

14. L. H. Little, *Infrared Spectra of Adsorbed Species*, Academic Press, New York, 1966, p. 265.

15. R. P. Eischens, W. A. Pliskin, and S. A. Francis, *J. Chem. Phys.*, 1954, **22**, 1786.

16. W. E. Osberg and D. F. Hornig, *J. Chem. Phys.*, 1952, **20**, 1345.

17. M. E. Jacox and D. E. Milligan, *Spectrochim. Acta*, 1961, **17**, 1196.

18. J. H. Taylor, W. S. Benedict, and J. Strong, *J. Chem. Phys.*, 1952, **20**, 1884.

19. L. H. Jones and E. McLaren, *J. Chem. Phys.*, 1958, **28**, 995.

20. A. H. Nelson and R. T. Lagemann, *J. Chem. Phys.*, 1954, **22**, 36.

21. R. P. Eischen and J. Jacknow, *Proceedings of the Third International Congress on Catalysis*, North–Holland, Amsterdam, Holand, 1984.

22. W. S. Benedict, N. Gailar, and E. K. Plyler, *J. Chem. Phys.*, 1956, **24**, 1139.

23. C. Hass and D. F. Hornig, *J. Chem. Phys.*, 1960, **32**, 1763.

24. G. Gompartez and W. J. Orville-Thomas, *J. Phys. Chem.*, 1959, **63**, 1331.

25. R. D. Waldron, *J. Chem. Phys.*, 1957, **26**, 809.

26. S. Pinchas and M. Halman, *J. Chem. Phys.*, 1959, **31**, 1692.

27. P. A. Staats, H. W. Morgan, and J. H. Goldstein, *J. Chem. Phys.*, 1956, **24**, 916.

28. H. C. Allen and E. K. Plyer, *J. Chem. Phys.*, 1956, **25**, 1132.

29. E. White, D. A. Dows, and G. C. Pimentel, *J. Chem. Phys.*, 1956, **25**, 224.

30. L. H. Jones, *J. Chem. Phys.*, 1954, **22**, 217.

31. K. A. Wickersheim, *J. Chem. Phys.*, 1959, **31**, 863.

32. R. M. Hexter, *J. Chem. Phys.*, 1961, **34**, 941.

33. R. A. Buchanan, H. H. Caspers, and H. R. Marlin, *J. Chem. Phys.*, 1964, **40**, 1125.

34. W. R. Busing, *J. Chem. Phys.*, 1955, **23**, 933.

35. J. A. Ibers, J. Kumamoto, and R. G. Snyder, *J. Chem. Phys.*, 1960, **33**, 1164; ibid, 1960, **33**, 1171.

36. R. A. Buchanan, *J. Chem. Phys.*, 1959, **31**, 870.

37. W. R. Busing and H. W. Morgan, *J. Chem. Phys.*, 1958, **28**, 998.

38. R. M. Hexter, *J. Opt. Soc. Am.*, 1958, **48**, 770.

39. H. R. Benesi, *J. Chem. Phys.*, 1959, **30**, 852.

40. R. T. Mara and G. B. B. M. Sutherland, *J. Opt. Soc. Am.*, 1953, **43**, 1100.

41. H. Ibach and D. L. Mills, *Electron Energy Loss Spectroscopy and Surface Vibrations*, Academic Press, Orlando, Fla., 1982.

42. M. A. Chester, *J. Electron Spectrosc. Relat. Phenom.*, 1986, **38**, 123.

43. M. Fleischmann, P. J. Hendra, and A. J. McQuillan, *Chem. Phys. Lett.*, 1974, **26**, 163.

44. S. Lehward and H. Ibach, *Surf. Sci.*, 1979, **89**, 1.

45. L. Hammer, T. Hertlain, and K. Müller, *Surf. Sci.*, 1986, **178**, 693.

46. S. Lehward, H. Ibach, and H. Steininger, *Surf. Sci.*, 1982, **117**, 342.

47. F. Zaera and R. B. Hall, *Surf. Sci.*, 1987, **180**, 1.

48. F. Zaera and R. B. Hall, *J. Phys. Chem.*, 1987, **91**, 4318.

49. C. E. Anson, B. J. Bandy, M. A. Chesters, B. Keiller, I. A. Oxton, and N. Sheppard, *J. Electron Spectrosc. Relat. Phenom.*, 1983, **29**, 315.

50. G. S. McDougall, Ph.D. thesis, University of East Anglia.

51. J. A. Stroscio, S. R. Bare and W. Ho, *Surf. Sci.*, 1984, **148**, 499.

52. M. A. Barteau, J. Q. Broughton, and D. Manzel. *Appl. Surf. Sci.*, **19**, 92.

53. M. A. Chesters and E. M. McCash, *Surf. Sci.*, 1987, **187**, L639.

54. I. J. Malik, M. E. Brubaker, S. B. Mohsin, and M. Trenary, *J. Chem. Phys.*, 1987, **87**, 5554.

55. H. Ibach and S. Lehward, *J. Vac. Sci. Techn.*, 1978, **15**, 407.

56. C. Nyberg, *Chem. Phys. Lett.*, 1982, **87**, 87.

57. P. Skiner, M. W. Howard, I. A. Oxton, S. F. A. Kettle, D. B. Powell, and N. Sheppard, *J. Chem. Soc. Faraday Trans.*, 1981, **277**, 1203.

58. N. R. Avery and N. Sheppard, *Proc. Roy. Soc*, 1986, **A405**, 1.

59. N. R. Avery and N. Sheppard, *Surf. Sci.*, 1986, **169**, L367.

60. N. R. Avery and N. Sheppard, *Proc. Roy. Soc.*, 1986, **A405**, 27.

61. R. J. Koestner, M. A. Van Howe, and G. A. Somorjai, *J. Phys. Chem.*, 1983, **87**, 203.

62. B. E. Bent, C. M. Mate, J. E. Crowell, B. E. Koel, and G. A. Somorjai, *J. Phys. Chem.*, 1987, **91**, 1493.

63. N. Sheppard, *J. Electron Spectrosc. Relat. Phenom.*, 1986, **38**, 175.

64. B. J. Bandy, M. A. Chester, M. E. Pemble, G. S. McDougall, and N. Sheppard, *Surf. Sci.*, 1984, **139**, 87.

65. M. A. Chester, and E. M. McCash, *J. Electron Spectrosc. Relat. Phenom.*, 1987, **44**, 99.

66. J. B. Gates and L. L. Kesmodel, *Surf. Sci.*, 1982, **120**, L461.

67. J. A. Gates and L. L. Kesmodel, *Surf. Sci.*, 1982, **124**, 68.

68. M. A. Chester, *J. Mol. Struct.*, 1987.

69. K. Horn and J. Pritchard. *Surf. Sci.*, 1975, **52**, 437.

70. M. A. Chester, in *Analytical Applications of Spectroscopy*, C. S. Creaser and A. M. C. Davies, Eds., Royal Society of Chemistry, London, 1988.

71. J. E. Demuth, H. Ibach, and S. Lehward, *Phys. Rev. Lett.*, 1978, **40**, 1044.

72. S. Lehwald and H. Ibach, *Surf. Sci.*, 1979, **89**, 425.

73. F. M. Hoffman and T. H. Upton, *J. Phys. Chem.*, 1984, **88**, 6209.

74. N. R. Avery, *Surf. Sci.*, 985, **163**, 357.

75. G. Herzberg, *Infrared and Raman Spectra*, Van Nostrand, New York, 1945.

76. J. C. Bertolini, J. Massardier, and G. Dalmai-Imelik, *J. Chem. Soc., Faraday Trans.*, 1978, **174**, 1720.

77. J. C. Bertolini and J. Rousseau, *Surf. Sci.*, 1979, **89**, 467.

78. S. Lehward, H. Ibach, and J. E. Demuth, *Surf. Sci.*, 1978, **78**, 577.

79. H. Jobic, B. Tardy, and J. C. Bertolini, *J. Electron Spectrosc. Relat. Phenom.*, 1986, **38**, 55.

80. J. C. Bertolini, J. Massardier, and B. Tardy, *J. Chem. Phys.*, 1981, **78**, 939.

81. M. Abon, J. C. Bertolini, J. Billy, J. Massardier, and B. Tardy, *Surf. Sci.*, 1985, **162**, 395.

82. M. Surman, S. R. Bare, P. Hofmann, and D. A. King, *Surf. Sci.*, 1983, **126**, 349.

83. B. Tardy, J. C. Bertolini, and R. Ducros, *Bull. Soc. Chim. Fr.*, 1985, **3**, 313.

84. B. E. Koel, J. E. Crowell, M. C. Mate, and G. A. Somorjai, *J. Phs. Chem.*, 1988, **88**, 1988.

85. M. C. Mate and G. A. Somorjai, *Surf. Sci.*, 1985, **160**, 542.

86. B. Tardy and J. C. Bertolini, *J. Chi. Phys.*, **82**, 407.

87. G. D. Waddill and L. L. Kesmodel, *Phys. Rev.*, 1985, **B32**, 2107.

88. G. D. Waddill and L. L. Kesmodel, *Phys. Rev.*, 1985, **B31**, 4940.

89. P. Avouris and J. E. Demuth, *J. Chem. Phys.*, 1981, **75**, 4783.

90. L. Hammer, T. Hertlein, and K. Muller, *Surf. Sci.*, 1986, **178**, 693.

91. S. Lehward, H. Ibach, and H. Steininger, *Surf. Sci.*, 1982, **117**, 342.

92. F. Zaera and R. B. Hall, *Surf. Sci.*, 1987, **180**, 1.

93. F. Zaera and R. B. Hall, *J. Phys. Chem.*, 1987, **91**, 4318.

94. H. Ibach and S. Lehward, *J. Vac. Sci. Techn.*, 1978, **15**, 407.

95. G. H. Hatzikos and R. I. Masel, *Surf. Sci.*, 1987, **185**, 479.

96. W. Erley, A. M. Baro, and H. Ibach, *Surf. Sci.*, 1982, **120**, 273.

97. M. A. Chesters, G. S. McDougall, M. E. Pemble, and N. Sheppard, *Appl. Surf. Sci.*, 1985, **22/23**, 369.

98. H. Steininger, H. Ibach, and S. Lehwald, *Surf. Sci.*, 1982, **117**, 685.

99. A. M. Baro and H. Ibach, *J. Chem. Phys.*, 1981, **74**, 4194.

100. L. H. Dubois, D. G. Castner, and G. A. Somorjai, *J. Chem. Phys.*, 1980, **72**, 5234.

101. B. E. Koel, B. E. Bent, and G. A. Somorjai, *Surf. Sci.*, 1984, **146**, 211.

102. T. S. Marinova and K. L. Kostov, *Surf. Sci.*, 1985, **181**, 573.

103. D. L. Jeanmaire and R. P. van Duyne, *J. Electroanal. Chem.*, 1977, **84**, 1.

104. F. W. King, R. P. van Duyne, and G. C. Schatz, *J. Chem. Phys.*, 1978, **69**, 4472.

105. M. Moskowits, *J. Chem. Phys.*, 1978, **69**, 4159.

106. T. E. Furtak and J. Reyes, *Surf. Sci.*, 1980, **93**, 351, and references cited therein.

CHAPTER 5

ADSORPTION ON METAL OXIDES

5.1 ALKALINE-EARTH OXIDES

Adsorption on metal oxide surfaces is a very important aspect of the surface characteristics since many catalytically controlled reactions involve and often require the presence of a metal oxide as an active ingredient of the reaction. Reactions such as Ziegler–Natta and other polymerization processes, including isomerization or alkylation, play a significant role in both commercial production and scientific research.

Typical substrates in the alkaline-earth family include MgO, CaO, SrO, and BaO. These oxides essentially exhibit a simple rock–salt structure and are stable in a wide range of temperatures. Figures 5.1 and 5.2 show fluoride-type and three distorted cubic peroxide structures in ABO_3-type oxides. In order to establish the surface activity, several models that define the nature of the surface-active sites more closely have seen proposed. For the most part, these models have been based on the results of diffuse reflectance[1-3] and photoluminescence[4] spectroscopy combined with electron microscopy.[5] It has been proposed that the surface defects and irregularities, such as that depicted in Fig. 5.3, containing ions with a reduced coordination number, may be responsible for the additional bands in UV–VIS spectra and, at the same time, may serve as the adsorption sites.

In general, there are three groups of UV–VIS bands that can be related to the three families of the surface sites: penta-coordinated O_p^{2-} and M_p^{2+} ions, which occur at the (001) planes, tetra-coordinated O_t^{2-} and M_t^{2+} ions on edges, and tri-coordinated O_{tr} and M_{tr} ions in corners. Such a surface, as that presented in Fig. 5.3, with various active sites, becomes deactivated on exposure to gaseous

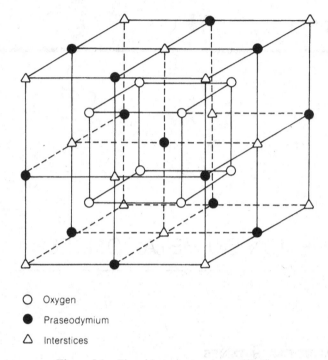

○   Oxygen

●   Praseodymium

△   Interstices

Figure 5.1   Fluoride-type structure of PrO_2.

vapors that are adsorbed in the surface. As a result, the absorption bands disappear or shift and new UV–VIS transitions can be detected.

Infrared spectroscopic studies of molecular hydrogen adsorbed on the oxide surfaces at liquid and solid nitrogen temperatures have shown the presence of the adsorbed molecular hydrogen bands. At this point it is appropriate to mention that the H_2 molecule in a nonadsorbed state is not "seen" by IR, similar to the O_2 molecule discussed in Chapter 2. While Table 5.1 lists the vibrational energies of the adsorbed species, it should be noted that for the surfaces with a lower acidity of the metal sides, such as MgO and CaO, the bands characteristic of adsorption has not been detected. In some cases, two, instead of one, IR bands are detected. This is due to the differences in the metal–hydrogen bond strengths, attributed to the higher coordination numbers of the surface metal atoms. Again, one should realize that the surface purity may be a key factor.

In contrast, hydrogen adsorption at room temperature on MgO, CaO, and SrO indicates the presence of IR bands in two regions: $3800-3000 \, cm^{-1}$ and

➤

Figure 5.3   A schematic representation of the surface alkaline-earth oxides the surface sites having various coordinations.

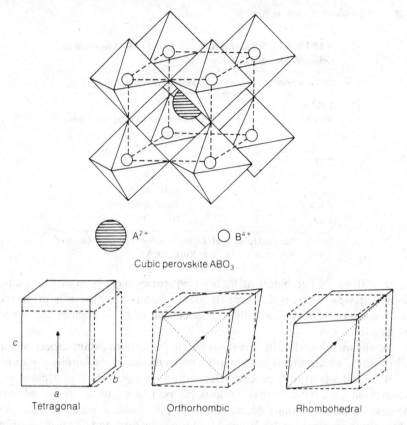

A²⁺   ◯ B⁴⁺

Cubic perovskite ABO₃

Tetragonal   Orthorhombic   Rhombohedral

Figure 5.2 Distortions of cubic perovskite structure in BaTiO₃ leading to other structures.

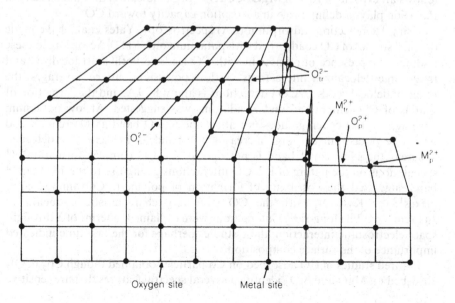

Oxygen site   Metal site

TABLE 5.1 **Vibrational Bands Due to H$_2$ Absorbed to Selected Metal Oxide Surfaces**

Metal Oxide	H—H Stretch (cm^{-1})
BeO	4110
MgO	Not observed because of low surface reactivity
CaO	"
SiO$_2$	4140
ZrO$_2$	4115
Al$_2$O$_3$	4110, 4060
ZnO	4040, 4080
TiO$_2$	4040, 4084

Source: Adopted from L. H. Little, *Infrared Spectra of Adsorbed Species*, Academic Press, New York, 1966.

1350–600 cm^{-1}. The bands at higher frequencies have been attributed to hydroxyl groups, whereas at lower frequency bands are due to hydride groups produced as a result of heterolytic dissociation of H$_2$ molecules above metal–oxygen ion pairs.

As was mentioned in the previous chapters, one important aspect of surface adsorption, which often causes many controversies, is the surface composition prior to the adsorption process. Depending on the surface pretreatment, the adsorption process may take various paths. For example, in the adsorption process of carbon monoxide, dehydration procedure is necessary in order to obtain reproducible results. When MgO is a substrate, and CO was adsorbed at 300 K, the structures presented in Fig. 5.4 were detected.[6] The uptake of CO follows an expected order of MgO < CaO < SrO, indicating that the basicity of the oxide plays a definite role in adsorption capacity toward CO.[7]

Using IR reflection–adsorption (RA) spectroscopy, Yates et al.[8] have made detailed studies of CO coadsorbed with potassium on Ni(111). According to these findings, the presence of potassium with CO leads to both short localized and long-range delocalized interactions. At low potassium surface coverages, the range of delocalized K ··· CO interactions is about 8–9 Å, and the formation of islands of CO centered around K adatoms was suggested. At low potassium coverages, a CO coverage does not affect the C—O band at 1750 cm^{-1}, and when the potassium coverage increases, the band at 1434 cm^{-1} is detected. Whereas the 1750-cm^{-1} band is not due to K—CO complex formation[9][11] speculations on the nature of K ··· CO interactions giving rise to the 1434-cm^{-1} band suggested the possibility of formation of polymeric CO anionic complexes[12] or K$_2$(CO)$_3$ with the CO stretching characteristic frequency at 1445 cm^{-1}. While long-range interactions were explained in terms of a through-space electrostatic interaction, these studies perhaps for the first time indicated importance of the surface composition.

Infrared studies of CO adsorbed on Co particles obtained though Co$_2$(CO)$_8$ adsorbed on MgO and SiO$_2$ identified several spectral features that are sensitive

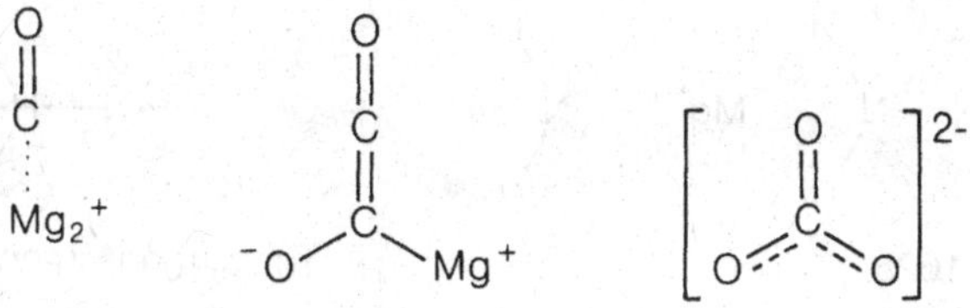

$$2200 \, cm^{-1}$$

$$\begin{array}{l} 2064\text{–}2108 \, cm^{-1} \\ 1318\text{–}1392 \, cm^{-1} \end{array}$$

$$\begin{array}{l} 1090\text{–}1100; \quad 880\text{–}850 \\ 1430\text{–}1505; \quad 705\text{–}725 \end{array}$$

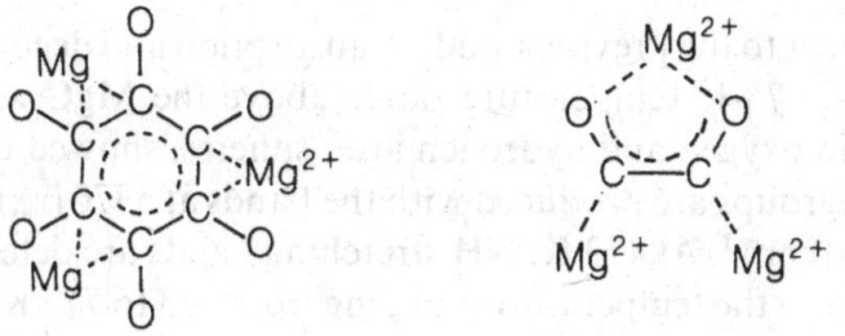

I   II   III

$$\begin{array}{l} 1480, \quad 1775, \quad 1197 \\ 1066 \, cm^{-1} \end{array}$$

$$1548\text{–}1582 \quad \text{and} \quad 1160 \, cm^{-1}$$

IV   V

Figure 5.4 Possible structures of C≡O adsorbed on MgO and vibrational frequencies detected on adsorption.

to the partial CO pressure: for example, at 0.4 torr, the bands at 2025 (the strongest band in the spectrum), 1940, 1880, 1855, 1665, 1610, 1520, 1370, 1320, and 1285 cm^{-1} were observed.[13] As the pressure is raised to 40 torr CO, the bands at 1940, 1880, and 1855 cm^{-1} become stronger and the 2025-cm^{-1} band shifts to 2040 cm^{-1}.

The situation becomes even more complex when ammonia is added to the oxide surface. Three structures have been identified and are depicted in Fig. 5.5, but it is possible that a small fraction of ammonia may dissociate on a cation–anion pair, producing OH$^-$ and NH$_2^+$ groups. The reaction that is believed to govern the formation of OH$^-$ and NH$_2^+$ species occurs on Mg and O, which have a coordination number equal to 3 (tr refers to tri-coordinated ions; Fig. 5.3).

$$Mg_{tr}^{2+} \, O_{tr}^{2-} + NH_3 \text{---} (Mg\text{---}NH_2)^+(OH)^-$$

The experiments of NH$_3$ adsorption on preheated MgO surfaces showed that when NH$_3$ is introduced at 350°C, two peaks at 2170 and 2150 cm^{-1} are detected. Both bands shift with the ^{15}N substitution to 2150 and 2135 cm^{-1} and were attributed to N$_2$ species formed during ammonia decomposition on the oxide

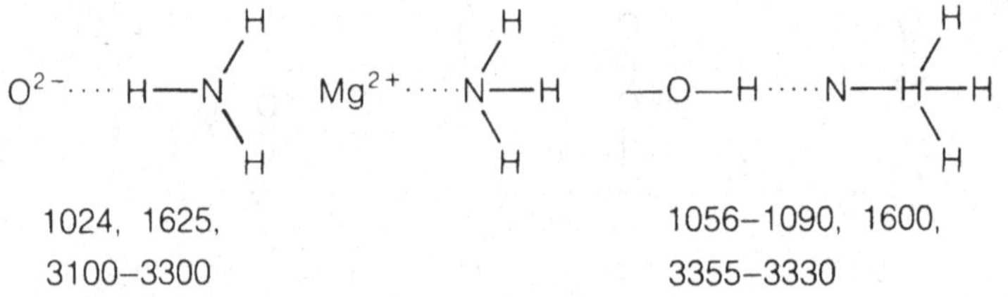

$$O^{2-}\cdots H—N\underset{H}{\overset{H}{<}}\qquad Mg^{2+}\cdots\underset{H}{\overset{H}{>}}N—H\qquad —O—H\cdots N—H\underset{H}{\overset{H}{<}}—H$$

1024, 1625,   1056–1090, 1600,
3100–3300   3355–3330

Figure 5.5 Different structures of NH_3 interacting with surface cation, anion, and hydroxyl groups.

surface.[14,15] In contrast to the previous studies, adsorption and decomposition of ammonia in the 298–773-K temperature range above the MgO surfaces, also preheated at 773 K in oxygen and hydrogen atmospheres, showed that at room temperatures amino groups are produced with the bands at 3370 (v_{as}), 3310 (v_{sym}), 1550 (δ), and 1110 (w) cm^{-1}. At 473 K, NH stretching bands are detected at 3370 (v) and 1410 (δ) cm^{-1}. In the temperatures ranging from 573 to 773 K, three types of $N\equiv N$ structures are represented by three different $N\equiv N$ stretching frequencies at 2220, 2195, and 2080 cm^{-1}.[16]

While the examples cited above generally illustrate the capability of alkaline-earth oxides of dissociative chemisorption, these oxides also behave as Bronsted bases. Using the existing experimental data, a correlation between pK_a of molecules in homogeneous media and reactivity on MgO was made.[17] Apparently, the weaker the acidity, the greater the degree of coordinative unsaturation required in order for molecules to dissociate at the ion-pair surface sites. For example, when propene is adsorbed to the MgO surface, the presence of non-dissociative molecules is accompanied by the ionic σ- and π-allyl species. This is illustrated in Fig. 5.6 and spectroscopically detected by the presence of $C=C$ and $C—C$ stretching modes of the σ-allyl species at 1625 and 950 cm^{-1}. The bands at 1555, 1250, 1055, and 1020 cm^{-1} are associated with the ionic π-allyl groups.

Catalytic properties of surfaces stimulated many model experimental studies that used mixtures of simple gases in order to follow the reactions on surfaces. For example, when CO and H_2 are simultaneously admitted over MgO, at higher

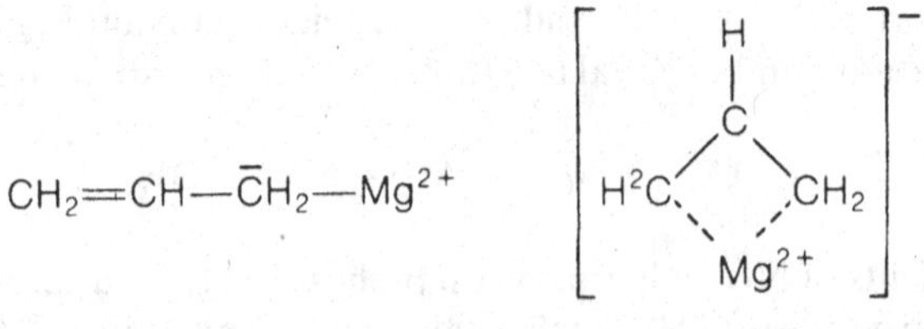

$$CH_2=CH—\overline{C}H_2—Mg^{2+}$$

σ-allyl   Ionic π-allyl

Figure 5.6 σ-allyl and ionic π-allyl on MgO surface.

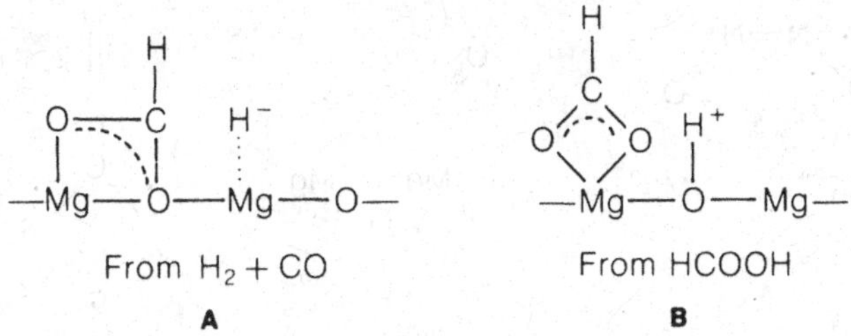

Figure 5.7 Formate obtained from H_2 + CO adsorption on MgO (structure A); formate structures resulting from HCOOH adsorption on MgO (structure **B**).

temperatures, one of the primary products are formyl groups that, on reaction with MgO, give formate ions with the characteristic frequencies at 2840 (C—H stretch), 1604 (OCO asymmetric stretch), and 1370 cm^{-1} (OCO symmetric stretch).[18] It should be noted that this structure is thought to be quite different from that of formic acid (HCOOH) adsorption; both are depicted in Fig. 5.7. These types of structures have been detected on other oxides including CaO, SrO, La_2O_3, ThO_2, and ZrO_2.

The reactions on MgO surfaces have been extended to NH_3 and CO mixtures and have led to the formation of several types of ions, which are listed in Table 5.2, along with the characteristic vibrational frequencies. Similar results were obtained on other oxides and supported metals[19,20] along with the adsorption of propene with carbon monoxide. In the latter case, it was proposed that carbon oxide attacks the allyl group, giving vinylacetyldehyde, which remains in equilibrium with croton–aldehyde. Both aldehydes undergo the hybrid transfer on O^{2-} of MgO, yielding corboxylates and/or alcohols adsorbed on the surface.

The studies of reactions of CO and NO or their mixtures over the alkaline-earth oxides have suggested the presence of appreciable amounts of the species in various structural arrangements, depicted in Fig. 5.8.[21,22] The studies of

TABLE 5.2 Species Resulting from Reactions of H_2 and NH_3 over the Surface of MgO and Their Characteristic Vibrational Features

Species	Temperature Range (K)	Characteristic Bands (cm^{-1})
HCONH$^-$ (formamine monoanion)	< 453	1630, 1587, 1395, 1335, 1150
HCON^{2-} (formamine dianion)	< 453	1604, 1380, 1330
HCO$_2^-$ (formates)	< 453	1615, 1375, 1320
O=C=N—COO—	453–603	2187, 1590
(NCO)$^-$	773	2190–2220

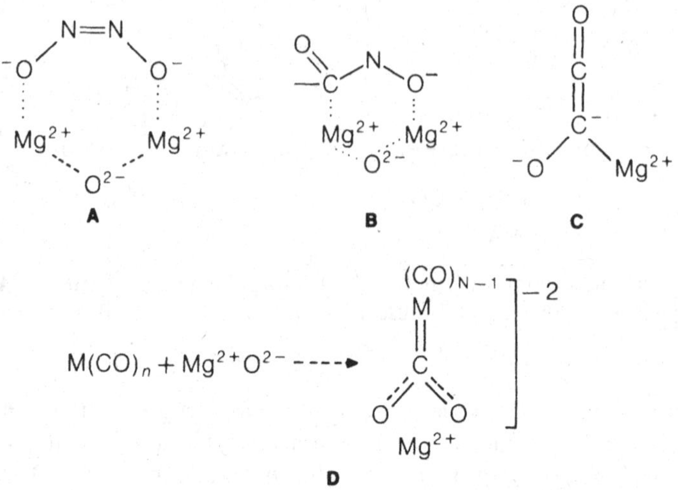

Figure 5.8   Surface structures resulting from adsorption of NO and CO: $N_2O_2^{2-}$ (structure **A**); CNO_2^- (structure **B**); $C_2O_2^-$ (structure **C**).

monometallic and dimetallic carbonyls have shown that the adsorption mechanism is similar to that illustrated in Fig. 5.4., leading to the formation of carboxylic–carbenoid groups.[23] [25] As one would expect for lithium alkyl and its derivatives Li^+R^-, similar structures were found, but the situation becomes more complex for dimetallic carbonyls such as $Co_2(CO)_8$.[26] Several structures in charged and chemical equilibrium may exist simultaneously.

Whereas fine powders of MgO and CaO exhibit strong surface basicity and moderate acidity[27] that gives rise to extremely high surface reactivity and terrific catalytic properties, these properties can be further enhanced by preparing ultrahigh surface area powders.[28] This issue is particularly relevant in adsorption of toxic organophosphrous compounds. When dimethyl methylphosphonate $OP(CH)_3(OCH_3)_2$ (DMMP) is adsorbed on the MgO surface, the main volatile product released on decomposition was formic acid[29] and the following surface species were produced:

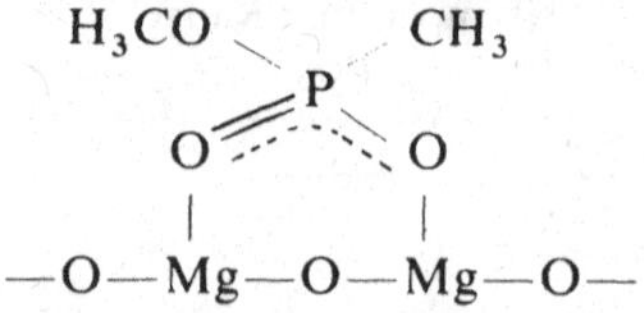

On adsorption, the $P{=}O$ stretching band at 1242 shifts to 1184. However, when triethyl phosphate (TEP) and trimethyl phosphate (TMP) were adsorbed on the MgO surface,[30] the $P{=}O$ stretch shifted from 1267 to $1216\,cm^{-1}$ for TEP, and from 1275 to $1220\,cm^{-1}$ for TMP.

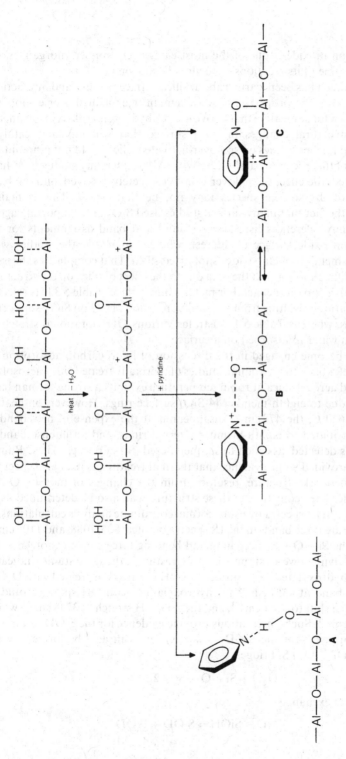

Figure 5.9 Adsorption model of pyridine on aluminium oxide: formation of pyridinium ion pyN⁺—H (structure **A**); formation of pyridinium oxide (structures **B** and **C**). (Reproduced with permission from ref. 56, copyright 1986, Society for Applied Spectroscopy.)

5.2 SILICA

Silica or silicon dioxide is one of the most extensively studied inorganic compounds, and some of its reactions were already shown.

Although much has been learned about silica surface features and interactions with adsorbates, some properties of silica remain unexplained. For example, it has been known for some time that above 425°C, hydrogen spillover from metals can induce structural changes in silica surface that will influence catalytic activity,[31] and further studies on spillover-activated silica could not pinpoint the exact nature of this effect on catalytic activity. Although many studies[32 35] have been conducted, the effect of spillover is not completely resolved, and the band assignments of the surface species may not be that trivial. This is mainly attributed to the fact that many surface studies used SiO_2 as a supporting agent. Therefore, many references list surface studies and band assignments for the metal or metal oxide surface of interest, whereas in reality the studies were conducted on metal- or metal oxide–supported silica. That complicates the issue because it is difficult to sort out the data due to the effect of silica on the structural features that develop on the metal or metal oxide surface. Table 5.3 lists selected vibrational features resulting from the surface reaction of CO on SiO_2-supported metal surfaces, whereas Table 5.4 illustrates various OH vibrational stretching energies when water is adsorbed on a surface.

Before we become engaged in the discussion of hydrocarbon adsorption on silica, let us refer back to the Raman studies of thermally treated silica aerosols.[36] The untreated aerogel silica at room temperature exhibits a strong Raman band at $478\,cm^{-1}$ due to eighth-member (4-SiO) surface rings. However, on heating from 320 to 620°C, the $478\,cm^{-1}$ disappeared at the expense of new band at $490\,cm^{-1}$, attributed to eighth-member internal rings, and an intense band at $600\,cm^{-1}$ was detected, assigned to six-membered (Si—O) rings. These Raman studies also provided some evidence that thermal treatments may also effect the equilibrium network structure, resulting from the changes of the Si—O—Si bridging angle. The geometries of these structures can also be determined using information from IR spectra by using normal coordinate analysis calculations. In essence, there are three bands in the IR spectra of silica: 805, 1088, and $1105\,cm^{-1}$ from which the Si—O—Si stretching and bending force constants of 4.6×10^5 and 1.1×10^5 dyn/cm were estimated.[37] Interestingly, the same studies indicated that when dehydroxylated silica surface (no OH groups) was treated with D_2O, a combination band at $4557\,cm^{-1}$ for hydroxylation with OH species would be detected, with a shift to $3361\,cm^{-1}$, and the SiO—H stretch at $3748\,cm^{-1}$ would shift to $2755\,cm^{-1}$. Such observations provide evidence for the SiOD formation. The latter suggests that the SiOD surface groups can also be formed by the reaction with Si—O—Si bridges

$$D_2O + Si—O—Si \Rightarrow 2SiOD$$

instead of H/D exchange:

$$D_2O + SiOH \Rightarrow SiOD + HOD$$

TABLE 5.3 Characteristic Vibrational Bands Detected as a Result of Adsorption of CO on SiO$_2$ Supported Metal Surfaces

Metal	Support	Band Frequency (cm^{-1})	Intensity	Assigned Structure	Ref.[a]
Fe	Cabosil silica powder	1960	Strong	—Fe···C≡O	A
Cu	Cabosil silica powder	2120	Strong	—Cu···C≡O	B
		~1830	Weak	Cu ＼ C=O ／ Cu	
Pt	Cabosil silica powder	2070	Strong	Pt···C≡O	B
Ni	Cabosil silica powder	2030	Weak	—Ni···C≡O	B
				—Ni ＼ C=O ／ —Ni	
Pd	Cabosil silica powder	2050	Weak	—Pd···C≡O	B
		1920	Strong	—Pd ＼ C=O ／	
		1827	Shoulder	—Pd	
Rh	Alumina powder	2110	Strong	⟋C≡O —Rh	C
		2040		＼C≡O	
		2045	Strong	—Rh···C≡O	
		1925	Medium	—Rh ＼ C=O ／	
Ni/SiO$_2$	CO + O$_2$	1570	Asymmetric	—Rh	
		1330	Symmetric	⟋O —C ＼O	D

REFERENCES:
A. R. P. M. Eischens and W. A. Pliskin, *Advances in Catalysis*, Vol. X, Academic Press, New York, 1958.
B. R. P. Eischens, W. A. Pliskin, and S. A. Francis, *J. Chem. Phys.*, 1954, **22**, 1786.
C. A. C. Yang and C. W. Garland, *J. Phys. Chem.*, 1957, **61**, 1504.
D. J. B. Peri, *Proceedings of the Second International Congress on Catalysis* (Paris, 1960), Edition Technip, Paris, 1961, p. 1333.

TABLE 5.4 Effect of Other Surface Species on O—H Stretching Energy of Hydroxyl Groups on Silica

Adsorbent	Adsorbate	OH Stretch of Unperturbed Adsorbent (cm^{-1})	OH Stretch in the Presence of Adsorbate (cm^{-1})	Nature of Interaction with Hydroxyl Groups	Ref.[a]
Silica powder	H_2O	~3750		No adsorption coeff. changes. OH is an adsorption site	D
Cabosil silica powder	H_2O	3750, ~3650	3750	Hydrogen bonding	E
Silica glass	H_2O	3750		No interaction with OH except at high coverage	F
Silica alumina	H_2O			No interaction with surface OH groups	G
Silica glass	H_2O	3750, ~3680	3680	Little interaction with free OH; adsorption of water occurs on H-bonded OH groups and on "sites of second kind"	H
Silica	H_2O			Adsorption on pair of OH groups on full hydroxylated surface	I, J / K, L

Material	Adsorbate			Assignment	Ref
Aerosil silica powder	tert-Butanol	3750, ~3650	3750	Hydrogen bonding to free OH groups on partly hydroxylated surface	M
Aerosil silica powder	Diethyl ether	3750, ~3650	3750	Hydrogen bonding to free OH groups on partly hydroxylated surface	N
Silica powder	Alkyl benzenes	3743, 3692	3743	Hydrogen bonding to free OH groups	O
Silica glass	NH_3	3750, 3650	3750	Hydrogen bonding to free groups	A, B
Silica glass	CH_3OH	3750, 3680	3750	OH replaced by—OCH	C

a REFERENCES

A. M. Folman and D. J. C. Yates, *Proc. Roy. Soc.*, 1958, **A246**, 32.

B. N. W. Cant and L. H. Little, *Can. J. Chem.*, 1964, **42**, 802.

C. A. N. Sidorov, *Zh. fiz. khim.*, 1956, **30**, 995.

D. G. J. Young, *J. Colloid Sci.* 1958, **13**, 67.

E. R. S. McDonald, *J. Phys. Chem.* 1958, **62**, 1168.

F. V. A. Nikitin, A. N. Sidorov, A. V. Karyakin, *Zh. fiz. khim.*, 1956, **30**, 117.

G. M. R. Basila, *J. Phys. Chem.*, 1962, **66**, 2223.

H. A. N. Sidorov, *Optika Spektrosk.*, 1960, **8**, 806. (Engl. transl. *Opt. Spectrosc.* **8**, 424).

I. A. V. Kiselev, in *Structure and Properties of Porous Materials*, D. H. Everett and F. S. Stone, Eds., Butterworth, London, 1958, p. 210.

J. A. V. Kiselev and V. I. Lygin, *Kolloid. Zh.*, 1959, **21**, 581.

K. A. V. Kiselev and V. I. Lygin, *Usp. Khim.*, 1962, **31**, 351 (Engl. transl. *Russ. Chem. Rev.* 1962, **31**, 175).

L. J. J. Fripiat, and J. Uytterhoeven, *J. Phys. Chem.*, 1962, **66**, 800.

M. V. Y. Davydov, A. V. Kiselev, V. I. Lygin, *Zh. fiz. khim.* 1963, **37**, 469 (Engl. transl. *Russ. J. Phys. Chem.*, 1963, **37**, 243).

N. V. Y. Davydov, A. V. Kiselev, and V. I. Lygin, *Kolloid. Zh.*, 1963, **25**, 152.

O. M. R. Basila, *J. Chem. Phys.*, 1961, **35**, 1151.

When ethyl alcohol is added to the silica surface containing hydroxyl groups, the following reaction may occur:[38]

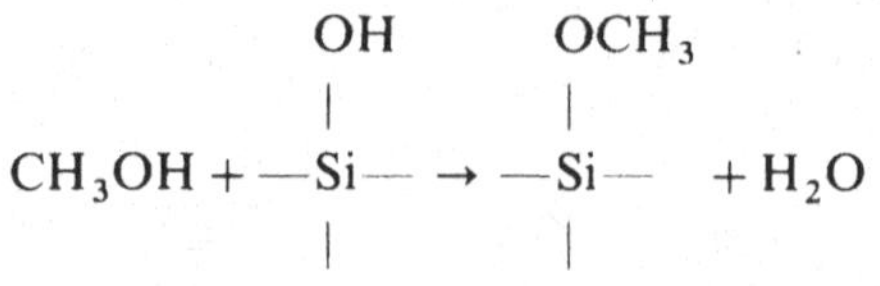

$$
\begin{array}{ccc}
 & OH & OCH_3 \\
 & | & | \\
CH_3OH + & -Si- \rightarrow & -Si- \quad + H_2O \\
 & | & |
\end{array}
$$

This reaction not only produces surface methoxy groups but also effectively eliminates the Si—OH functionalities.[39] This reaction is, however, reversible when the sample is heated under vacuum and exposed to H_2O.

The IR spectrum of gaseous acethylene shows the band at 3287 cm^{-1} due to asymmetric C—H vibrations, which are shown below:

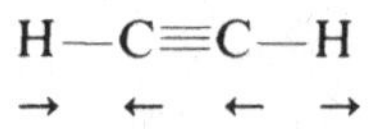

$$
H-C\equiv C-H
$$
$$
\rightarrow \quad \leftarrow \quad \leftarrow \quad \rightarrow
$$

As we recall from the previous chapters, selection rules make symmetric stretching vibrations forbidden in the IR spectrum; instead, they are Raman-active because of the polarizibility tensor changes during the vibrations, giving the vibrational energy at 3373 cm^{-1} with the atoms moving in the directions depicted below:

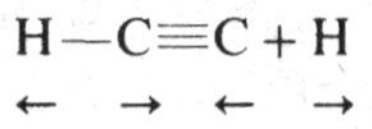

$$
H-C\equiv C + H
$$
$$
\leftarrow \quad \rightarrow \quad \leftarrow \quad \rightarrow
$$

When an acetylene mocelule becomes attached to the surface, the situation changes. This change will depend, however, upon the structures that develop during the adsorption process which, in the past, has been the subject of various controversies. For the most part, it was believed that an HCCH molecule is normal to the surface; that is, only one end of the molecule would be attached to the surface, drastically changing the selection rules making CC stretching vibration IR-active. Table 5.5 provides a comparison of the C—H asymmetric frequencies for C_2H_2 and its deuterated counterpart along with the effect of H—D substitution on the C—H or C—D stretching vibrational frequencies. Apparently, on the basis of the same trends detected in both types of species, it was concluded that one end of the surface adsorbed molecule may be attached (lower frequency) giving a similar effect to that for the C—H to C—D exchange. However, this issue does not seem to indicate a horizontal orientation of the molecule with respect to the surface. While the argument of orientation may be an important one, the above observations and correlations do not conclusively address the issue, especially in a view that the majority of oxide surfaces are covered with the terminal hydroxyl groups.

Although the use of lanthana and other rare earth oxides as supports has added more complexity to the surface analysis, such supports have been found to

TABLE 5.5   Effect of Isotopic Substitution in Acetylene on C≡C Stretching Vibrational Frequency

Molecule	Frequency (cm^{-1})
HC≡·CH	1974
HC≡CD	1851
DC≡CD	1762
HC≡CCH$_3$	2142
CH$_3$C≡CCH$_3$	2233
Strongly adsorbed	
HC≡CH(or Hc≡C—surface) on Al$_2$O$_3$	

enhance the formation of oxygenated products during CO hydrogenation over Rh.[40–42] Underwood and Bell[43] also postulated that the presence of promoting lanthana on Rh/SiO$_2$ surfaces inhibits CO adsorption on Rh sites. The 1725-cm^{-1} surface band was attributed to the CO adjacent to LaO$_x$ islands. These studies also postulated that the interaction of CO with lanthana is responsible for dissociation and the formation of CH$_4$ products. Infrared studies also provided evidence for the presence of acyl, formate, and acetate groups.

TiO$_2$ (7 w/w%)/SiO$_2$ can be prepared using chemical modification of the SiO$_2$ surfaces by reacting it with Ti[OCH(CH$_3$)$_2$]$_4$.[44] Among other techniques, DRIFT spectra at 80 K showed better sensitivity for adsorption of CO on these surfaces. It is most likely that under high-vacuum conditions of XPS (X-ray photoemission spectroscopy) spectra, the surface was altered such that CO were removed from the surface. Ti^{3+} can be formed by either reduction in H$_2$ or evacuation at high temperatures.

5.3   ALUMINAS

When ethylene is introduced on the alumina surface, a few observations became very apparent. Firstly, no exchange with the present on the surface OH (or OD) groups was detected; instead, the adsorbed molecules showed hydrogen bonding with the OH groups, as demonstrated by a shift to sharp OH stretch at 3785 cm^{-1} to a broadband at ∼3500 cm^{-1}.[45] Other OH stretching bands at ∼3740 and ∼3710 cm^{-1} were not affected. Table 5.6 provides the C—H stretching for ethylene adsorbed on Al$_2$O$_3$. Considering the intensity changes of the C—H bands for CH$_2$/CH$_3$, the following structures were proposed:

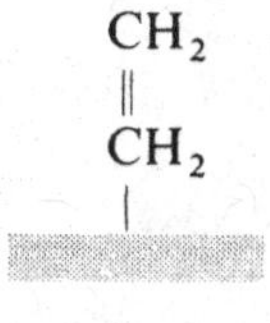

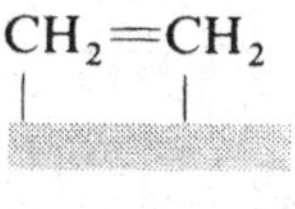

TABLE 5.6 Effect of C_2H_2 and CH_3CCCH_3 on O—H and O—D Stretching Modes on Al_2O_3 Surface

Sample Pretreatment	Adsorbate	Adsorption Bands (cm^{-1})	Assignment	Structure	Desorption
Al_2O_3 deuterated with D_2	C_2H_2	3300	≡C—H stretch		Strongly adsorbed; desorbed by evacuation at 300°C for 10 min
		2007	C≡C stretch		
	C_2H_2	3220	Asymmetric CH stretch of C_2H_2	HC≡CH	Weakly adsorbed; desorbed by evacuation at 20°C for 5 sec
		1950	C≡C stretch		
	$CH_3C≡CH$	~3300	≡CH stretch		Strongly adsorbed; band became weaker on standing; band removed by 200°C evacuation
		2155	C≡C stretch		
	C_2H_2	3320	≡CH stretch	$H_3CC≡CH$	Weakly adsorbed; desorbed by evacuation at 20°C
		2120	C≡C stretch		
	$CH_3C≡CCH_3$	Adsorption bands below 3000 cm^{-1}	CH_3 stretching and bending bands	$CH_3C≡CCH_3$	Some weakly and some strongly adsorbed
		2145	C≡C stretch	$CH_3C≡CH$ impurity	

Source: D. J. C. Yates, and P. J. Lucchesi, J. Chem. Phys., 1961, **35**, 243.

When acetylene is adsorbed on alumina, the band at $3300\,cm^{-1}$ is detected. This band was originally assigned by Yates and Lucchisi [46] and two structures, such as that below, were proposed:

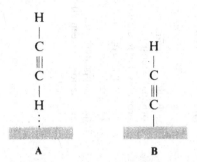

The proposed structures indicate that, unlike the surface on SiO_2, which may also produce a parallel orientation of the C_2H_2 molecule, the Al_2O_3 surface gives vertical orientation. Tables 5.6–5.8 summarize the results of these studies and indicate that the Al_2O_3 surface exhibits quite different activity, which also may be influenced by the presence of other species.

It is almost impossible to realize the adsorption process of an organic molecule on a metal oxide surface without considering the presence of other surface species. It is well known that adsorption of hydroxyl groups on oxide surfaces occurs quite readily, as usually tetra or more valent metal ions will easily bond oxygen of accessively available hydroxyl or water molecules.

In many adsorption studies, especially on oxides that exhibit strong hydroscopic properties, the presence of various forms of water is unavoidable and therefore should be taken into an account. For example, when butene molecules are adsorbed on alumina containing hydroxyl groups, the hydroxyl bands at 3785, 3740, and $3710\,cm^{-1}$, due to the stretching vibrations of hydroxyl groups in various surface environments, decrease in intensity on the expense of a broadband centered at about $3500\,cm^{-1}$.[47] Their OD stretching counterparts are observed in 2803, 2759, and $2733\,cm^{-1}$. The effect of C_2H_2 and NH_3 on the OH stretching modes of OH adsorbed on Al_2O_3 have been a subject of numerous studies, and Table. 5.9 lists the major effects.

When butene is deposited on the surface, interestingly, all hydroxyl bands are restored on evacuation of butene from the alumina surface, indicating that only weak, physical adsorption between the hydroxyl groups and the adsorbate molecules has occurred. These studies suggested that as the surface was evacuated, the adjacent hydroxyl ions became condensed and eliminated as water, leaving oxide ions on the surface. Reduction of hydroxyl groups below the 33% level resulted in either formation of adjacent aluminium ions exposed to the surface or oxide ions with the four nearest neighbors of the same kind in the outermost layer. Although the exact understanding of the spectral changes is still an open question, defect surface structures may possibly accommodate olefin adsorption by coordination of their π-electron-filled orbitals in a way similar to

TABLE 5.7 Effect of C_2H_2, CH_3CCH, CH_3CCCH_3 Adsorption on a Deuterated Al_2O_3 Surface on CC and CH Stretching Vibrations

Adsorbate	Bands of Adsorbent (cm^{-1})	Assignment	Effect of Adsorption	Effect of Desorption of Adsorbate
C_2H_2, $CH_3C{\equiv}CH$, or $CH_3C{\equiv}CCH_3$	2790	OD stretch of free groups	Band disappeared	Band did not reappear on desorption, probably due to exchange of the type 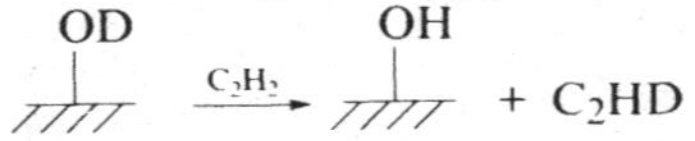
	2755 2720	OD stretch of interacting OD groups	Bands weakened	Bands did not regain original intensity, probably due to exchange as shown above, occurring to slight extent
	2650	New band which appeared during adsorption OD stretch of 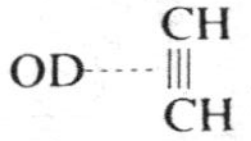 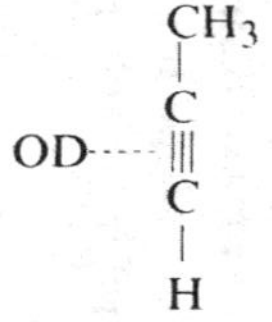	Band increased progressively	Band disappeared with evacuation between 20°C and 300°C

Source: D. J. C. Yates and P. J. Lucchesi. *J. Chem. Phys.*, 1961, **35**, 243.

TABLE 5.8 Comparison of Asymmetric CC Stretching Frequencies in C_2H_2 and C_2D_2 of Strongly and Weakly Adsorbed Acetylenes on Al_2O_3 and Gas Phase

	C_2H_2 CH Asymmetric Stretching		C_2D_2 CD Asymmetric Stretching	
	Frequency (cm^{-1})	Frequency Difference (cm^{-1})	Frequency (cm^{-1})	Frequency Difference (cm^{-1})
Strongly adsorbed molecules on Al_2O_3	3300	13	2570	143
Gas phase	3287	67	2427	17
Weakly adsorbed molecules on Al_2O_3	3220	$C\equiv C$ stretching	2410	vibration
Strongly adsorbed molecules on Al_2O_3	2007	33	1890	128
Gas phase	1947 (only in Raman spectrum)	24	1762 (only in Raman spectrum)	22
Weakly adsorbed molecules on Al_2O_3	1950		1740	
$H-C\equiv C-D$ (gas phase) $H-C\equiv C-Na^+$ (solid)*[a]	CH stretching 3335	CD stretching 2584		$C\equiv C$ stretching 1851

[a] Data from J. Goubeau, and D. Beurer, Z. anorg. allg. Chem., 1961, **310**, 110.

Source: D. J. C. Yates and P. J. Lucchesi, J. Chem. Phys., 1961, **25**, 243.

TABLE 5.9 Characteristic Vibrational Bands of C_2H_2, 1-butene, D_2, and NH_3 Detected as a Result of Adsorption on Al_2O_3

Adsorbent	Adsorbate	Hydroxyl Frequencies of Unperturbed Adsorbent (cm^{-1})	Frequency of Hydroxyl that Interacts Most Readily with Adsorbate (cm^{-1})	Nature of Interaction with Hydroxyl Groups	Ref.
Al_2O_3 deuterated	C_2H_2	2790 ⎫ 2755 ⎬ OD bands 2720 ⎭	2790	Exchange with C_2H_2 OD—OH	1
Al_2O_3	C_2H_2	3785 3740 3710	3785	Hydrogen bonding	2
Al_2O_3 deuterated	1-Butene	2803 ⎫ 2759 ⎬ OD bands 2733 ⎭	2733	Hydrogen bonding exchange reaction OD—OH. Only 2733 cm^{-1} band exchanged at room temperature and 200°C; at higher temperatures the 2759-cm^{-1} band was slowest to exchange OH/OD;	3,4
Al_2O_3	D_2	3795 3737 3598	3698	exchange OH—OD was most rapid with the 3698- cm^{-1} band at 250°C; the 3737-cm^{-1} band was slowest to exchange	3
Al_2O_3	NH_3	3795 3737 3698	3795 3737 3698	Bands disappeared Probably Hydrogen bonding	3,5

Source: D. J. C. Yates and P. J. Lucchesi, *J. Chem. Phys.*, 1961, **25**, 243.

how metal–olefin coordination complexes are formed. If that were the case, the C—C double bond should remain only slightly affected because only small perturbations of the C—H stretching frequencies were detected. These studies also suggested that none of the three OH cited were responsible for the isomerization process of butenes because the rate of isomerization was much faster than the rate of exchange, and the isomerization increased with the decreasing hydroxyl concentration.

In the isotope OH/OD exchange experiments prior to the butene adsorption, in addition to the OH stretching bands, three OD counterparts at 2800, 2750, and 2730 cm^{-1} were observed.[48] As a result of butene adsorption, the 2730 cm^{-1} band weakened at the expense of the increasing 3698-cm^{-1} band. It should be kept in mind that when several OH stretching vibrations are detected, the most reactive species are usually associated with the band at 3780 cm^{-1}, likely adjacent to the strong Lewis acid groups.[49] On the basis of these spectroscopic observations, the following mechanism for butene isomerization was proposed:

$$OD + CH_3CH_2CH{=}CH_2 \quad CH_3CH_2\overset{+}{C}HCH_2D$$

$$\longrightarrow$$

$$OH + CH_3CH{=}CHCH_2D$$

The pioneering work of Peri,[50] that the peripherial crystal termination appears to be a key factor in the structures that develop on the oxide surfaces, still appears to be one of the major considerations in assigning the observed OH bands. Although this approach provided the initial assessments of the vibrational frequencies, other more sophisticated approaches included catalytic surface activity[51] in the frequency considerations. These, combined with the experimental observations, have led to frequency–structure relationships shown in Table. 5.10.

Because alumina surfaces exhibit different structures from that of the bulk, the Al—O surface stretching band at 1025–1050 cm^{-1} was isolated as a shoulder on the high-frequency side of a much stronger Al—O stretch. With that in mind let us consider how other species adsorbed to the Al$_2$O$_3$ surface can be detected.

When ethanol is added to the surface of alumina, the following fragmental functionalities are detected: —OH, —CH$_2$CH$_2$—and ethoxyl groups. The latter, however, are found to be bonded to an aluminum surface giving Al—O—CH$_2$CH$_3$. Whereas a random surface orientation would be expected for short aliphatic species, unless the adsorbing species contain preferential chemical entities leading to ordered structures (Langmuir–Blodgett films), the situation may be different for such species as pyridine. The complexity arises because (1) the nature of the acidity of the aluminum oxide catalysts has been and still is a matter of controversy and (2), adsorption of pyridine onto alumina

TABLE 5.10 Effect of Coordination Number of Alumina on OH Stretching Frequency

Al Coordination Number	Approximate OH Stretching (cm^{-1})
Tetrahedral (example: Al^{IV} in $AlPO_4$; ref. 24)	3800
Al ions occupy only octahedral coordination (example: Al ^{VI} in α-Al_2O_3 or; $MgAl_2O_4$; refs. 25, 26,	3730
Al ions oocupy two different coordination sites, or defective α-Al_2O_3 (example: Al^{IV} and Al^{VI} in transition compounds; ref. 27)	3760, 3738, 3690
Nonstoichiometric solid solutions (example: MgO/Al_2O_3; ref. 28)	3740, 3690
Transition aluminas	Up to five bands can be detected ranging from 3680 to 3800

and silica has been used to determine acidic sites on these as well as other catalysts.[52,53] In all cases in which pyridine was used to measure the strength of acid surface sites, two bands have been of particular interest: at $1540 \, cm^{-1}$ due to formation of a pyridinium ion $(pyN^+ —H)$ and at $1450 \, cm^{-1}$, due to coordinately bonded pyridine.[54,55] Table. 5.11 summarizes observed IR bands of pyridine, pyridine/Al_2O_3, and pyridine oxide and indicates that if pyridine is H-bonded to the surface OH groups present on the surface of Al_2O_3, the band at $1580 \, cm^{-1}$ is due to the $N^+ —H$ deformation modes and the corresponding $N^+ —H$ stretching is found at $2450 \, cm^{-1}$. Another band at $1261 \, cm^{-1}$ is a characteristic feature of the pyridine molecule adsorbed on the surface of Al_2O_3 and is attributed to the $N—O$ stretching mode.[56] In pyridine oxide itself, the $N—O$ stretching frequency appears at $1250 \, cm^{-1}$ in nonpolar solvents.[57] Thus, there are two distinctly

TABLE 5.11 Infrared Bands of Pyridine, Pyridine/Al_2O_3, and Pyridine Oxide

Pyridine		Py/Al_2O_3	PyN–Oxide
Transmittance[a] (cm^{-1})	Drift[b] (cm^{-1})	PA/FT-IR[b] (cm^{-1})	Transmittance[a] (cm^{-1})
991	991	801	841
		1030	1025
1069	1068	1072	
1180	1147	1214	1175
1217	1217	1261 (N—O)	1250
1448	1438	1447	
1482	1452	1461	1464
1582		1580	1488
1600	1601	1592	1608
1633	1633	1620	

[a] N. B. Colthup, L. H. Daly, and S. E. Wiberley, *Introduction to Infrared and Raman Spectroscopy*, 2nd ed., Academic Press, New York, 1975.
[b] M. W. Urban and J. L. Koenig, *Appl. Spectrosc.*, 1986, **40**(6), 851.

different sorption processes of pyridine on Al_2O_3: (1) formation of pyridinium ion N^+—H with surface hydroxyl groups; and (2) oxidation of pyridine on the alumina surface when OH groups are not present. Furthermore, the orientation effect of pyridine on the alumina surface was also addressed and is schematically depicted in Fig. 5.9.

Eyring et al.,[58] using photoacoustic FT-IR spectroscopy, identified a relative number of Bronsted and Lewis acid sites on silica–alumina when pyridine was used as a probe of the surface acidity. In contrast to the previous studies that identified only one Lewis acid site,[59] two types of Lewis acid sites were identified by the presence of two bands at 1624 and 1614 cm^{-1}. The latter appears to be in agreement with the previous studies.[60]

Using temperature program dissorption (TPD) techniques, Soled et al.[61] examined pyridine adsorption on WO_3/γ-Al_2O_3 surfaces and indicated that the adsorbed pyridine DRIFT spectra provide evidence for coordinatively bound Lewis-type and protonated Bronsted-type pyridinium ions.[62,63] By monitoring intensity changes of pyridinium bands as a function of temperature, acid site strengths can be determined.

Preparation of Pt, Ir, and Pt-Ir on Al_2O_3 under different conditions can be followed using Raman spectroscopy.[64] For example, for Pt/Al_2O_3 thermally treated at 110°C in air the following Raman bands were reported: 174, 204, 242, 322, 580 cm^{-1}. As the temperature was raised to 270°C, the 198-, 206-, 234-, 335-, and 580-cm^{-1} bands were detected, and at 500°C, only four bands were detected: 198, 208, 332, and 580 cm^{-1}. For Ir/Al_2O_3, the following bands were detected: for 110°C—316, 480, and 775 cm^{-1}; for 270°C—196, 229, 315, 480, and 745 cm^{-1}; and for 500°C—556 and 745 cm^{-1}. The formation of $PtCl_6^{2-}$ and ICl_6^{2-} salts (K^+)[65] on the surfaces was postulated.

Methanation of CO/H_2 mixtures over 10% Ni/Al_2O_3 surfaces has been studied extensively[66] and showed that both linear and bridged CO surface configurations are possible. In other studies, Young et al.[67] proposed that the surface coverage of CH_x species resulting from the CO and H_2 reactions is independent on the CO/H_2 ratio. In contrast, other studies indicated that this is not the case.[68,69]

CO_2 and NO chemisorption experiments on Mo/γ-Al_2O_3 indicated that the formation of surface species definitely depends on Mo loading.[70] Four characteristic bands at 3610, 1640, 1485, and 1235 cm^{-1} were found when CO_2 was adsorbed on the catalyst and were attributed to OH stretch, asymmetric C—O stretch, symmetric C—O stretch, and C—O—H bending modes of bicarbonate.[71,72] When NO was added to the surface alone, two bands at 1807 and 1700 cm^{-1} were observed and assigned to symmetric and asymmetric NO stretching modes in $Mo(NO)_2$.[73,74]

5.4   ZINC OXIDE

It was believed that when hydrogen is adsorbed on ZnO, depending on the adsorption temperature, the surface of ZnO may exhibit two types of structures. In fact, Table 5.1 lists two bands, possibly confirming this theory; however, this issue appears to be more complex. These structural differences are supposedly reflected in electrical conductivity differences: below 100°C, the conductivity remains the same as that of nonadsorbed ZnO, but at higher temperatures it increases. In the earlier studies, the presence of 3500- and 1710-cm^{-1} bands on ZnO exposure to 1/2 atm of H_2 and 30°C was reported.[75] When D_2 was admitted to ZnO, the bands at 2600 and 1230 cm^{-1}, the counterparts of the 3500- and 1710-cm^{-1}, bands, were detected. From these isotopic substitution experiments, it was concluded that the following reaction for H_2 and D_2 occurred:

$$ZnO + H_2 \Rightarrow OH + ZnH \quad \text{or} \quad ZnO + D_2 \Rightarrow OD + ZnH$$

Although one could correctly argue that such assessments represent only wishful thinking, there is more to that than just a wish. The conductivity data suggest that this is not really the surface phenomenon but reduction of ZnO in the presence of reducing H_2 atmosphere, which causes the removal of O^{2-} ions from the lattice. To maintain charge balance, two electrons are trapped in the oxygen vacancy, thus enhancing electrical conductivity. It should be remembered that reactions such as that listed below are also possible, but they do not necessarily occur on the surface:

$$ZnO + H_2 \Rightarrow Zn + H_2O$$

Above 130°C, however, ZnO may be reduced along with the liberation of H_2O molecules:

$$2ZnOH \Rightarrow ZnO + Zn^0 + H_2O$$

leading to

$$Zn^0 = Zn^+ + e$$

Another possiblity is that

$$H_2 + 2O^{2-} = 2OH^- + 2e$$
$$H_2 + O^{2-} + Zn^{2+} + O^{2-} = Zn^0 + 2OH^-$$

The reactions shown above represent a perfect example of how bulk phenomenon can be confused with the surface studies. This is why it is usually good practice to conduct independent experiments that indeed provide further evidence of the surface effect.

In the early 1980s, however, the research on molecular hydrogen chemisorption revealed new features.[76] It was established that hydrides and hydroxylic groups are formed as a result of chemisorption on the surface $Zn^{2+}O^{2-}$ pairs. On the basis of H_2 and D_2 isotopic substitutions and the frequency shifts due to the surface coverage observed for ZnH and ZnD ($\Delta v = 5$ and 3.5 cm^{-1}, respectively), the dipole–dipole interactions were proposed, and agreed well with the Hammaker model. On the other hand, the ZnOH and ZnOD shifts revealed that the nature of this bonding is primarily static.

The complexity of the surface science comes not only from the detection limits and sensitivity of detection but also from the difficulties associated with the surface preparation. In essence, it is often difficult to prepare the surface that has only one type of species deposited on it. For that reason, it is often necessary to consider the presence of mutually existing species. In the case of H$_2$, carbon monoxide is often present. Griffin and Yates[77] have studied the mutual CO and H$_2$ existence on ZnO surface and have shown that the presence of CO induces a marked shift of the Zn—H and OH stretching modes, attributed to the electron-donating and/or —withdrawing properties of each adsorbate molecule. Furthermore, both molecules interact chemically, giving formyl groups:

$$
\begin{array}{ccc}
 & & \text{H} \quad \text{O} \\
 & \text{H} & \backslash \; /\!/ \\
 & | & \text{C} \\
\text{Zn}^{2+}\text{O}^{2-} \xrightarrow{\;\text{H}_2\;} & \text{Zn—OH} & \rightarrow \text{ZnOH}
\end{array}
$$

This mechanism has received a significant amount of attention since it helped further understanding of the synthesis of methanol.

Catalytic surfaces of Zn mixed oxides have been found useful in many heterogeneous reactions. For example, surfaces of solid solutions of Zn—Cr—O may serve as an attractive synthetic site of alcohols from carbon oxides and hydrogen.[78] The commercial importance of this process stimulated an immense amount of research in this area, and because of the complexity of possible reactions on the surface, usually several methods have been used for characterization. Lietti et al.[79] combined GC (gas chromatography), GC-FT-IR, and GC-MS (MS—mass spectrometry) techniques in an effort to identify a tremendous number of desorption products from the surfaces of Zn—Cr—O, including 1-butanol, C$_8$ and C$_{12}$ aldehydes, C$_7$ and C$_8$ ketones, and C$_3$, C$_4$, and C$_7$ olefins. Other products were dienes, trienes, aromatics, and other hydrocarbons.

5.5 TITANIUM DIOXIDE

It is well accepted that the outer surface layer of TiO$_2$ contains hydroxyl groups that usually absorb in the 3800–3300-cm^{-1} range and dissociated water with the bending modes around 1600–1650 cm^{-1}. The vibrational frequency as well as the intensities of OH groups vary depending on the crystal structure of the oxide[80] as well as the preparation procedures.[81,82] The Raman spectra of Ti^{16}O$_2$ and Ti^{18}O$_2$ are shown in Fig. 5.10[83] and indicate that the A_{1g}, B_{1g}, and E_g modes are particularly sensitive to the isotopic substitution. Whereas these spectra are given here for reference purposes, a great deal of work has been accomplished on the

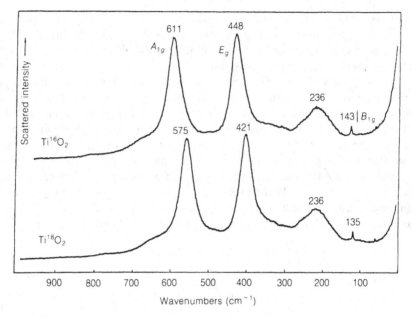

Figure 5.10 Raman spectra of a rutile form of $Ti^{16}O_2$ and $Ti^{18}O_2$.

formation and spectroscopic characterization of the surface species on TiO_2. Considering the immense number of often controversial studies, only a short summary will be provided in the following section:

Hydrated titanias contain OH groups and water; this is reflected in the IR spectra by the appearance of the bands in the $3800-3300$-cm^{-1} region and undissociated water with a characteristic bending mode in the $1650-1600$-cm^{-1} region. In identifying the structural surface features, one should be aware that the frequency and the intensity changes vary and depends primarily on the crystal phase and the excess to the oxide surface. In fact, certain crystal phases can carry only hydroxyl groups, whereas others carry undissociated H_2O and/or both.[84,85]

Surface of TiO_2 can be reduced by vacuum–thermal and chemical treatment using reducing agents.

It appears that the presence of impurities (e.g., sulfates on the surface) often induces surface acidity of the protonic type (Bronsted acidity), whereas a lack of impurities may result in the presence of Lewis acidity. The latter is usually associated with the unsaturated surface Ti^{4+}.[86]

The adsorption of hydrocarbons on rutile results in the surface interaction that perturbs hydroxyl groups. When aliphatic hydrocarbons are on the surface, the extent of perturbation is not as pronounced as for aromatic hydrocarbons. Such perturbations are usually reflected in the vibrational frequency shifts; for aliphatics, OH may shift as much as $400\,cm^{-1}$ down from $3700\,cm^{-1}$, and its magnitude will depend on the ability of inter- and/or intramolecular bonding; specifically, there are a few possibilities. For example, if the concentration of the OH groups is high, or there are preferential areas of higher concentration, the OH groups will have a tendency to self-assemble for $-OH \cdots OH-$associations. However, if an acid COOH group is nearby, because of a strong affinity toward OH to form H bonding the $-OH \cdots HO-$ associations may break and new structures $-COOH \cdots HO$ can form. Again, all that will depend on concentration and conformational ability of both species.

When OH and CO are present on the rutile surface and alkyl acetate and acetic acid are deposited from n-heptane, two adsorbed species having characteristics of Lewis acid–base types may be formed. This is indirectly reflected by a shift of the C—O stretching mode by about $60-90\,cm^{-1}$ as well as a complete elimination of the OH bands because acetate anions may be produced.[87,88]

When anatase contains a certain amount of sulfates covalently bonded to the surface and ethylene is introduced, polymerization of ethylene to form polyethyl-ene is detected.[89]

Infrared spectroscopic studies of CO and H_2 on various surfaces is highly attractive and unfortunately provided a significant amount of contradictory data. For example, in one study on TiO_2-supported Rh, the surface bands characteristic of surface hydrocarbons, formate, and carbonates were detected and were not believed as reaction intermmediates,[90,91] whereas others indicated that these species play a role in further synthesis in TiO_2-supported Rh surfaces.

5.6 IRON OXIDE

When Fe^{3+} salts are hydrolyzed, a dark precipitate α-FeOOH, known as *geothite*, can be obtained. However, its precipitation depends on the precipitating agent, and the starting iron salt.[92,93] The hydroxyl groups in α-FeOOH are usually detected at $3140-3160\,cm^{-1}$, whereas the same species on the surface have been shown to absorb at higher wavenumbers; for example, Perfitt et al.[94] reported 3660- and 3486-cm^{-1} bands, Rochester and Topham[95] reported 3660 and $3500\,cm^{-1}$, whereas Morterra et al.[96] suggested the presence of three bands at 3675, 3630, and $3480\,cm^{-1}$. Although there is general agreement that at least three types of OH groups are present at the surface, many discrepancies are caused not by spectroscopy, but by sample preparation and its reproducubility, such as surface area, surface acidity, and surface purity. Because of this, many OH groups in the $3760-3320$-cm^{-1} range, with many vibrational characteristics, can be found. As many as 11 structures were suggested, and they are not listed here because these data do not reveal the surface conditions that would result in that

TABLE 5.12 Vibrational Energies and Structures Resulting From Adsorption of CO, CO_2, O_2, and NO on Surfaces of Various Oxides

Adsorbent	Adsorbate	Frequency	Assignment	
			Vibration	Structure
NiO/Al_2O_3	CO	1450	Asymmetric stretch	$\overset{\cdot\cdot}{C}O_3^{\cdots}$
	$CO_2 + O_2$	1650	Asymmetric stretch	
		1560	Asymmetric stretch	
		1380	Symmetric stretch	
CrO_3/Al_2O_3	CO_2	1630–1610	Asymmetric stretch	
	CO	1610–1580	Asymmetric stretch	HO—C
	CO_2	1750	C=O stretch	
	CO_2, CO	1430	Asymmetric stretch	CO_3^{2-}
Fe	CO_2(180°C)	1510	Asymmetric stretch	
	CO + O_2 (180°C)	1530	Asymmetric stretch	
	CO + H_2(180°C)	1560	Carbide or oxide	
		1435		
ZnO	CO_2	1430	Asymmetric stretch	
		1640	Symmetric stretch	
		1570		
		1380		

Substrate	Gas	Wavenumber (cm⁻¹)	Assignment	Structure
NiO	CO	1620		
		1575		
		1440		
	CO₂	1620		
		1360		
Ni	CO₂	1560	Asymmetric stretch	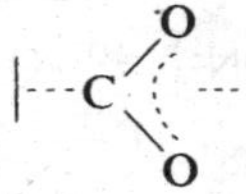
		1400	Symmetric stretch	
NiO	CO₂	1640	Asymmetric stretch	
		1390	Symmetric stretch	
		1640	Asymmetric stretch	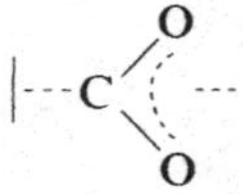
		1390	Symmetric stretch	
Ni/NiO	CO + O₂	1540	Asymmetric stretch	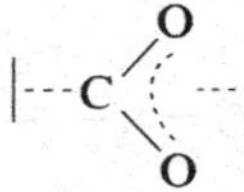
		1390	Symmetric stretch	
TiO₂	CO₂	1570	Asymmetric stretch	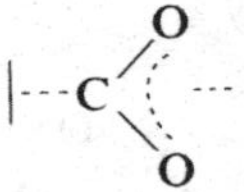
		1330	Symmetric stretch	
	NO	2010	NO stretch	^+NO or $\;^-|\;N{\equiv}O^+$
		1905	NO stretch	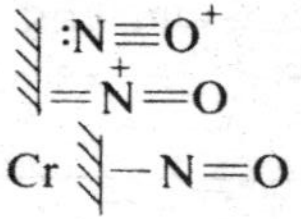 $:N{\equiv}O^+$
		1830	NO stretch	$=N^+{=}O$
		1735	NO stretch	
		1698		$Cr{-}N{=}O$
		1660		
		1625	NO stretch	$O{-}|{-}N{=}O$ (NO on Al₂O₃)

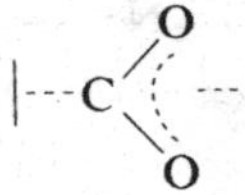

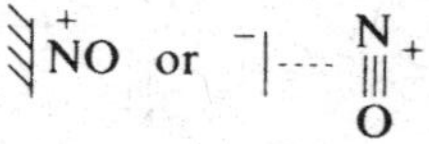

(Carbonate/carboxylate surface structures shown at right as bidentate C–O species bound to the metal surface.)

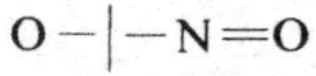

TABLE 5.12 (*Continued*)

Adsorbent	Adsorbate	Assignment		
		Vibration	Frequency	Structure
Fe on alumina gel	NO		2008	$^-|N\equiv N \overset{+}{\cdots} O$ or $^-|NO$
			1805	$^-|{-}N{=}O$ or $|:NO$
			1735	$^-Fe|{-}{-}N{=}O$
			1698 1660 1625	(NO on Al_2O_3) $O{-}|{-}N{=}O$ and $-|\overset{+}{-}N{=}O$
Ni/Al_2O_3	NO		1850	$Ni|{-}{-}N{=}O$
			1735 1698 1660 1625	(NO on Al_2O_3) $O{-}|{-}N{=}O$
Fe_2O_3 gel	NO		1927 1865	NO dimer capillary condensed
			1806 1770	$|{-}NO_2$ $-|\overset{+}{-}N{=}O$ or $|:NO$, NO dimer capillary condensed

		Structure	Wavenumber (cm⁻¹)
		$Fe-|-N=O$	1738
		(NO on Fe_2O_3) $\quad O-|-N=O$	
O_2 added	NO	$|-NO_2$	1700
			1665
			1625
		$-|-N=O$ or $|:NO$	1927
			1806
			1620
			1738
			1700
NiO	NO	$Ni-|-N=O$	1805
			1735
			1695
			1660
Cr_2O_3	NO	$-|NO+$	1625
		$-Cr-|-N=O$	2093
		$Cr-|-N=O$	2028
		$O-|-N=O$	1842
			1810
			1734
			1698
			1663
			1625

Source: Adopted from L. H. Little, *Infrared Spectra of Adsorbed Molecules*, Academic Press, New York, 1966.

many species. As far as other species are concerned, the situation is very similar with a lot of controversies and discrepancies. Therefore, the readers are referred to the original literature and advised to use their own judgment.

Another product of the Fe^{3+} salt hydrolysis is so-called ferrigel which, depending on pH and the nature of the salt, will also precipitate from the solution. When these species are heated in oxygen at 300°C, they are coverted to α-Fe_2O_3, known as *hematite*. If hematite is further heated to 600°C, it forms magnetite λ-Fe_2O_3 with a spinal structure. Table. 5.12 lists selected vibrational frequencies for iron as well as other oxides. The relatively low cost of iron and its use in the ammonia process stimulated not only industrial[97] and practical developments in the Fischer–Tropsch process[98] but also further research in this area. For example, IR spectroscopic studies[99] of interactions of CO, NO, and NO + CO with Fe_2O_3/Al_2O_3 catalysts indicated that depending on the temperature of the catalyst preparation and the gas partial pressure, various species may be generated, such as CO, CNO, and NO.

When Cr^{3+} ions are incorporated in γ-Fe_2O_3 in amounts not exceeding 15 w/w%, the catalytic activity of γ-Fe_2O_3 increases.[100] IR analysis indicated a splitting of the bands in the 800–500- and 500–350- cm^{-1} regions into 720, 640, 575, and 440, 410 cm^{-1}, respectively. According to Tarte,[101] the bands in these regions are due to the modes in FeO_6. In contrast, Ishii et al.[102] attributed the bands in the 800–500-cm^{-1} region to octahedral and tetrahedral units, whereas the lower-frequency region was assigned to octahedral vibrations only.

5.7 MOLYBDENUM OXIDE

Because the most widely applied transmission IR measurements may impose experimental difficulties in the studies of metal oxides and, more importantly, do not provide surface selective information, other approaches have been taken. They include reflection absorption, diffuse reflectance, photoacoustic and emission measurements. Li et al.[103] have chosen emission spectroscopy to study IR spectra of molybdenum oxide and molybdena-supported Al_2O_3, SiO_2, ZrO_2, and TiO_2. Three emission bands at 990, 880, and 820 cm^{-1} were detected in MoO_3. As a result of reduction, the 880-cm^{-1} band is suppressed and a new band at 995 cm^{-1} due to Mo=O stretching was detected. When MoO_3 was supported on Al_2O_3 and SiO_2, no bands due to MoO_3 were detected. In contrast, however, MoO_3 on ZrO_2 and TiO_2 showed the bands at 990 and 890 cm^{-1} after heating above 250 °C. At low concentrations of MoO_2 (2 wt%), instead of MoO_3 phase, MoO_4^{2-} tetrahedral species were detected. A variety of molybdenum oxides included isolated and polymeric species. In the Raman spectra,[104,105] the Mo—O vibrational bands of the MoO_4^{2-} are observed at 930–880 cm^{-1} and 880–800 cm^{-1}, due to MoO_6^{6-} in the 990–930- and 860–800-cm^{-1} regions. As far as the band assignments are concerned, there is a discrepancy as the regions listed above are attributed to Mo=O or Mo—O—M (M is the cation in the supporting metal oxide).

In other studies, using Raman spectroscopy, Kasztelan[106] showed that molybdosilicic acid can be formed on the surface of Mo/SiO_2, with characteristic bands at 995, 969, 911, 636, and 252 cm^{-1} that have been assigned to $Mo—O$, $Mo—O—Mo$, and $Mo_3—O$ symmetric stretching modes and $Mo—O$ bending, respectively.[107,108]

5.8   OTHER METAL OXIDES

As was eluded to earlier, as a result of catalytic properties, aluminas have been extensively studied and, unlike other oxides, exhibit several (some claim up to five) non-hydrogen-bonded OH stretching bands ranging from 3800 to 3690 cm^{-1}. It appears that, depending on the crystal surface termination,[109] the number of bands may vary. However, the issue is further complicated because, similar to silica, alumina was often used as a supporting material in the studies of transition-metal or other metal oxides adsorptions. To illustrate the effect of

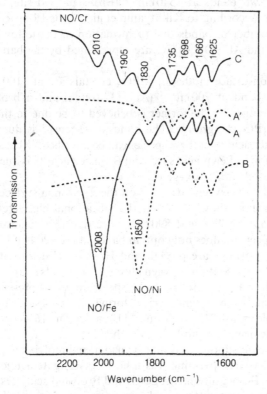

Figure 5.11   Infrared spectra of NO adsorbed on: (A) Fe at 20°C; (A') Fe at 20°C after evacuation; (B) Ni; (C) Cr. (Reproduced with permission from ref. 109, copyright 1965, American Chemical Society.)

alumina-supported transition metal on iron oxide, Figure 5.11 illustrates two IR spectra of NO adsorbed on Fe (tracing 1) and Ni (tracing 2). As can be clearly seen, the N=O stretching vibration is significantly different for different acceptor metal. Table 5.11 illustrates how a combination of metal and metal oxide may affect vibrational bands of adsorbed O_2, CO, CO_2, and NO molecules and what structures may be expected.

Other oxides with rather unique properties are aluminophosphates of the $AlPO_4$ type. They can be synthesized by hydrothermal treatments of aluminophosphates gels formed from mixing aluminium and phosphorus in an organic medium.[110] The surfaces of these species may have several characteristic surface bands, depending on the surface pretreatment, and are sensitive to the $Al_2O_3 : P_2O_5$ ratio. For example, the 3675-cm^{-1} band was attributed to the PO—H stretching normal mode,[111] whereas the 3790 cm^{-1} band is due to AlO—H stretch in tetrahedral coordination.[112] When Al and O atoms are in octahedral coordination, a broadband at 3730–3740 is detected. The dual nature of the surface cations (Al and P) renders the surface both Bronsted-and Lewis-acidic. The studies of pyridine adsorption on aluminophosphates[113] indicated the presence of two bands at 1540 cm^{-1} (Bronsted acid site) and 1450 cm^{-1} (Lewis acid site). According to Kikhtyanin et al.,[114] the IR spectra of adsorbed CO exhibit a number of bands due to formation of the following complexes: P—OH ··· CO and Al—OH ··· CO are represented by the bands at 2172 and 2163 cm^{-1}.

In general, nonsurface modes in $AlPO_4$ crystals are at 1100 cm^{-1} (strong), narrow, intense band at 700 cm^{-1} (Al—O stretch) and a broadband in the 530–400-cm^{-1} range.[115] The former is believed to be due to the triply degenerated band of $(PO_4)^{3-}$ groups. The band at 620 cm^{-1} is due to O—P—O vibrations. Surface acidity of these species can be modified[116] because phosphorous atoms function as Lewis acid sites, and Bronsted acidity is due to the protons attached to exposed surface oxygens.

Infrared studies of oxygen adsorbed on the α-chromia surfaces preheated to 500°C under vacuum showed[117] that the vibrational characteristic bands at 1235, 1145, 1045, 975, 920, and 890 cm^{-1} are due to unoxidized Cr_2O_3 surfaces.[118] If the spectrum does not contain bands in the 1200–1700-cm^{-1} region, usually no CO_3^{2-} groups are present and H—O—H deformation bands are detected. If the surface contains oxygen atoms, three bands at approximately 980, 995, and 1015 cm^{-1} are detected in the IR spectrum. In the presence of ammonia the spectrum contains three symmetric deformation modes due to three types of M—NH_3 complexes at 1215, 1150, and 1110 cm^{-1}. The 1610-cm^{-1} band is due to asymmetric deformation modes of M—NH_3.

In the studies of unsupported titania, vanadia, and molybdena, and their SiO_2 supported counterparts, pyridine was used to investigate acidic sites on these species.[119] Using IR measurements, Lewis and Bronsted acid sites were found on the unsupported surfaces of V_2O_5 and molybdena, whereas only Lewis sites were detected on TiO_2. The characteristic absorption bands of pyridine on Lewis acid sites are at 1448 and 1610 cm^{-1}, whereas Bronsted acid sites exhibit characteristic bands at $1531 \, cm^{-1}$. Vanadium oxide layers on silica were also investigated using Raman spectroscopy.[120]

It should be clear by now that based on the cited reports, it is very difficult to establish precise origins or mechanisms of formation of many surface species deposited on metal oxides. This issue becomes even more complex in so-called heteropolycompounds. A heteropolycompound is one of the strongly active acid catalysts that can be used for various acid-catalyzed reactions.[121] For example, when 12-tungstophosphoric acid (HTP—$H_3PW_{12}O_{40}$) and metal carbonates are decomposed, various polycompounds can be formed, including CaTP, SrTP, PbTP, BiTP, MgTP, CoTP, FeTP, NiTP, and CuTP.[122] The characteristic bands of the heteropolycompounds are P—O stretch at $1080 \, cm^{-1}$, W=O at $980 \, cm^{-1}$, W—O—W at $895 \, cm^{-1}$, and W—O—W at $805 \, cm^{-1}$.[123] However, when ammonia is adsorbed, the situation changes since these species give new bands in the IR spectra that can be classified in the following spectral regions: 1600, 1400, 1190–1240, 1030–1070, and $500–950 \, cm^{-1}$.[124]

5.9 ZEOLITES

5.9.1 General Features of Zeolites

Chemically, zeolites can be represented by the empirical formula

$$M_{2/n}O \cdot Al_2O_3 \cdot SiO_2 \cdot wH_2O$$

where $y > 2$, n is the cation valence state, and w represents the number of moles of water in the voids of the zeolite. From this formula, it is quite obvious that these species have complex structures containing AlO_4 and SiO_4 tetrahedra linked to each other through oxygen bridges. Table 5.13 provides several commonly known zeolites.

As in any characterization process, the first step is to identify the type of bulk structures present, which is followed by surface characterization. The utility of such detailed characterization will formulate the basis for further surface studies, including adsorption and other surface properties. Structural determination of catalytic activity, protonic acid sites, Lewis acid sites, covalently bonded centers, or oxidazing–reducing environments can be determined as described in the previous chapters.

Vibrational spectroscopy of zeolites goes back to the early 1960s[125] with the studies of variety crystalline structures and Si : Al composition ratios. In essence, the primary spectroscopic features are similar to those observed for Si—O and Al—O bonds. Such bands as 405, 465, $683 \, cm^{-1}$ and strong features above $900 \, cm^{-1}$ are due to SiO_2, whereas 568 and $758\text{-}cm^{-1}$ bands are sensitive to composition changes, particularly the Al^{3+}—Si^{4+} exchange as the $758 \, cm^{-1}$ is attributed to the Al—O stretching modes. When zeolites are doped with other species or the species are adsorbed on the zeolite surfaces, other characteristic bands can be present. For example, the presence of NH_4^+ ion will be manifested

TABLE 5.13   Typical Formulas for Natural and Synthetic Zeolites

Zeolite	Typical Formula[a]
Natural	
Chabazite	$Ca_2[(AlO_2)_4(SiO_2)_8].13H_2O$
Mordenite	$Na_8[(AlO_2)_8(SiO_2)_{40}].24H_2O$
Erionite	$(Ca, Mg, Na_2, K_2)_{4.5}[(AlO_2)_9(SiO_2)_{27}].27H_2O$
Faujasite	$(Ca, Mg, Na_2, K_2)_{29.5}[(AlO_2)_{59}(SiO_2)_{133}].235H_2O$
Clinoptilolite	$Na_6[(AlO_2)_6(SiO_2)_{30}].24H_2O$
Synthetic	
Zeolite A	$Na_{12}[(AlO_2)_{12}(SiO_2)_{12}].27H_2O$
Zeolite X	$Na_{86}[(AlO_2)_{86}(SiO_2)_{106}].264H_2O$
Zeolite Y	$Na_{56}[(AlO_2)_{56}(SiO_2)_{136}].250H_2O$
Zeolite L	$K_9[(AlO_2)_9(SiO_2)_{27}].22H_2O$
Zeolite omega	$Na_{6.8}TMA_{1.6}[(AlO_2)_8(SiO_2)_{28}].21H_2O$[a]
ZSM-5	$(Na, TPA)_3[(AlO_2)_3(SiO_2)_{93}].16H_2O$[b]

[a] TMA = tetramethylammonium; TPA = tetrapropylammonium.

by the NH deformation bands at around $1450 \, cm^{-1}$, whereas OH bending due to water is expected near $1640 \, cm^{-1}$. Similarly with metal oxides, the OH stretching bands are expected in the $3600–3750$-cm^{-1} region. Although many IR studies have been conducted on zeolites, Raman spectroscopy has not been that popular.[126,127] Perhaps one interesting aspect is the Si/Al ratio in zeolites that can be correlated to the $500 \, cm^{-1}$ Raman mode.[128]

5.9.2   Adsorption on Zeolites

Normal vibrations of triatomic species such as water have been described in the previous chapters. However, when water is adsorbed on the surface, additional bands may be detected and usually result from symmetric and asymmetric perturbations on the surface. Figure 5.12 illustrates atomic displacements for both perturbations and usually symmetric perturbations are characteristic for adsorption on zeolites.

When NH_4^+ is attached to a zeolite surface, the intensity of the band at $1450 \, cm^{-1}$ due to NH bending decreases on sample evacuation. The presence of this band parallels the band due to water at $1640 \, cm^{-1}$. As was proposed by Kiselev et al.,[129] and later confirmed by Fripiat,[130] the desorption mechanism shown below was proposed:

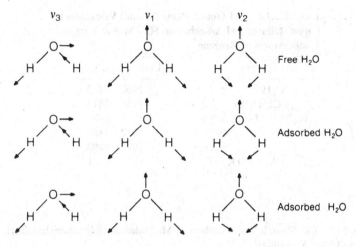

Figure 5.12 Atomic displacements of atoms for asymmetrically and symmetrically perturbed water molecule.

Understanding adsorption of hydrocarbons on zeolites is important as it may reveal insights not only about selectivity of catalytic reactions but also molecular-sieve effects.[131] For that purpose adsorption characteristics of benzene was used to model the behaviour of hydrocarbons on zeolites.[132-134] The behavior of benzene was investigated by Eyring et al.[135] using DRIFT. On the basis of analysis of the CH out-of-plane normal vibrations of benzene that shift on benzene adsorption (see Table 5.14), the authors concluded that benzene molecules may selectively adsorb on the most energetic Bronsted acid ZSM-5 sites, and at least two benzene adsorption species can be identified.

As a result of available space in the zeolite structures, many species can be trapped in it. When $Mo(CO)_6$ was introduced into the faujasite-type zeolites $Na_{85}X$, $Na_{56}Y$, and Y was investigated using DRIFT as a function of the Si/Al ratio and $Mo(CO)_6$ loading level it was found that a reversible reaction, occurring in sequence decomposition of $Mo(CO)_6$ to $Mo(CO)_3$ followed by further decomposition to Mo–aggregates can occur.[136] Table 5.15 provides vibrational frequencies for C=O stretch for the case when $Mo(CO)_5(Oz)$ takes C_{4v} symmetry where $2A_1 + B_1 + E$ IR modes are active, C_{2v} symmetry with $2A_1 + B_1 + B_2$ modes IR-active for $Mo(CO)_4(O_z)$ species, and C_{3v} symmetry for $Mo(CO)_3(O_z$-oxygen atom of zeolite framework) species with $A_1 + E$ modes IR allowed.

Spectroscopic features resulting from adsorption of pyridine on zeolites appear to be very similar to that on metal oxides. In the case of zeolites, the bands due to chemisorbed pyridine are detected at approximately 1546 and 1457 cm^{-1}. Whereas these bands are believed to indicate both Bronsted and Lewis acid sites, respectively, the presence of the 1491 cm^{-1} is due to both. However, the band at 3610 cm^{-1} attributed to Al—OH becomes weaker and, at the same time, the

TABLE 5.14 CH Out-of-Plane Normal Vibrations in Liquid, Diluted and Adsorbed on HZSM-5 at Various Temperatures of Benzene

Benzene Species	Frequencies (cm^{-1})
Liquid C_6H_6	1960, 1815
C_6H_6/CCl_4 (1:50)	1956, 1813
C_6H_6/HZSM-5 at 300 K	1977, 1880
403 K	1977, 1880
553 K	1977, 1880
C_6H_6/HY at 300 K	1977, 1841

TABLE 5.15 CO Stretch Wavenumbers of Molybdenum Subcarbonyls Adsorbed in $Na_{56}Y$ and $Na_{85}X$ Zeolites

Compound	$A_1{}^a$	$A_1{}^a$	$B_1{}^a$	$B_2{}^a$	E^a
$Mo(CO)_5(NC_5H_5)^b$	2074 w^c	1922 sh	1988 w		1943 s
$[Mo(CO)_5.Na_{56}Y]$	2075 w	N.O.	N.O.		1940 br
$[Mo(CO)_5.Na_{85}X]$	2075 w	N.O.	N.O.		1940 br
$Mo(CO)_4(NC_5H_5)_2{}^b$	2025 m	1881 sh	1839 s	1907 s	
$[Mo(CO)_4.Na_{56}Y]$	2040 m	1900 sh	1840 s	1910 s	
$[Mo(CO)_4.Na_{85}X]$	2035 m	1900 sh	1840 s	1910 s	
$Mo(CO)_3(NC_5H_5)_3{}^b$		1908 s			1777 s
$[Mo(CO)_3.Na_{56}Y]$		1907 s			1770 br
$[Mo(CO)_3.Na_{85}X]$		1912 s			1775 s-1750 sh

^{a} Symmetry of the CO stretching modes according to the molecular geometry reported in C. S. Kraihanzel and F. A. Cotton, *Inorg. Chem.* **2**, 533 (1963).
^{b} From C. Barbeau and J. Turcotte, *Canad. J. Chem.* **54**, 1903 (1976).
^{c} Relative intensity: w (weak), sh (shoulder), s (strong), br (broad), N.O. (not observed).

Si—OH counterpart remains unchanged.[137] Simultaneous presence of phosphorus results in attenuation of the bands at 3610 and 3740 cm^{-1}, and a new band at 3685 cm^{-1} is detected. Modification of H-ZMS-5 zeolites by impregnation with H_3PO_4 is a commonly used technique for enhancement of activity and selectivity.[138] In essence, the primary controversy is whether, on surface treatment, structure A[139,140] or structure B[141,142] dominates.

Although zeolite OH groups can serve as active sites for various acid-catalyzed reactions, they create a lot of controversy. Interestingly, spectroscopic evidence seems to indicate that several kinds of OH groups that may play different roles in surface catalytic activity. This was certainly illustrated for the NaH–ZMS-5 zeolite,[143] and does not seem to agree with the theoretical studies[144,145] that suggested existence of only one type of OH group and that the catalytic activity would depend on the overall composition of zeolite. The existence of different vibrational OH stretching frequencies has also been accounted for by different chemical composition of the nearest neighbouring group and the coordination number.[146] Table 5.16 provides a comparison of the OH stretching vibrations in different environments. When structural and chemical characteristics of bridging OH groups are considered,[147] the acid strength of the OH groups is affected, whereas OH stretching energies are influenced by the changes in surrounding structures. Furthermore, doping of zeolites with various metals has a tremendous effect on the intensity and a number of OH bands

TABLE 5.16   Calculated and Observed OH Stretching Vibrations

Surface OH Group		Calculated	Observed
$Si(OH)_4$	OH—Si	3750	$3750^{a,b}$
$OP(OH)_3$	OH—P	3676	3680^{c-e}
$B(OH)_3$	OH—B	3684	3695^f
$H_2OAl(OH)_3$	OH—Al	3803	$3800^{c,d}$

[a] R. E. Sempels and P. G. Rouxhet, *J. Colloid. Interface Sci.*, 1976, **55**, 263.
[b] L. M. Kustov, V. B. Kazansky, S. Beran, L. Kubelkova, and P. Jiru, *J. Phys. Chem.*, 1987, **91**, 5247.
[c] J. B. Moffat, *Catal. Rev. Sci. Eng.*, 1978, **18**, 199.
[d] J. B. Peri, *Discuss. Faraday Soc.*, 1971, **52**, 55.
[e] O. V. Kithtyanim, E. A. Paukshtis, K. G. Ione, and V. M. Mastikhin, *J. Catal.*, 1990, **126**, 1.
[f] V. M. Bermudez, *J. Phys. Chem.*, 1971, **75**, 3249.

observed in IR. For example, the presence of Eu^{3+} may result in an additional band in the proximity of the Si–OH region.[148] In other studies, when $Os_3(CO)_{12}$ was adsorbed on an acid zeolite (HX), the carbonyl bands characteristic to the osmium adsorption were found,[149] and on pressure and temperature treatments ($>140\,°C$, $100\,kPa$ of CO), two bands at 2132 and $2046\,cm^{-1}$ were detected. These bands arise from $Os(CO)_x$ ($x = 2$). However, in the absence of CO, this species transforms to a new form, giving three CO bands at 2128, 2045, and $1960\,cm^{-1}$ that are presumably due to a mixture of $Os(CO)_2$[150] and $Os(CO)_3$.[151]

Rhodium-containing Y-type zeolites are known catalysts for olefin hydroformylation[152,153] and two types of surface species are detected when 1:1 ratio of CO/H_2 is injected. The $Rh_6(CO)_{16}$ species has characteristic IR bands at 2098 and $1763\,cm^{-1}$, whereas the presence of the 2115, 2098, 2044, and $2019\,cm^{-1}$ bands results from the formation of dicarbonyl species $RH^I(CO)_2$ resulting from the following reaction:[154,156]

$$Rh^{III} + 3CO + H_2O \rightarrow Rh^I(CO)_2 + CO_2 + 2H^+$$

The doublet bands at 2098 and $2019\,cm^{-1}$ are attributed to $Rh^I(CO)_2(O_z)_2$, whereas the bands at 2115 and $2044\,cm^{-1}$ are due to $Rh^I(CO)_2(O_z)_2(H_2O)$ (O_z represents an oxygen atom being a part of the zeolite network). The bands at 2115, 2098, 2044, and $2015\,cm^{-1}$ shift to 2065, 2050, 2002, and $1975\,cm^{-1}$ on ^{12}CO–^{13}CO isotopic substitution.[156] Whereas catalytic conversion of aqueous ethanol to polyethylene on silicon-rich H-ZMS-5 is sensitive to the Si/Al ratio,[157] the alkylation of toluene with methanol over H-ZMS-5 zeolites provides another interesting example.[158] As indicated earlier, usually two active sites are demonstrated by the presence of two hydroxyl groups. We will use the bands at 3745 and $3610\,cm^{-1}$ as an example. When organic reactant or reactants are introduced to the surface, the band intensities diminish and new bands usually show up. In the case of toluene, the new bands are detected at 3600 and $3460\,cm^{-1}$. While the former is attributed to silanol groups interacting with toluene, the latter is due to OH groups of methanol molecules protonated at the strong Bronsted acid sites interacting with the aromatic rings of toluene.[159] Other bands in the spectra would include the CH stretching bands at 3086, 3060, 3030, 2925, and $2877\,cm^{-1}$ due to toluene and at 2958 and $2856\,cm^{-1}$ due to methanol. Other detected bands in the spectrum include a broad OH deformation band due to $CH_3OH_2^+$ ion and 1604- and $1495\text{-}cm^{-1}$ ring vibrations in toluene. The structure responsible for these spectral features is shown in Fig. 5.13. Other aromatic species can be detected by the presence of their characteristic bands at 1485 and $1620\,cm^{-1}$ for m-xylene, 1496 and $1466\,cm^{-1}$ for o-xylane, 1474 and $1447\,cm^{-1}$ for 1, 2, 3-trimethylbenzene with additional bands at $1606\,cm^{-1}$ for 1, 2, 5-trimethylbenzene, and $1506\,cm^{-1}$ for 1, 2, 4-trimethylbenzene.

The surface of ZSM-5 catalyst can serve as a proton-exchange site employed during the process of conversion of methanol to gasoline.[160] Therefore, mechanistic and structural aspects of the conversion of methanol to dimethylether[161] and then to higher hydrocarbons at 350°C are particularly important. It seems

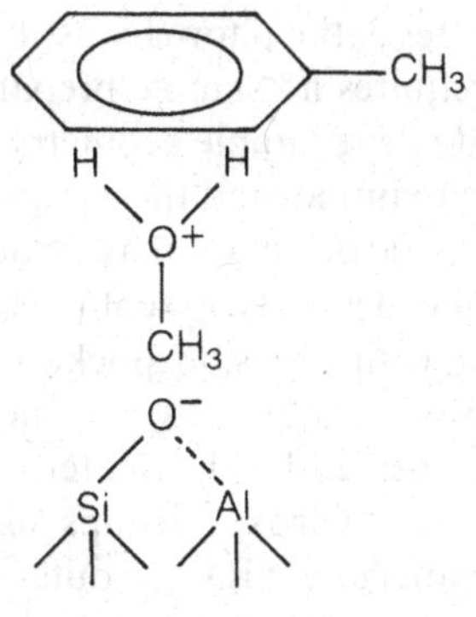

Figure 5.13 Orientation resulting from the coadsorption process of toluene and methanol at the strong Bronsted acid sites.

that the most rationable approach is to analyze three spectral regions: examine the zeolite OH stretching region, the organic fragments C—H stretching region, and the changes resulting from zeolite framework vibrations, and parallel these measurements with the temperature studies. The OH stretching band due to terminal silanols at approximately $3740 \, cm^{-1}$ is usually undisturbed by methanol adsorption. However, the $3676\text{-}cm^{-1}$ band shifts to $3700 \, cm^{-1}$ as a result of desorption of water at increased temperature. In contrast, the weakest band at $3605 \, cm^{-1}$ at room temperature becomes strong at 200°C as the acidic OH bridged structures Al—OH—Si are formed.[162] At room temperature, adsorption of methanol will be domonstrated by the bands due to methoxide at 2997, 2954, and $2845 \, cm^{-1}$.[163,164] At elevated temperatures, the $2997\text{-}cm^{-1}$ band disappears at the expense of a new band at $3011 \, cm^{-1}$, whereas other bands become broader. Adsorption of dimethylether on alumina above 75°C resulted in presence of the $3036\text{-}cm^{-1}$ band. This band was attributed to the C—H stretching normal modes of methoxy groups and results from the reduced symmetry on adsorption.[165]

Whereas spectroscopic adsorption studies provide a significant amount of information about catalytic surface properties, polymerization on the surfaces of zeolites adds another dimension to their use. Hendra et al.[166] reported on the formation of polyacetylene on X-zeolite and Al_2O_3 activated at higher temperatures. Of course, the most extensively used catalysts for polymerization of acetylene are the Ziegler–Natta catalysts, based on titanium alkoxides and alkylaluminum compounds.[167] However, another catalytic active center may be achieved by using cobalt and nickel salts in the presence of borohydride reducing agents.[168] A similar activity can be achieved on Y zeolites when doped with selected transition metals. For example, no acetylene polymerization reactions were observed for MnY, CuY, and ZnY, whereas CoY and NiY resulted in the formation of *trans*-polyacetylene. Raman bands due to *trans*-polyacetylene are detected at 1015, 1140, 1300, and $1560 \, cm^{-1}$.[169]

As was indicated in Chapter 3, the photoacoustic mode of detection provides several unique features; it requires no sample preparation and does not impose special requirements related to the sample geometry, color, or shape. Since little or no sample preparation is required, and the sample can be exposed to various environments, the photoacoustic detection may be advantageous for adsorption studies of zeolites. Since a mordenite-type zeolite is built of negatively charged SiO_4 and AlO_4 tetrahedra with the side pockets of 0.57×0.29 nm, usually single-valent cations serve as neutralizers. In such systems, three types of hydroxyl groups can be substituted by the deuterium isotope exchange. When these cations are protons, the hydroxyl groups usually are detected around 3740 cm^{-1} (hydroxyl on amorphous SiO_2 or outer zeolite crystal surface), the 3650 cm^{-1} attributed to the nonacidic hydroxyls, most likely bonded to $[AlO]^+$ structures and at around 3600 cm^{-1} due to acidic hydroxyl groups.[170] Of course, OH groups can be replaced by OD groups by deuterating an outgassed sample with D_2O. As a result, new bands due to the OD corresponding to the order for OH species will be detected at around 2760, 2700, and 2660 cm^{-1}.[171] In addition, Si—O—D vibration may occur at approximately 890 cm^{-1}. Such pretreated surfaces, on further treatment with diboranes, show new spectral features at 2535 and 1368 cm^{-1} due to BH and BO normal vibrations, indicating that the following reaction may have occurred:

$$2Si—OH + B_2H_6 \rightarrow 2Si—O—BH_2 + 2H_2$$

On further chemisorption of ammonia on such boronated surface, NH stretching and bending modes at 3311 and 1617 cm^{-1} are detected. Along with that, the BH normal vibrations are shifted from 2535 to 2483 and 2355 cm^{-1}.

Because IR spectra of oxide surfaces often contain unique information regarding the nature of adsorption sites, considerable interest has been devoted to the OH stretching frequencies. In fact, even quantum-chemical molecular models designed to reproduce the OH stretching vibrational energies of Si—OH, P—OH, B—OH, and Al—OH were developed[172] and a comparison with the experimental results is presented in Table 5.15.

5.10   CLAY MINERALS

Clays are usually anisotropic microcrystals that display large specific surface areas. The lattices of clays have a common feature, that is they are built of continuous planes containing oxygen atoms linked to such cations as magnesium, aluminum, lithium, or iron in octahedral coordination. A schematic sketch of the kaolinite layered structure is depicted in Fig. 5.14. As a result of isomorphic substitutions of Si^{4+} by Al^{3+} in tetrahedral coordination, the overall charge of these lattices is negative. This negative charge is usually compensated by adsorption of exchangeable Na^+, K^+, and Ca^{2+} cations. Many surface-catalyzed reactions have been described in the literature.[173,174] Perhaps the most

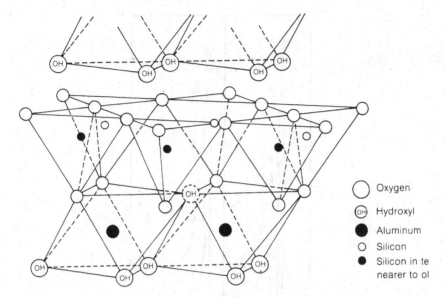

Oxygen

Hydroxyl

Aluminum

Silicon

Silicon in te
nearer to ol

Figure 5.14   Schematic representation of the kaolinite layer structures.

typical catalytical reaction involves transalkylation reaction:[175]

$$RNH_3^+ + H_2O \rightarrow ROH + NH_4^+$$

$$ROH + RNH_3^+ \rightarrow R_2NH_2^+ + H_2O$$

Interestingly, in the absence of water interlayer this reaction is not observed and, on more severe conditions, deamination or dehydration of H_3O^+ may result in the exchange at the surface sites. Furthermore, the presence of negative charges arising from the $Al^{3+} - Si^{4+}$ exchange will result in opening of the Si—O—Al bridges.[176] This process can be monitored using IR, as demonstrated by the symmetry changes from T_g to C_{3v}, making the A_1 mode IR-active.[177] Formation of surface Si—OH—Al acidic groups on heating NH_4^+ beidellite is demonstrated in Fig. 5.15, and leaching Na beidellite with an acid solution, followed by calcination at the indicated temperature, is reflected in the disappearance of the NH stretching bands at $3440\,cm^{-1}$.

A significant effort has been devoted to understanding interactions between organic molecules with mineral surfaces. In situ Raman and FT-IR analysis of kaolinate–hydrozine (KH)-type intercalation complexes has demonstrated that the inner-surface hydroxyl stretching bands of kaolinate are strongly influenced by the presence of hydrazine in the intermellar region.[178] This is not very surprising as the interlamellar region of kaolinite consists of two molecular sheets: one contains siloxane surface with a hexagonal network of ditrigonal cavities, whereas the attached sheet is of that of aluminum hydroxyl groups. As a

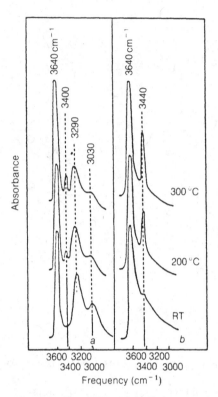

Figure 5.15   Formation of surface Si — OH — Al acidic groups on heating NH$_4^+$ beidellite (a) and leaching Na beidellite with an acid solution (b), followed by calcination at the indicated temperature (RT = room temperature). (Reproduced with permission from ref. 177.)

result of local environmental changes between the layers, hydroxyl groups are expected to be mostly affected, causing their vibrational energy to change.[179] In particular, the band intensities of the inner surface at 3652, 3668, 3688, and 3695 cm^{-1} are strongly reduced on intercalation. Furthermore, as these bands diminish in intensity, a set of new bands at lower energies is detected. When hydrozine is introduced between the Si and Al layers, the story outlined above changes quite a bit. For example, Ledoux and White[180] observed that in the inner surface layer (often called "ISu") hydroxyl groups are strongly perturbed. Furthermore, the N—H stretching bands of kaolinate are observed at higher frequencies. Table 5.17 provides FT-IR and Raman data for the kaolinite–hydrazine complexes.

Silane coupling agents are used to enhance bonding between silicones and polymer matrices. DRIFT can be also used to quantify adsorption and polymerization processes of silane coupling agents on kaolin clays.[181] On a similar note, adsorption of silanes on mica can be analyzed quantitatively.[182] In this case, γ-APS was monitored using the internal standard band at 3625 cm^{-1}. Because

TABLE 5.17   FT-IR and Raman Bands Observed in the Kaolinite–Hydrazine Complex

Kaolinite–Hydrazine Intercalate			Hydrazine		Kaolinite	
Raman	FT-IR(1 atm)	FT-IR(evac.)	Raman	IR(liquid)	Raman	IR
140	n/a[a]	n/a		130	130	
					141	
				201	190	
				244		
				271	268	
336	n/a	n/a			338	
					352	
					366	
400	n/a	n/a			397	405
436	n/a	n/a			431	440
467	n/a	n/a			461	470
514	n/a	n/a			512	
					555	
636					637	638
				700	692	
739		750			750	754
794		796			790	792
					799	
	871					
903						
903			903			
903		914			915	918
948		945			940	941
	1014				1018	
1033						
	1044		1042		1040	
1095			1112			1101
1131		1127				1120
1273	1265	1286	1283			
1313	1339		1335	1324		
		1628	1608			
3185						
3200	3196	3225	3195	3189		
3283			3280			
3302						
3310						
3344	3336	3361	3346	3310		
3362						
3620	3620	3628			3621	3620
	3656			3652	3652	
	3673			3668	3668	
				3688		
3692	3694			3696	3695	

[a] Not applicable.

Source: C. T. Johnson, in *Spectroscopic Characterization of Minerals and Their Surfaces,* ACS Symposium Series No. 415, L. M. Coyne, S. W. S. McKeever, and D. F. Blake, Eds., American Chemical Society, Washington, D.C., 1990, Chapter 22.

mica can be cleaved into very thin sheets giving tremendous interference fringes, transmission FT-IR can be used to monitor adsorbed monolayers of octadecyl-trichlorosilane (OTS).[183]

REFERENCES AND NOTE

1. R. L. Nelson and J. W. Hale, *Discuss. Faraday Soc.*, 1971, **52**, 77.
2. E. Garrone, A. Zecchina, and F. S. Stone, *Philos. Mag., B*, 1980,**42**, 683.
3. M. W. Urban and J. L. Koenig, *Appl. Spectrosc.*, 1986.
4. S. Coluccia, A. M. Dean, and A. J. Tench, *J. Chem. Soc., Faraday Trans. 1*, 1978, **74**, 29.
5. S. Coluccia, A. J. Tench, and R. L. Segall, *J. Chem. Soc., Faraday Trans, 1*, 1979, **75**, 1769.
6. A. Zecchina, E. Garrone, and E. Guglielminotti, *Specialist Periodical Reports, Catalysis*, Vol. 6., Royal Society of Chemistry, London, 1983, p. 90, and references cited therein.
7. S. Coluccia, E. Garrone, E. Guglielminotti, and A. Zecchina, *J. Chem. Soc., Faraday Trans. 1*, 1981, **77**, 1063.
8. K. J. Uram, L. Ng, and J. T. Yates, *Surf. Sci.*, 1976, **177**, 253.
9. M. Ito and R. West, *J. Am. Chem. Soc.*, 1963, **85**, 2580.
10. M. Takahashi, K. Kaya, and M. Ito, *Chem. Phys.*, 1978, **35**, 293.
11. R. West, D. Eggerding, J. Perkins, D. Hardy, and E. C. Tuazon, *J. Am. Chem. Soc.*, 1979, **101**, 1710.
12. D. Lackey, M. Surman, S. Jacobs, D. Grinder, and D. A. King, *Surf. Sci.*, 1985, **152/153**, 513.
13. K. Mohana Rao, G. Spoto, and A. Zecchina, *J. Catal.*, 1988, **113**, 466.
14. K. Aida, H. Midorikawa, K. Alika, and A. Ozaki, *J. Catal.*, 1982, **78**, 147.
15. K. Aida, H. Midorikawa, K. Aika, and A. Ozaki, *J. Phys. Chem.*, 1982, **86**, 3263.
16. S. Kagami, T. Onishi, and K. Tamaru, *J. Chem. Soc. Faraday Trans. 1*, 1984, **80**, 29.
17. E. Garrone and F. S. Stone, *Proceedings of the Eighth International Congress on Catalysis*, Vol. 3, Verlag Chemie, Dechema, Berlin, 1984, p. 441.
18. G. Wang and H. Hattori, *J. Chem.Soc., Faraday Trans. 1*, 1984, **80**, 1039.
19. F. Solymosi, and T. Bansagi, *J. Phys. Chem.*, 1979, **83**, 552.
20. F. Solymosi and L. Volgyesi, J. Rasko, *Z. Phys. Chem. N. F.*, 1980, **120**, 79.
21. L. Cerruti, E. Modone, E. Guglielminotti, and E. Borello, *J. Chem. Soc., Faraday Trans. 1*, 1974, **70**, 729.
22. E. Giamello, E. Garrone, E. Guglielminotti, and A. Zecchina, *J. Chem. Soc., Faraday Trans. 1*, 1984, **80**, 2723.
23. E. Guglielminotti, A. Zecchina, F. Boccizi, and E. Borello, in *Growth and Properties of Metal Clusters*, J. Bourdon, Ed., Elsevier, Amsterdam, 1980.
24. E. Guglielminotti and A. Zecchina, *J. Chem. Phys.*, 1981, **78**, 891.
25. E. Guglielminotti and A. Zecchina, *J. Mol. Catal.*, 1984, **24**, 331.
26. P. G. Gopal and K. L. Watters, *Proceedings of the Eighth International Conference on Catalysis*, Verlag Chemie, Berlin, 1984, p. 75.

27. K. Tanabe, *Solid Acids and Bases*, Academic Press, New York, 1970.

28. S. Utamapanya, K. J. Klabunde, and J. R. Schlup, *Chem. Mater.*, 1991, **3**, 175.

29. Y.-Xi. Li and K. J. Klabunde, *Langmuir*, 1991, **7**, 1388.

30. Y.-Xi. Li, J. R. Schlup, and K. J. Klabunde, *Langmuir*, 1991, **7**, 1394.

31. J. M. Sinfelt and P. J. Lucchisi, *J. Am. Chem. Soc.*, 1963, **85**, 3365.

32. G. E. E. Gardes, G. M. Pajonk, and S. J. Teichner, *Hebd. Seances Acad. Sci.*, 1973, **227C**, 191.

33. D. Bianchi, G. E. E. Gardes, G. M. Pajonk, and S. J. Teichner, *J. Catal.*, 1975, **38**, 135.

34. T. C. Sheng and I. D. Gay, *J. Catal.* 1981, **71**, 119.

35. D. H. Lenz and W. C. Conner, *J. Catal.*, 1987, **104**, 288.

36. G. E. Walrafen and M. S. Hokmabadi, *J. Chem. Phys.*, 1986, **85**, 771.

37. R. L. White and A. Nair, *Appl. Spectrosc.*, 1990, **44**(1), 89.

38. S. Hennings and L. Svenson, Report No. LUNFDD6/(NFFL-7001)/1-434/1979, University of Lund, Sweden; *Phys. Sci.*, 1981, **23**, 697.

39. G. E. Walrafen, M. S. Hokmabadi, N. C. Holmes, W. J. Nellis, and S. Henning, *J. Chem. Phys.*, 1985, **82**, 2472.

40. M. Ichikawa, *Bull. Chem. Soc. Jpn.*, 1978, **51**, 2273.

41. J. R. Katzer, A. W. Sleight, P. Gajardo, J. B. Michel, E. F. Gleason, and S. McMillan, *Faraday Discuss Chem. Soc.*, 1981, **72**, 121.

42. R. P. Underwood and A. T. Bell, *Appl. Catal.*, 1986, **21**, 157.

43. R. P. Underwood and A. T. Bell *J. Catal.*, 1988, **111**, 325.

44. A. Fernandez, J. Leyer, A. R. Gonzalez-Elipe, G. Munuera, and H. Knozinger, *J. Catal.*, 1988, **112**, 489.

45. P. J. Lucchisi, J. L. Carter, and D. J. C. Yates, *J. Phys. Chem.*, 1962, **66**, 1451.

46. D. J. C. Yates and P. J. Lucchisi, *J. Phys. Chem.*, 1961, **35**, 243.

47. J. B. Peri, *J. Phys. Chem.*, 1965, **69**, 220, and references cited therein.

48. J. B. Peri, *Proceedings of the Second International Congress on Catalysis*, Paris, Ed.Technip, 1960, p. 1333

49. C. Morterra, A. Chiorino, G. Ghiotti, and E. Garrone, *J. Chem. Soc., Faraday Trans. 1*, 1979, **75**, 271.

50. J. B. Peri, *J. Phys. Chem.*, 1965, **69**, 220.

51. H. Knozinger and P. Ratnasamy, *Catal. Rev. Sci. Eng.*, 1978, **17**, 31.

52. G. Della Gatta, B. Fubini, G. Ghiotti, and C. Morterra, *J. Catal.*, 1976, **43**, 90.

53. A. Zecchina, S. Coluccia, and C. Morterra, *Appl. Spectrosc. Rev.*, 1985, **21**, 259.

54. C. Morterra, A. Zacchina, S. Coluccia, and A. Chiorino, *J. Chem. Soc., Faraday Trans. 1*, 1977, **77**, 1544.

55. S. M. Riseman, F. E. Massoth, G. M. Dhar, and E. M. Eyring, *J. Chem. Phys.*, 1982, **86**, 1760.

56. M. W. Urban and J. L. Koenig, *Appl. Spectrosc.*, 1986, **40**(6), 851.

57. N. B. Colthup, H. L. Daly, and S. E. Wiberlay, *Introduction to Infrared and Raman Spectroscopy*, Academic Press, New York, 1975.

58. S. M. Riseman, F. E. Massoth, G. Murral Dhar, and E. M. Eyring, *J. Phys. Chem.*, 1982, **86**, 1760.

59. T. R. Hughes and H. M. White, *J. Phys. Chem.*, 1967, **71**, 2192.

60. R. Mone, in *Preparation of Catalysts*, B. Delmon, P. A. Jacobs, and G. Poncelet, Eds., Elsevier, Amsterdam, 1976.

61. S. L. Soled, G. B. McVicker, L. L. Murrell, L. G. Sherman, N. C. Dispenziere, S. L. Hsu, and D. Waldman, *J. Catal.*, 1988, **111**, 286.

62. E. P. Parry, *J. Catal.*, 1963, **2**, 371.

63. M. R. Basilo and T. R. Kantner, *J. Phys. Chem.*, 1966, **70**, 1681.

64. S. C. Chan, S. C. Fung, and J. H. Sinfelt, *J. Catal.*, 1988, **113**, 164.

65. P. Hendra, *Spectrochim. Acta, A*, 1967, **23**, 2871.

66. D. M. Stockwell, J. S. Chung, and C. O. Bennet, *J. Catal.*, 1988, **112**, 135.

67. C. H. Young, Y. Soong, and P. Biloen, *J. Catal.*, 1985, **94**, 306.

68. D. W. Goodman, R. D. Kelley, T. E. Madey, and J. M. White, *J. Catal.*, 1980, **64**, 479.

69. R. P. Underwood and C. O. Bennett, *J. Catal.*, 1984, **86**, 245.

70. C. H. O'Young, C. H. Young, S. J. DeCanio, M. S. Patel, and D. A. Storm, *J. Catal.*, 1988, **113**, 307.

71. K. I. Segawa and W. K. Hall, *J. Catal.*, 1982, **77**, 221.

72. N. D. Darkyns, *J. Chem. Soc., A*, 1969, **410**; *J. Phys. Chem.*, 1971, **75**, 526.

73. J. B. Peri, *J. Phys. Chem.*, 1982, **86**, 1615.

74. W. S. Millmann and W. K. Hall, *J. Phys. Chem.*, 1979, **83**, 427.

75. R. P. Eischens, W. A. Pliskin, and M. J. D. Low, *J. Catal.*, 1962, **1**, 180.

76. G. L. Griffin and J. T. Yates, *J. Chem. Phys.*, 1982, **77**, 3744.

77. G. L. Griffin and J. T. Yates, *J. Chem. Phys.*, 1982, **77**, 3752.

78. N. Natta, in *Catalysis*, Vol. III, P. H. Emmett, Ed., Reinhold, New York, 1955, p. 349.

79. L. Lietti, D. Botta, P. Forzatti, E. Mantica, E. Tronconi, and I. Pasquon, *J. Catal.*, 1988, **111**, 360.

80. Crystalline phases of TiO_2: rutile—tetragonal with a space group D_{4h}^{14} ($P4_2/mnm$); brookite—orthorhombic with complex crystal structure containing eight molecules in one unit cell, but all atoms occupy positions of a simple orthorhombic space group D_{2h}^{15} (Pbca); anatase with a space group D_{4h}^{19} ($I4_1/amd$)—tetragonal. In addition, there are two high-pressure phases (372 kbar: TiO-II with the α-PbO_2 structure and TiO_2-III).

81. C. Mortera, A. Chirina, A. Zecchina, and E. Fisicaro, *Gazz. Chim. Ital.*, 1979, **109**, 683.

82. C. Mortera, A. Chirina, A. Zecchina, and E. Fisicaro, *Gazz. Chim. Ital.*, 1979, **109**, 691.

83. B. C. Cornilsen and M. W. Urban, unpublished results.

84. C. Mortera, A. Chirina, A. Zecchina, and E. Fisicaro, *Gazz. Chim. Ital.*, 1979, **109**, 693.

85. C. Mortera, A. Chirina, A. Zecchina, and E. Fisicaro, *Gazz. Chim. Ital.*, 1979, **109**, 697.

86. C. Mortera, G. Ghiotti, E. Garrone, and E. Fisicaro, *J. Chem. Soc., Faraday Trans. 1* 1980, **76**, 2102.

87. J. Graham, C. H. Rachester, and R. Rudham, *J. Chem. Soc., Faraday Trans. 1*, 1981, **77**, 2735.

88. A. D. Buckland, J. Graham, R. Rudham, and C. H. Rochester, *J. Chem. Soc., Faraday Trans. 1*, 1981, **77**, 2845.

89. F. Al-Mashta, C. U. Davanzo, and N. Sheppard, *J. Chem. Soc., Chem. Commun.*, 1983, **78**, 2649.

90. R. A. Dalla Betta and M. Shelef, *J. Catal.*, 1977, **48**, 111.

91. J. G. Ekerdt and A. T. Bell, *J. Catal.*, 1979, **58**, 170.

92. C. H. Rochester and S. A. Topham, *J. Chem. Soc., Faraday Trans. 1*, 1979, **75**, 1073.

93. C. H. Rochester and S. A. Topham, *J. Chem. Soc., Faraday Trans. 1*, 1979, **75**, 591.

94. R. L. Parfitt, J. D. Russell, and V. C. Farmer, *J. Chem. Soc., Faraday Trans. 1*, 1976, **72**, 1082.

95. C. H. Rochester and S. A. Topham, *J. Chem. Soc. Faraday Trans. 1*, 1979, **75**, 591.

96. C. Morterra, A. Chirino, and E. Borello, *Mater. Chem. Phys.*, 1984, **10**, 199.

97. H. Pichler, *Adv. Catal.*, 1952, **4**, 272.

98. M. E. Dry, *Ind. Eng. Chem. Prod. Res. Dev.*, 1976, **15**, 282.

99. H. Batis, *J. Soc. Chem., Tun.*, 1986, **II**, 19.

100. M. L. Kundu, A. C. Sengupta, G. C. Maiti, B. Sen, S. K. Ghoshi, V. I. Kuznetow, G. N. Kustova, and E. N. Yurchenko, *J. Catal.*, 1988, **112**, 375.

101. J. Preudhomme and P. Tarte, *Spectrochim. Acta A*, 1971, **27**, 1817.

102. M. Ishii, M. Nakahira, and T. Yamanaka, *Solid State Commun.*, 1972, **11**(1), 209.

103. C. Li, Q. Xin, K. L. Wang, and X. Guo, *Appl. Spectrosc.*, 1991, **45**(5), 874.

104. D. S. Kim, Y. Kurusu, I. E. Wachs, F. D. Hardcastle, and K. Segawa, *J. Catal.*, 1989, **120**, 325.

105. S. Kasztelan, E. Payen, H. Toulhoat, J. Grimblot, and J. P. Bonnelle, *Polyhydron*, 1986, **5**, 157.

106. S. Kasztelan, E. Payen, and J. B. Moffat, *J. Catal.*, 1988, **112**, 320.

107. R. Thouvenot, M. Fournier, R. Frank, and C. Ricchiccioli-Deltcheff, *Inorg. Chem.*, 1982, **23**, 598.

108. M. S. Kasprzak, G. E. Leroi, and S. R. Crouch, *Appl. Spectrosc.*, 1982, **36**, 285.

109. J. B. Peri, *J. Phys. Chem.*, 1965, **69**, 220.

110. U.S. Patent, 4310440, 1982.

111. J. B. Moffat, *Catal. Rev. Sci. Eng.*, 1978, **18**, 199.

112. E. Baumgarten and F. Weinstrauch, *Spectrochim. Acta*, 1979, **35**, Part A, 1315.

113. L. H. Little, *Infrared Spectra of Adsorbed Species*, Academic Press, New York, 1966.

114. O. V. Kikhtyanin, E. A. Paukshtis, K. G. Ione, and V. M. Mastikhin, *J. Catal.*, 1990, **126**, 1.

115. J. M. Campelo, A. Garcia, D. Luna, and J. M. Marinas, *J. Catal.*, 1988, **111**, 106.

116. J. B. Moffat, R. Vetrivel, and B. Viswanathan, *J. Mol. Catal.*, 1985, **30**, 171.

117. D. Klissurski and K. Hadjiivanov, *J. Catal.*, 1988, **111**, 421.

118. A. Zecchina, F. Coluccia, E. Guglielminotti, and G. Ghiotti, *J. Phys. Chem.*, 1971, **75**, 2774.

119. T. Takaoka, and J. A. Dumesic, *J. Catal.*, 1988, **112**, 66.

120. H. Myiata, S. Tokuda, and T. Yashida, *Appl. Spectrsoc.*, 1989, **43**(3), 522.

121. P. J. Gelings, in *Catalysis*, G. C. Bond and G. Webb, Eds., *Specialist Periodical Reports*, Vol. 7, p. 105, Royal Society of Chemistry, London, 1985.

122. G. A. Tsigdinos, *Ind. Eng. Chem. Prod. Res. Dev.*, 1974, **13**, 267.

123. K. Nakamota, *Infrared and Raman Spectra of Inorganic and Coordination Compounds*, 3rd ed. Wiley, New York, 1978.

124. G. Seo, J. W. Lim, and J.T. Kim, *J. Catal.*, 1988, **114**, 469.

125. S. P. Zhdanow, A. V. Kiselev, V. I. Lygin, and T. I. Titova, *Zh. Eiz. Khim.*, 1964, **38**, 2408; 1965, **39**, 2554.

126. C. L. Angell, *J. Phys. Chem.*, 1973, **77**, 222.

127. P. K. Dutta, R. E. Zaykoski, and M. A. Thomson, *Zeolites*, 1986, **6**, 423.

128. P. K. Dutta, D. C. Shieh, and M. Puri, *Zeolites*, 1988, **8**, 306.

129. A. V. Kiselev and V. I. Lygin, in *Infrared Spectra of Adsorbed Species*, L. H. Little, Ed., Academic Press, London, 1966.

130. J. J. Fripiat, in *Spectroscopic Characterization of Minerals and Their Surfaces*, ACS Symposium Series No. 415, L. M. Coyne, S. W. S. McKeever, and D. F. Blake, Eds., American Chemical Society, Washington, D.C. 1990.

131. S. M. Csicsery, *Zeolites*, 1984, **4**, 202.

132. G. Busca, T. Zerlia, V. Lorenzelli, and A. Girelli, *J. Catal.*, 1984, **88**, 131.

133. A. Jentys and J. A. Lercher, in *Zeolites as Catalysts, Sorbents and Detergent Builders*, H. G. Krge and J. Weitkamp, Eds., Elsevier, New York, 1989.

134. A. N. Fitch, H. Jobic, and A. Renouprez, *J. Phys. Chem.*, 1986, **90**, 1311.

135. H. P. Wang, T. Yu, B. A. Garland, and E. M. Eyring, *Appl. Spectrosc.*, 1990, **44**(6), 1070.

136. C. Bremard, C. Depecker, H. Des Grousillier, and P. Legrand, *Appl. Spectrosc.*, 1991, **45**(8), 1278.

137. A. Rahman, G. Lemay, A. Adnot, and S. Kallaguine, *J. Catal.*, 1988, **112**, 453.

138. D. H. Olson, G. T. Kokotailo, S. L. Lawton, and W. M. Meier, *J. Phys. Chem.*, 1981, **85**, 1981.

139. J. A. Lercher, G. Rumplmayr, *Appl. Catal.*, 1986, **25**, 215.

140. H. Vinek, G. Rumplmayr, and J. A. Lercher, *J. Catal.*, 1989, **61**, 115.

141. J. C. Vedrine, A. Auroux, P. Dejaifve, V. Ducarme, H. Hoser, and S. Zhou, *J. Catal.*, 1982, **73**, 147.

142. W. W. Kaeding and S. A. Buttler, *J. Catal.*, **61**, 155.

143. J. Datka, M. Boczar, and P. Rymarowicz, *J. Catal.*, 1988, **114**, 368.

144. W. J. Molier, *J. Catal.*, 1978, **55**, 138.

145. P. A. Jacobs, *Catal. Rev. Sci. Eng.*, 1982, **24**, 415.

146. A. A. Tsyganenko and V. N. Felimov, *J. Mol. Struct.*, 1973, **19**, 579.

147. A. G. Pelmenshchikov, E. V. Paukshtis, V. G. Stepanov, V. I. Pavlov, E. N. Yurchenko, K. G. Ione, and G. M. Zhidomirov, *J. Phys. Chem.*, 1989, **93**, 6725.

148. J. R. Bartlett, R. P. Cooney, and R. A. Kydd, *J. Catal.*, 1988, **114**, 53.

149. M. Lenarda, J. Kaspar, R. Ganzerla, A. Trovarelli, and M. Graziani, *J. Catal.*, 1988, **112**, 1.

150. R. Psaro, R. Ugo, and J. M. Basset, *J. Organmetal. Chem.*, 1981, **213**, 215.

151. H. Knozinger and Y. Zhao, *J. Catal.*, 1981, **71**, 337.

152. E. Mantovani, N. Palladino, and A. Zanobi, *J. Mol. Catal.*, 1977, **3**, 285.

153. E. J. Davis, E. J. Rode, D. Taylor, and B. E. Hanson, *J. Catal.*, 1984, **86**, 67.

154. M. Primet, J. C. Vedrine, and C. Nacchache, *J. Mol. Catal.*, 1978, **4**, 411.

155. M. Primet, *J. Chem. Soc., Faraday Trans. 1*, 1980, **74**, 2570.

156. N. Takahashi, A. Mijin, H. Suematsu, S. Shinohara, and H. Matsuoka, *J. Catal.*, 1989, **117**, 348.

157. W. R. Moser, C. C. Chiang, and R. W. Thomson, *J. Catal.*, 1989, **115**, 532.

158. G. Mirth and J. A. Lercher, *J. Catal.*, 1991, **132**, 244.

159. N. B. Colthup, L. H. Daly, and S. E. Wiberley, *Introduction to Infrared and Raman Spectroscopy*, Academic Press, New York, 1975.

160. S. L. Meisel, *Philos. Trans. Roy. Soc.* (London), 1981, **A330**, 157.

161. E. D. Derouane, J. B. Nagy, P. Dejaifve, H. C. van Hoof, B. P. Spekman, J. C. Vedrine, and C. Naccache, *J. Catal.*, 1982, **75**, 284.

162. K. A. Martin and R. F. Zabransky, *Appl. Spectrosc.*, 1991, **45**(1), 68.

163. R. G. Greenler, *J. Chem. Phys.*, 1962, **37**, 2094.

164. P. Salvador and W. Kladnig, *J. Chem. Soc., Faraday Trans. 1*, 1977, **73**, 1153.

165. J. G. Chen, P. Basu, T. H. Ballinger, and J. T. Yates, *Langmuir*, 1989, **5**, 352.

166. J. Heaviside, P. J. Hendra, P. Tsai, and R. P. Cooney, *J. Chem. Soc., Faraday Trans. 1*, 1978, **74**, 2542.

167. T. Ito, H. Shirakawa, and S. Ikeda, *J. Polym. Sci., Polym. Chem. Ed.*, 1974, **12**, 11.

168. L. B. Lutinger, *J. Org. Chem.*, 1962, **27**, 1591.

169. P. K. Dutta, M. Puri, *J. Catal.*, 1988, **111**, 453.

170. H. Karge, *Z. Phys. Chem., Neue Folge*, 1980, **122**, 103.

171. J. Philippaerts, E. F. Vansant, G. Peeters, and E. Vanderheyden, *Anal. Chim. Acta*, 1987, **195**, 2237.

172. A. G. Pelmenschikov, G. Morosi, and A. Gamba, 1991, **95**(24), 10037.

173. J. M. Thomas, in *Intercalation Chemistry*, M. S. Wittingham and A. J. Jacobson, Eds., Academic Press, New York, 1982.

174. J. A. Balantine, in *Chemical Reaction in Organic and Inorganic Constrained Systems*, R. Setton, Ed., NATO, ASI Series C165: Reidel, Dordrecht, 1986.

175. B. Durand, R. Pelet, and J. Fripiat, *J. Clays Clay Min.*, 1972, **20**, 21.

176. A. Schutz, W. E. E. Stone, G. Poncelet, and J. J. Fripiat, *Clays Clay Min.*, 1987, **35**, 251.

177. B. Chourabi and J. J. Fripiat, *Clay Min.*, 1981, **29**, 260.

178. C. T. Johnston, in *Spectroscopic Characterization of Minerals and Their Surfaces*, L. M. Coyne, S. W. S. McKeever, and D. F. Blake, Eds., ACS Symposium Series, No. 145, American Chemical Society, 1990, Washington D.C.

179. P. M. Costanzo, R. F. Giese, and M. Lipsicas, *Clays Clay Min.*, 1984, **32**, 419.

180. R. L. Ledoux and J. L. White, *Science*, 1964, **145**, 47.

181. T. J. Porro and S. C. Pattacini, *Appl. Spectrosc.*, 1990, **44**(7), 1170.

182. L. Vagberg, P. DePotocki, and P. Stenius, *Appl. Spectrosc.*, 1989, **43**(7), 1240.

183. D. A. Guzonas, M. L. Hair, and C. P. Tripp, *Appl. Spectrsc.*, 1990, **44**(2), 290.

CHAPTER 6

VIBRATIONAL FEATURES OF INORGANIC MACROMOLECULES

6.1 SILICON-CONTAINING COMPOUNDS

The analysis of silanes and and their derivatives was shown in Chapter 6. Here, we will focus on Si-containing functional groups and their sensitivity to the environment. Because of high polarity of the Si atoms, the Si involved vibrations are fairly sensitive to the neighboring substituents. Table 6.1 provides tentative band assignments for the silicon-containing polymers and their selective derivative, along with several sources discussing the fundamental vibrational frequencies of linear siloxane polymers.[1][4] Because, for the most part, these polymers are used in a crosslinked form deposited as a film, the effects of crosslinking and silica filler on vibrational frequencies of siloxanes are essential. Table 6.2 provides a comparison between the vibrational frequencies of crosslinked and linear silicone polymers.

6.1.1 Water and Silane Coupling Agents

As was eluded to earlier, water is adsorbed on oxides in the form of hydroxyl groups (M—OH) and as molecular water held by hydrogen bonding to the surface hydroxyl groups. In the case of surface adsorbed water on amorphous silica, although it was claimed that physically or chemically adsorbed water can be removed by heating to 800°C in vacuum for a few hours,[5] the presence of a low-intensity band at $3750 \, cm^{-1}$ was attributed to non-hydrogen-bonded hydroxyl groups (Si—OH). This process is reversible; exposure to water leads to a significant increase of the $3750 \, cm^{-1}$ band and a rise of a new band at around $3600 \, cm^{-1}$. It should be kept in mind that while the $3750 \, cm^{-1}$ band remains quite steady above $3700 \, cm^{-1}$, the bands due to hydrogen-bonded water vary significantly in

TABLE 6.1 Band Assignments of Si-Containing Macromolecules

Group	Frequency (cm^{-1})	Assignment
Si—OH	3700–3200	General O—H stretching region
	3750	Isolated O—H stretching of silica
	3690	Free O—H stretching
	3400–3200	Hydrogen bonded O—H stretching, often broad
	~1030	Si—OH deformation
	910–830	Si—asymmetric stretching
Si—H	2250–2100	General Si—H str. region
	950–800	General Si—H deformation region
R—SiH$_3$	2160–2140	Si—H stretching
	945–930	Asymmetric Si—H deformation
	930–910	Symmetric Si—H deformation
	680–540	Si—H rock
RR'SiH$_2$	2150–2115	Si—H stretching
	945–920	Si—H deformation
	895–840	Si—H wagging
	745–625	Si—H twisting
	600–460	Si—H rock
RR'R"SiH	2135–2095	Si—H stretching
	845–800	Si—H wagging
Ph—SiH$_3$	2157–2152	Si—H stretching
	945–930	Asymmetric Si—H deformation
	930–910	Symmetric Si—H deformation
Ph$_2$SiH$_2$	2147–2130	Si—H stretching
	940–928	Si—H deformation
	870–843	Si—H wagging
Ph$_3$SiH	2132–2112	Si—H stretching
	842–800	Si—H wagging

TABLE 6.1 (*Continued*)

Group	Frequency (cm^{-1})	Assignment
PhRSiH$_2$	2142–2128	Si—H stretching
	938–923	Si—H deformation
	870–843	Si—H wagging
Ph$_2$RSiH	2125–2115	Si—Hstretching
	842–800	Si—H wagging
PhRR′SiH	2115–2103	Si—H stretching
	842–800	Si—H wagging
F$_3$SiH	2282	Si—H stretching
Br$_3$SiH	2236	Si—H stretching
(CH$_3$O)$_3$SiH	2203	Si—H stretching
Si—R		
Si(CH$_3$)$_n$ (n = 1, 2, 3, or 4)	1410–1390	General asymmetric CH–deformation region for methyl-substituted Si
	1280–1240	General symmetric CH–deformation region for methyl-substituted Si
Si—CH$_3$	800–735	Si—C stretching, general range, very sensitive to other Si groups and goes above 800 in some cases
	815–790	Si—C stretching (cyclic siloxanes)
	~765	Si—CH$_3$ rock
Si(CH$_3$)$_2$	~1259	Symmetric CH$_3$ deformation
	~855	Si—CH$_3$ rock
	815–802	Si—CH$_3$ rock (cyclic systems)
	~800	Si—CH$_3$ rock (open-chain systems)
Si(CH$_3$)$_3$	~840	Si—CH$_3$ rock
	~765	Si—CH$_3$ rock
	ˉ715–680	Asymmetric Si—C stretching
	660–485	Symmetric Si—C stretching
	330–240	Si(CH$_3$)$_3$ rock

Group	Frequency (cm⁻¹)	Assignment
—OSiCH$_3$	850–840	Si—C stretching, linear siloxane end group
—OSiCH$_3$	820–780	Si—C stretching, cyclic siloxanes
—OSiCH$_2$CH$_3$	810–800	Si—C stretching, linear siloxane end group
SiCH$_2$CH$_3$	1250–1220	Symmetric CH$_2$ deformation
?	1020–1000	
?	970–945	
SiCH$_2$(CH$_2$)$_X$CH$_3$	∼1410	Asymmetric CH$_2$ deformation
	1220–1200	Symmetric CH$_2$ stretching ($X = 1$)
	1200–1190	Symmetric CH$_2$ stretching ($X = 2$)
	1190–1170	Symmetric CH$_2$ stretching ($X \geqslant 3$)
	760–670	CH$_2$ rock
SiCH=CH$_2$	∼1925	Overtone
	1615–1590	C=C stretching
	1410–1390	CH$_2$ in-plane deformation
	1020–1000	CH wagging
	980–940	CH$_2$ wagging
	580–515	H out-of-plane deformation
Si—C≡N	2222–2174	C≡N stretching
Si—C≡CH	∼3300	CH stretching
$\overset{O}{\underset{\parallel}{\text{RCSiR}_3}}$	∼2040	C≡C stretching
	∼1618	C=O stretching
Si—Ph	3080–3030	C—H stretching
	2000–1660	Conventional benzene substitution patterns
	1610–1590	C=C stretching
	1480–1425	Ring vibration, band generally appears at ∼1430 cm⁻¹
	1130–1090	"X-sensitive" planar ring vibration
	1030	Phenyl ring deformation
	1000	Phenyl ring deformation
	760–710	C—H out-of-plane vibration, most often appears at ∼730 cm⁻¹
	700–690	C—H out-of-plane vibration

TABLE 6.1 (*Continued*)

Group	Frequency (cm^{-1})	Assignment
R$_3$SiPh	670–625	In-plane ring bending
	490–445	Si—C out-of-plane bending
	405–345	Si—C stretching and in-plane ring deformation
	~290	Si—Ph in-plane deformation
R$_2$SiPh$_2$	1130–1090	"X-sensitive" planar ring vibration, often split into two bands
	635–605	In-plane ring bending
	495–470	Si—C out-of-plane bending
	445–400	Asymmetric Si—C stretching
	380–305	Symmetric Si—C stretching
RSiPh$_3$	625–605	In-plane ring bending
	515–485	Si—C out-of-plane bending
	445–420	Asymmetric Si—C stretching
	~330	Symmetric Si—C stretching
	1135–1110	Planar ring vibration
	~1543	Ring stretching mode
	~1350	Ring stretching mode
	~1009	C—C stretching between phenyl rings
	~833	
Si—O—R		
SiOCH$_3$	2860–2820	Asymmetric CH-stretching
	~1190	CH-rock
	1100–1080	Asymmetric Si—O—C stretching
	850–800	Symmetric Si—O—C stretching
Si(OCH$_3$)$_2$	390–360	Asymmetric Si—O—C deformation
Si(OCH$_3$)$_3$	480–440	Asymmetric Si—O—C deformation

Group	Frequency	Assignment
SiOCH$_2$CH$_3$	1175–1160	Asymmetric Si—O—C stretching, usually a doublet
	1100–1075	Symmetric Si—O—C stretching
Si(OCH$_3$)$_2$	970–940	Asymmetric Si—O—C deformation
Si(OCH$_3$)$_3$	475–405	Asymmetric Si—O—C deformation
SiOCH$_2$CH$_2$CH$_3$	500–440	
	~1155	Asymmetric Si—O—C stretching
	1100, 1085	
SiOCH(CH$_3$)$_2$	~1020	
	1385, 1370	
	1175	
	1140–1110, 1537,	
	1541,1543, 1542	
	1055–1030, 1544	
	~890	
$\overset{\|}{\underset{\|}{\text{SiOCR}_3}}$	1750–1695	C=O stretching
	1190–1135	Asymmetric Si—O—C stretching
	970–925	Symmetric Si—O—C stretching
SiOC$_6$H$_5$	1135–1090	Asymmetric Si—O—C stretching, often several bands
	975–920	
Si—O—Si		
Si—O—Si	1130–1000	Asymmetric Si—O—Si stretching, general region, as siloxane chain lengthens—the band splits into two bands and broadens
	625–480	Symmetric Si—O—Si stretching, general region
R(R$_2$SiO)$_x$SiR$_3$	1093–1076	Asymmetric Si—O—Si stretching, frequency goes from the low end of the range of high as X increases
	1055–1024	Asymmetric Si—O—Si stretching, frequency goes from the high end of the range to low as X increases
(R$_2$SiO)$_3$	1020–1010	Asymmetric Si—O—Si stretching
(R$_2$SiO)$_{4-5}$	1085–1076	Asymmetric Si—O—Si stretching, frequency goes from the high end of the range to the low as ring size increases

TABLE 6.1 (*Continued*)

Group	Frequency (cm^{-1})	Assignment
(R$_2$SiO)$_{6-8}$	1100–1097	Asymmetric Si—O—Si stretching
	1070–1055	Asymmetric Si—O—Si stretching, frequency goes from the high end of the range to low as ring size increases
Si—R—Si		
SiCH$_2$Si	~1351	CH$_2$ twisting
	1110–1000	Si—C—Si stretching
SiCH$_2$CH$_2$Si	1250–1000	Two bands
Si—N		
Si—NH$_2$	3570–3475	N—H stretching
	3410–3390	N—H stretching
	1550–1530	NH$_2$ deformation
Si—NH—Si	3400–3390	N—H stretching
	1200–1150	N—H bending
(Si)$_3$N	950–920	Asymmetric Si—N—Si stretching
	~1000	Si—N stretching
H$_2$NSiNH$_2$	800–785	Symmetric N—Si—N stretching
Si-N=C=O	~2280	Asymmetric N=C=O stretching
Si-N=N=N	2180–2120	Asymmetric N=N=N stretching
Si—(halogen)		
SiF	920–820	Si—F stretching
	425–265	Si—F deformation, general region
SiF$_2$	945–915	Asymmetric Si—F stretching
	910–870	Symmetric Si—F stretching
SiF$_3$	980–945	Asymmetric Si—F stretching
	910–860	Symmetric Si—F stretching
SiCl	550–470	Si—Cl stretching
	250–150	Si—Cl deformation, general region

SiCl₂	595–535	Asymmetric Si—Cl stretching
	540–460	Symmetric Si—Cl stretching
	~241	Si—Cl₂ rock
	~232	Si—Cl₂ rock
	~177	Si—Cl₂ twisting
	~168	Si—Cl₂ deformation
SiCl₃	625–570	Asymmetric Si—Cl stretching
	535–450	Symmetric Si—Cl stretching
	~229	Si—Cl₃ deformation
	~164	Si—Cl₃ rock
SiBr	430–360	Si—Br stretching
SiBr₂	460–425	Asymmetric Si—Br stretching
	395–330	Symmetric Si—Br stretching
SiBr₃	480–450	Asymmetric Si—Br stretching
	360–300	Symmetric Si—Br stretching
SiI	365–280	Si—I stretching
SiI₂	390–330	Asymmetric Si—I stretching
	325–275	Symmetric Si—I stretching
SiI₃	410–365	Asymmetric Si—I stretching
	280–220	Symmetric Si—I stretching
Miscellaneous		
Si—S	515–450	Si—S stretching; this range is based on a limited amount of information
SiOP	~1031	
SiP	~455	
SiO—Mt	1000–900	Where Mt is a metal atom such as Na, K, Sn, Hg, Tl
Si-D	1690–1570	Si—D stretching
	710–665	Si—D deformation
	600–530	Si—D rock

Source: Courtesy S. R. Gaboury, PhD Dissertation 1993, North Dakota State University.

TABLE 6.2 Band Assignments of Si-Containing Linear and Crosslinked Macromolecules

Vibrational Mode	Linear	Crosslinked	S and A[b]
C—H stretch	2965	2963	2965
C—H stretch	2907	2907	2905
1260 + 1414 combination	2664	2664	—
1260 overtone	2500	2502	—
	2127	2127	—
	2052	2052	—
	1946	1946	—
SiCH=CH	1599	—	—
C—H bend	1445	1445	—
C—H bend	1414	1414	1411
C—H bend	1261	1258	1260
Si—O—Si v stretch	1097	1076	1087
Si—O—Si v stretch	1024	1007	1022
CH wag	866	864	864
CH wag + Si—C bend	806	785	802
Si—C bend	700	704	704
	687	689	688
Si—CH$_2$	662	662	—
	—	—	633
Si—O—Si	500	505	500
Si—C$_2$	—	—	407[a], 395[a]
Si—C$_2$	—	—	363[a], 328[a] 277[a]

[a] Low-temperature spectra–crystalline form.
[b] From A. L. Smith and D. R. Anderson, *Appl. Spectroscopy*, 1984, **38**, 822.

energy, depending on the surface coverage and the strength of the bonding. The latter is greatly affected by the composition of the surface. As an example, the surface chemistry of glass departs considerably from that observed in pure silica or pure oxides because silicate glasses are mixtures of metal oxides that are not evenly distributed in silica network. E-glass contains 55% of SiO_2, and the remaining oxides are CaO (16%), Al_2O_3 (14.5%), B_2O_3 (9.5%), and MgO (5%). Such a composition makes E-glass hydroscopic so that the water adsorption on such surfaces leads to heterogeneous hydration.

This observation leads us to the adhesion aspect of bonding, which was recognized many years ago as a principal operation for corrosion protection of metals. In the case of glasses, one of the driving forces for the development of chemical bonding characteristics was dictated by fiber-glass-reinforced composites. Because water may hydrolyze on the glass surface and stresses may develop across the polymer–fiber interfaces resulting from the mismatch in thermal expansion coefficients, organometallic coupling agents are used across the organic–inorganic interface to help overcome these obstacles. The entire chemistry of silane coupling agents has been well treated in the literature, and the reader is referred to an excellent monograph by Pluddemann.[6] A brief description of the hydrolysis process is provided in Chapter 8.

6.2 PHOSPHATES

Table 6.3 provides the band assignments for the most common-phosphorous-containing groups. The bond that is of particular interest with respect to both chemistry and vibrational energy is that of $P=O$. Because of the tremendous changes of dipole moment, the $P=O$ appears to be strong in the IR spectrum and, interestingly because of the size of the phosphorous atom, is almost independent of the type of compound in which the group is present. It is, however, influenced by electronegativity of the neighboring atoms. This is related to the fact that highly electronegative groups withdraw electrons from the phosphorous atom, thus competing with the oxygen, which would otherwise have a tendency to form $P^+ - O^-$. In fact, the $P=O$ stretch may split as a result of the tendency of these species to form various isomers. Table 6.4 provides vibrational frequencies for $P=O$ stretching vibrations with selected substituents.[7]

Although the OH vibrations are relatively independent of the nature of attached groups, for acid compounds containing $P=OH$ groups the hydrogen-bonding effects become more significant than those in carboxylic acids. Perhaps the most convincing evidence was found by Daasch and Smith[8] and Belamy and Beecher,[9] who determined that the OH streching vibrations usually observed above 3000 cm^{-1} were detected in the 2560–2700-cm^{-1} range. Inorganic phosphoric acids and related species exhibit similar behavior, although according to Thomas and Chittenden,[7] the structures such as PS—OH lead to higher OH stretching frequencies, reflecting the weaker association forces. This is also reflected in the P—O and P—S stretching frequences that occur in the 1350–1080- and 940–910-cm^{-1} regions, respectively.

TABLE 6.3 Spectral Regions for Phosphate-Containing Molecules

Phosphorus-Containing Bond	Vibrational Frequency (cm^{-1})
$P=O$	1300–1100
$P-OH$	2700–2560
$P-H$	2460–2270
$P-CH_2$	1440–1405
$P-CH_3$	1320–1280; 960–860
$P-C_6H_5$	1450–1425; 1130–1090; 1010–990
$P-F$	940–740
$P-Cl$	580–440
$P-Br$	400–380
$P=S$	865–655; 730–550

TABLE 6.4 Band Assignments for Various Phosphate-Containing Molecules

Funtional Groups	Spectral Region (cm^{-1})
P=O	
P=O (unassociated)	1350–1175
P = O (associated)	1250–1150
$(RO)_3$ P=O	1285–1255
$(ArO)_3$ P=O	1315–1290
P—H	
Akyl phosphine (P—H)	2285–2265
Aryl phosphine (P—H)	2285–2270
Phosphonates, $(RO)_2$ HP=O	2455–2400
Phosphonates, R_2 HP=O	2340–2280
P—C	
P—CH_2—	780–760
P—CH_3—	790–770
P—CH_2—Ar	795–740
P—X	
R_2P(O) F	835–805
RP(O) F_2	930–895
P—Cl	605–435
P—Cl_2	590–485
P—Br	485–400
P—S or P=S	
$(RO)_3$P=S	845–800
$(RO)_2$RP=S	800–770
$(RO)R_2$P=O	790–765
R_3P=S	770–685
R_2ClP=S	775–750
RCl_2P=S	780–775

One of the most appealing and friendly regions in the IR spectrum is the spectral range from 2000 to 2800 cm^{-1}, the region where there are fewer species that absorb. One species is the P—H bond, which, depending on the neighboring substituents, may have vibrational energies different from those of typical P—H stretching bands. These variations are listed in Table 6.5. The vibrational energy of the PH is also affected by the presence of inductive effects resulting from the neighboring groups. Although this behavior more or less parallels the effects detected for P=O vibrational frequency, the presence of two P—H bands in the phosphinic acid esters[9] was attributed to the presence of rotational isomers. Another useful region of identification for P—CH_3 compounds is related to the symmetric deformational bonds of CH_3 absorbing in the 1330–1270-cm^{-1} region. Aliphatic substituents attached to such species as trimethyl phosphine oxide appear in the 750–650-cm^{-1} region, common for many organophosphorous species.

TABLE 6.5 IR Tentative Band Assignments for Silicone Phthalocyanine Polymers and Monomers

A Photoacoustic FT-IR Spectral Data of $H_2(Pc)$, $Ni(Pc)$, $Cu(Pc)$, and $Zn(Pc)$ Monomers[a,b]

Vibrational Mode		Band	$H_2(Pc)$ β	Ni(Pc)	Cu(Pc)	Zn(Pc)
C—C	Def. re	1	(beyond detection limit)			
C—C	Def. re	2	487, 493 (m)	518 (m)		
C—C	Def. re	3	555 (m)	572 (m)	572 (m)	572 (m)
C—C	Def. re	4	613 (s) 684 ([t]s)	642 (w)	632 (w)	621 (s)
C—H	Def. e	5	719 (vs)	732 (vs)	725 (vs)	727 (vs)
C—H	Def. e	6	729, 734 (vs)	756 (s)	754 (s)	752 (s)
N—H	Def. e	7	748 (vs)	—	—	—
C—H α	Def. e	8	767 (s)	773 (s)	773 (s)	773 (m)
C—H β	Def. e	9	779 (s)	781 (m)	781 (m)	781 (w)
C—H	Def. e	10	—	804 (w)	802 (w)	799 (w)
C—H α	Def. e	11	—	867 (sh)	867 (sh)	867 (sh)
C—H β	Def. e	12	880 (sh)	873 (w)	873 (w)	873 (w)
C—H	Def. e	13	873 (vs)	916 (s)	901 (s)	887 (s)
		14	947 (vs)	949 (m)	949 (m)	949 (m)
	β	15	956 (m)	955 (w)	955 (w)	955 (sh)
N—H	Def. i	16	1006 (vs)	—	—	—
C—H	Def. i	17	—	1078 (sh)	1069 (s)	1061 (s)
C—N	Str.	18	1093 (vs)	1090 (s)	1090 (s)	1090 (s)
		19	—	—	—	—
C—H	Def. i	20	1118 (vs)	1124 (s)	1121 (s)	1119 (s) 1134 (s)
C—H	Def. i	21	1156 (m)	1167 (s)	1167 (s)	1167 (w)
C—H β	Def. i	22	1163 (m)	1191 (sh)	1191 (sh)	1191 (sh)
C=N	Def.		1249 (m)			
C—C	Str.	23	1276 (s)	1290 (s)	1287 (s)	1285 (s)
N—H	Def.	24	1303 (s)	—	—	—
N—H	Def.	25	1323 (s)	—	—	—
		26	1334 (vs) 1342 ([t]sh)	1335 (s)	1335 (s)	1335 (vs)
C—C	Str.	27	1436 (s)	1429 (s)	1421 (s)	1410 (m)
		28	1458 (w)	1470 (m)	1466 (m)	1454 (m)
		29	1477 (w)	1491 (w)	1479 (m)	1467 (sh)
C=N	Str.	30	1502 (m)	1533 (s)	1508 (s)	1485 (s)
C=C	Str.	31	1597 (sh)	1597 (sh)	1589 (sh)	1585 (sh)
C=C	Str.	32	1606, 1614 (m)	1610 (m)	1610 (m)	1610 (m)
N—H	Str.	33	3277 (s)	—	—	—

[a] Band numbering scheme (1–33) is in accordance with ref. A in (below) Str. = stretching, Def. = deformation, Def. re = out-of-plane ring deformation, Def. e = out-of-plane deformation (e.g., wagging and twisting), Def. i = in-plane deformation (e.g., scissoring and rocking), s = strong, m = medium, w = weak, bd = broad, sh = shoulder, v = very, α = alpha crystal form, β = beta crystal form, [t] = additional peaks found and reported by Kobayashi (ref. B, below).

[b] REFERENCES:

A. A. N. Sidorov and I. P. Kotlyer, *Opt. Spectrosc.*, 1961, **11**, 92.

B. T. Kobayashi F. Kurokawa, N. Uyeda, and E. Suito, *Spectrochim. Acta (Part A)*, 1970, **26A**, 1305–1311.

Source: Courtesy B. J. Exsted, PhD Dissertation, 1991, North Dakota State University.

TABLE 6.5 (*Continued*)

B Photoacoustic FT-IR Spectral Data of Ni(Pc), Ni(Pc)(COOH)$_4$, and [Ni(Pc)]$_n$(COOH)$_n$ [a]

Vibrational Mode			Band	Ni(Pc)	Ni(Pc)(COOH)$_4$	[Ni(Pc)]$_n$(COOH)$_n$
C—C		Def.re	1	(beyond detection limit)		
C—C		Def.re	2	518 (m)		
C—C		Def. re	3	572 (m)	563 (w)	594 (w)
C—C		Def. re	4	642 (w)	646 (w)	630 (m)
C—H		Def. e	5	732 (vs)	725 (sh)	734 (s)
C—H		Def. e	6	756 (s)	745 (s)	—
N—H		Def. e	7	—	—	—
C—H	α	Def. e	8	773 (sh)	770 (sh)	—
C—H	β	Def. e	9	781 (m)	—	794 (w)
C—H		Def. e	10	804 (w)	833 (s)	—
C—H	α	Def. e	11	867 (sh)	—	—
C—H	β	Def. e	12	873 (w)	901 (w)	904 (s)
C—H		Def. e	13	916 (s)	918 (w)	—
			14	949 (m)	949 (m)	—
	β		15	955 (w)	972 (w)	991 (w)
N—H		Def. i	16	—	—	—
C—H		Def. i	17	1078 (sh)	1045 (sh)	1039 (sh)
C—N		Str.	18	1090 (s)	1097 (vs)	1068 (m)
			19	—	—	—
C—H		Def. i	20	1124 (s)	—	1124 (w)
C—H		Def. i	21	1167 (s)	1165 (s)	1172 (w)
C—H	β	Def. i	22	1191 (sh)	1200 (w)	1209 (w)
C—C		Str.	23	1290 (s)	1292 (m)	1263 (s)
N—H		Def.	24	—	—	—
N—H		Def.	25	—	—	—
			26	1335 (s)	1329 (s)	1321 (w)
				1385 (s)		1370 (w)
C—C		Str.	27	1429 (s)	1404 (sh)	—
			28	1470 (m)	1474 (m)	—
			29	1491 (w)	1491 (sh)	—
C=N		Str.	30	1533 (s)	1533 (s)	1530 (bd)
C=C		Str.	31	1597 (sh)	1572 (m)	—
C=C		Str.	32	1610 (m)	1616 (s)	1612 (bd)
C=O		Str.		—	1720 (s)	1730 (s)
C=O		Str.		—	1792 (s)	1782 (vs)
C=O		Str.				1848 (vs)
N—H		Str.	33	—	—	—
O—H		Str.		—	3400 (bd)	3300 (bd)

[a] Band numbering scheme (1–33) is in accordance with ref. A in Table 6.5*A*. Str; = stretching, Def. = deformation; Def. re = out-of-plane ring deformation; Def. e = out-of-plane deformation (e.g., wagging and twisting); Def. i = in-plane deformation (e.g., scissoring and rocking); s = strong, m = medium, w = weak, bd = broad, sh = shoulder, v = very, α = alpha crystal form, β = beta crystal form.

TABLE 6.5 (*Continued*)

C Photoacoustic FT-IR Spectral Data of Cu(Pc), Cu(Pc)(COOH)$_4$, and [Cu(Pc)]$_n$(COOH)$_n$[a]

Vibrational Mode		Band	Cu(Pc)	Cu(Pc) (COOH)$_4$	[Cu(Pc)]$_n$ (COOH)$_n$
C—C	Def. re	1	(beyond detection limit)		
C—C	Def. re	2			
C—C	Def. re	3	572 (m)	560 (m)	594 (m)
C—C	Def. re	4	632 (w)	650 (m)	630 (m)
					658 (sh)
C—H	Def. e	5	725 (vs)	—	733 (s)
C—H	Def. e	6	754 (s)	743 (vs)	—
N—H	Def. e	7	—	—	—
C—H α	Def. e	8	773 (s)	773 (sh)	—
C—H β	Def. e	9	781 (m)	794 (sh)	794 (w)
C—H	Def. e	10	802 (w)	839 (s)	—
C—H α	Def. e	11	867 (sh)	—	—
C—H β	Def. e	12	873 (w)	899 (m)	864 (w)
C—H	Def. e	13	901 (s)	908 (w)	910 (s)
		14	949 (m)	937 (s)	—
β		15	955 (w)	964 (w)	—
N—H	Def. i	16	—	—	—
C—H	Def. i	17	1069 (s)	1055 (sh)	1024 (m)
C—N	Str.	18	1090 (s)	1097 (vs)	1066 (m)
		19	—	—	—
C—H	Def. i	20	1121 (s)	—	1124 (w)
C—H	Def. i	21	1167 (s)	1158 (s)	1177 (w)
C—H β	Def. i	22	1191 (sh)	—	—
C—C	Str.	23	1287 (s)	1294 (w)	1265 (s)
N—H	Def.	24	—	—	—
N—H	Def.	25	—	—	—
		26	1335 (s)	1335 (s)	1315 (m)
				1391 (s)	1375 (w)
C—C	Str.	27	1421 (s)	—	—
		28	1466 (m)	1463 (sh)	—
		29	1479 (m)	1490 (sh)	—
C≡N	Str.	30	1508 (s)	1506 (s)	1510 (m)
C=C	Str.	31	1589 (sh)	1576 (s)	1535 (sh)
C=C	Str.	32	1610 (m)	1616 (s)	1620 (bd)
				1690 (s)	
C=O	Str.		1722 (sh)	1719 (s)	1730 (s)
C=O	Str.		—	1792 (s)	1782 (vs)
C=O	Str.				1850 (vs)
N—H	Str.	33	—	—	—
O—H	Str.		3400 (bd)	3400 (bd)	3300 (bd)

[a] Band numbering scheme (1–33) is in accordance with ref. A in Table 6.5*A*. Str. = stretching, Def. = deformation; Def. re = out-of-plane ring deformation; Def. e = out-of-plane deformation (e.g., wagging and twisting); Def. i = in-plane deformation (e.g., scissoring and rocking); s = strong, m = medium, w = weak, bd = broad, sh = shoulder, v = very, α = alpha crystal form, β = beta crystal form.

TABLE 6.5 (*Continued*)

D Photoacoustic FT-IR Spectral Data of Zn(Pc), Zn(Pc)(COOH)$_4$, and [Zn(Pc)]$_n$(COOH)$_n^a$

Vibrational Mode			Band	Zn(Pc)	Zn(Pc) (COOH)$_4$	[Zn(Pc)]$_n$ (COOH)$_n$
C—C		Def. re	1	(beyond detection limit)		
C—C		Def. re	2	—	527 (m)	530 (w)
C—C		Def. re	3	572 (m)	560 (m)	596 (w)
C—C		Def. re	4	621 (s)	656 (w)	646 (w)
C—H		Def. e	5	727 (vs)	—	737 (m)
C—H		Def. e	6	752 (s)	746 (s)	—
N—H		Def. e	7	—	—	—
C—H	α	Def. e	8	773 (m)	763 (m)	—
C—H	β	Def. e	9	781 (w)	781 (m)	793 (w)
C—H		Def. e	10	799 (w)	839 (s)	—
C—H	α	Def. e	11	867 (sh)	—	—
C—H	β	Def. e	12	873 (w)	—	864 (w)
C—H		Def. e	13	887 (s)	908 (sh)	—
			14	949 (m)	932 (s)	905 (s)
	β		15	955 (sh)	—	—
N—H		Def. i	16	—	—	—
C—H		Def. i	17	1061 (s)	1053 (s)	1015 (m)
C—N		Str.	18	1090 (s)	1094 (s)	1065 (m)
			19	—	—	—
C—H		Def. i	20	1119 (s)	1101 (s)	1124 (w)
				1134 (s)		
C—H		Def. i	21	1167 (w)	1163 (m)	1175 (w)
C—H	β	Def. i	22	1191 (sh)	1201 (w)	—
C—C		Str.	23	1285 (s)	1269 (sh)	1269 (s)
N—H		Def.	24	—	—	—
N—H		Def.	25	—	—	—
			26	1335 (vs)	1335 (bd)	1313 (m)
					1398 (w)	1373 (w)
C—C		Str.	27	1410 (m)	—	—
			28	1454 (m)	—	—
			29	1467 (sh)	—	—
C=N		Str.	30	1485 (s)	1489 (s)	1489 (bd)
C=C		Str.	31	1585 (sh)	1528 (s)	—
C=C		Str.	32	1610 (m)	1578 (s)	1636 (bd)
				—	1714 (s)	1734 (s)
C=O		Str.		—	1757 (s)	1784 (vs)
C=O		Str.				
C=O		Str.				1846 (vs)
N—H		Str.	33	—	—	—
O—H		Str.		—	3400 (bd)	3300 (bd)

^{a} Band numbering scheme (1–33) is in accordance with ref. A in Table 6.5A. Str. = stretching, Def. = deformation; Def. re = out-of-plane ring deformation; Def. e = out-of-plane deformation (e.g., wagging and twisting), Def. i = in-plane deformation (e.g., scissoring and rocking); s = strong, m = medium, w = weak, bd = broad, sh = shoulder, v = very, α = alpha crystal form, β = beta crystal form.

TABLE 6.5 (*Continued*)

E Photoacoustic (FT-IR) Spectral Data of $Si(Pc)Cl_2$, $Si(Pc)(OH)_2$, and $[Si(Pc)O]_n{}^a$

Vibrational Mode			Band	$Si(Pc)Cl_2$	$Si(Pc)(OH)_2$	$[Si(Pc)O]_n{}^a$
C—C		Def. re	1		450 (w)	
Si—Cl		Str.		464 (‡vs)		
C—C		Def. re	2	507 (w)	—	—
				532 (‡s)	528 (‡m)	528 (‡m)
C—C		Def. re	3	574 (m)	574 (m)	574 (m)
				609 (‡w)	615 (‡w)	
C—C		Def. re	4	648 (m)	644 (m)	644 (w)
					673 (‡w)	
C—H		Def. e	5	694 (‡m)	702 (‡w)	690 (w)
C—H		Def. e	6	734 (s)	732 (vs)	727 (vs)
N—H		Def. e	7	—	—	—
C—H	α	Def. e	8	761 (s)	759 (vs)	758 (s)
C—H	β	Def. e	9	775 (w)	779 (m)	773 (w)
				785 (m)		
C—H		Def. e	10	808 (‡m)	806 (sh)	804 (‡w)
					831 (‡sh)	
Si—O		Str.			842 (vs)	
C—H	α	Def. e	11	866 (‡m)	871 (w)	866 (‡w)
C—H	β	Def. e	12	883 (m)	—	
C—H		Def. e	13	914 (s)	912 (m)	910 (s)
			14	947 (w)	950 (w)	947 (sh)
	β		15			
				960 (‡m)	974 (‡w)	
				987 (‡w)	993 (‡w)	
						999 (‡bd)
				1004 (‡w)	1004 (‡w)	
N—H		Def.	16	—	—	—
C—H		Def. i			1022 (‡w)	—
C—H		Def. i		1051 (‡sh)		
C—H		Def. i	17	1062 (vs)	1068 (vs)	
C—N		Str.	18	1082 (vs)	1078, 1089 (vs)	1082 (s)
			19	—	—	—
C—H		Def. i	20	1124 (vs)	1122, 1134 (s)	1122 (s)
C—H		Def. i	21	1165 (s)	1168 (m)	1165 (m)
C—H	β	Def. i	22	—	1175 (w)	—
				1240 (‡w)	1222 (‡w)	
					1249 (w)	
C—C		Str.	23	1292 (s)	1292 (s)	1288 (s)
N—H		Def.	24	—	—	—
N—H		Def.	25	—	—	—
			26	1338 (vs)	1336 (vs)	1334 (vs)
				1342 (‡sh)		
				1352 (‡sh)	1352 (‡sh)	1350 (‡sh)

Table 6.5   (*Continued*)

E   Photoacoustic (FT-IR) Spectral Data of Si(Pc)Cl$_2$, Si(Pc)(OH)$_2$, and [Si(Pc)O]$_n^a$

Vibrational Mode		Band	Si(Pc)Cl$_2$	Si(Pc)(OH)$_2$	[Si(Pc)O]$_n^a$
C—C	Str.	27	1431 (s)	1431 (s)	1427 (vs)
		28	1473 (s)	1471 (m)	1471 (m)
		29	1486 (w)	1487 (w)	—
C=N	Str.	30	1533 (s)	1519 (s)	1516 (s)
C=C	Str.	31	1596 (sh)	1595 (sh)	1597 (w)
C=C	Str.	32	1610 (s)	1608 (m)	1614 (m)
N—H	Str.	33	—	—	—
O—H	Str.		—	3500 (bd)	3500 (bd)

^{a} Band numbering scheme (1–33) is in accordance with ref. A in Table 6.5*A*. Str. = stretching; Def. = deformation; Def.re = out-of-plane ring deformation; Def.e = out-of-plane deformation (e.g., wagging and twisting); Def.i = in-plane deformation (e.g., scissoring and rocking); s = strong, m = medium, w = weak, bd = broad, sh = shoulder, v = very, α = alpha crystal form, β = beta crystal form, ‡ = additional peaks found and reported by Marks (C. W. Dirk, T. I. Inabe, T. J. Marks, K. F. Schoch, *J. Amer. Chem. Soc.*, 1983, **105**, 1539.).

The presence of halide atoms attached to P gives other spectral features detected below 1000 cm^{-1}. These bands are affected, however, by the oxidation state of the phosphorous atom and usually give rise to two symmetric and asymmetric vibrations. The presence of difluorides of the type —P(O)F$_2$ also leads to two bands at 930–895 and 890–870 cm^{-1}.

The band assignments for phosphorous-containing species are important because many such species serve a variety of functions from adhesion promoters and corrosion inhibitors to valuable insecticides. The latter also includes P=S- and P—S-containing groups, and Table 6.5 lists the characteristic regions of species containing these groups. Two separate bands in the regions listed in the table have been identified due to P=S adsorptions, and both vanish when phosphorothionates are converted to phosphorothiolates. Similarly to the P=O groups, the position of the P=S band is affected by electronegativity of adjacent groups.

6.3   PHTHALOCYANINES

Cofacial assembly of metallomacrocycles was first pioneered in the late 1950s by Lever[10] and Kenney et al.[11] for phthalocyaninogermanium and phthalocyanomanganese(IV) complexes. Soon thereafter it was discovered that Si(Pc)(OH)$_2$ can also form a stacked, planar phthalocyanine moiety on thermal dehydration.[12] The resulting poly(phthalocyanato) siloxane, [Si(Pc)O]$_n$, on iodine doping, has since gained wide attention as the cornerstone of a new class of electrically conductive macromolecules.[13 16]

In the traditional two-step thermal polymerization of poly(phthalocyanato) siloxane extreme parameters are required, including high vacuum (10^{-3} torr) and

TABLE 6.5 (Continued)

F Infrared Bands of $H_2(Pc)$, $Si(Pc)Cl_2$, $Si(Pc)(OH)_2$, and Both Thermally and Sonically Polymerized $[Si(Pc)O]_n$[a]

Vibrational Mode	Band	β H_2—Pc	$Si(Pc)Cl_2$	$Si(Pc)(CH)_2$	Thermal $[Si(Pc)O]_n$	Sonic $[Si(Pc)O]_n$
C—C Def. re	1		464(‡vs)	450 (W)	—	—
Si—Cl Str.			507 (w)			
C—C Def. re	2	487, 493 (m)	532 (‡s)	528 (‡m)	528 (‡m)	530 (‡m)
C—C Def. re	3	555 (m)	574 (m)	574 (m)	574 (m)	574 (m)
C—C Def. re	4	613 (s)	609 (‡w)	615 (‡w)	—	—
C—H Def. e	5	684(‡s)	648 (m)	644 (m)	644 (w)	646 (w)
				673 (‡w)		
C—H Def. e	6	719 (vs)	694 (m)	702 (w)	700 (w)	692 (w)
N—H Def. e	7	729, 734 (vs)	734 (s)	732 (vs)	727 (vs)	734 (vs)
C—H α Def. e	8	748 (vs)	761 (s)	759 (vs)	758 (s)	759 (s)
C—H β Def. e	9	767 (s)	775 (w)	779 (m)	773 (w)	773(w)
		779 (s)	785 (m)			
C—H Def. e	10		808 (‡m)	806 (sh)	804 (‡w)	804 (‡w)
Si—O Str.	11			831 (s)	831 (w)	831 (w)
C—H α Def. e	12	880 (sh)	868 (‡m)	871 (w)	866 (‡w)	869 (‡w)
C—H β Def. e	13	873 (vs)	883 (m)	912 (m)	910 (s)	910 (s)
C—H Def. e	14	947 (vs)	914 (s)	950 (w)	947 (sh)	947 (sh)
			947 (w)			
β	15	958 (m)	960 (‡m)	974 (‡w)		
			987 (‡w)	993 (‡w)		
Si—O—Si Str.					999 (‡bd)	999 (‡bd)
N—H Def. i	16	1006 (vs)	1004 (‡w)	1004 (‡w)		
C—H Def. i				1022 (‡w)		

231

TABLE 6.5 (Continued)

F Infrared Bands of $H_2(Pc)$, $Si(Pc)Cl_2$, $Si(Pc)(OH)_2$, and Both Thermally and Sonically Polymerized $[Si(Pc)O]_n$[a]

Vibrational Mode		Band	β H_2—Pc	$Si(Pc)Cl_2$	$Si(Pc)(CH)_2$	Thermal $[Si(Pc)O]_n$	Sonic $[Si(Pc)O]_n$
C—H	Def.i	17		1051 (‡sh)			
C—H	Def.i	18	—	1062 (vs)	1068 (vs)		
C—N	Str.	19	1093 (vs)	1082 (vs)	1078, 1089 (vs)	1082 (s)	1082 (s)
C—H	Def.i	20	1118 (vs)	1124 (vs)	1122, 1134 (‡s)	1122 (s)	1122 (s)
C—H	Def.i	21	1158 (m)	1165 (s)	1166 (m)	1165 (m)	1165 (m)
C—H	Def.i	22	1163 (m)	—	1175 (w)		
	β			1240 (‡w)	1222 (‡w)		
					1249 (‡w)		
C=N	Def.	23	1249 (m)			1288 (s)	1288 (s)
C—C	Str.	24	1276 (s)	1292 (s)	1292 (s)		
N—H	Def.	25	1303 (s)				
N—H	Def.	26	1323 (s)				
			1334 (vs)	1338 (vs)	1336 (vs)	1336 (vs)	1336 (vs)
			1342 (‡sh)	1342 (‡sh)			
				1352 (‡sh)	1352 (‡sh)	1352 (‡sh)	1352 (‡sh)
C—C	Str.	27	1436 (s)	1431 (s)	1431 (s)	1427 (vs)	1427 (vs)
		28	1458 (w)	1473 (s)	1471 (m)	1471 (m)	1471 (m)
		29	1477 (w)	1486 (w)	1487 (w)		1485 (w)
C=N	Str.	30	1502 (m)	1533 (s)	1519 (s)	1518 (s)	1517 (s)
C=C	Str.	31	1597 (sh)	1596 (sh)	1595 (sh)	1595 (w)	1595 (w)
C=C	Str.	32	1606, 1614 (m)	1610 (s)	1608 (m)	1614 (m)	1614 (m)
N—H	Str.	33	3277 (s)				
O—H	Str.			—	3500 (bd)	3500 (bd)	3500 (bd)

[a]Band numbering scheme (1–33) is in accordance with ref. A in Table 6.5A. Str. = stretching; Def. = deformation, Def.re = out-of-plane ring deformation; Def.e = out-of-plane deformation (e.g. wagging and twisting); Def.i = in-plane deformation (e.g. scissoring and rocking); s = strong, m = medium, w = weak, bd = broad, sh = shoulder, v = very, † = additional peaks found and reported by Kobayashi (ref. B in Table 6.5A); ‡ = additional peaks found and reported by Marks (ref. 13); e = $Si(Pc)Cl_2/Na_2Te$ sonicated in THF for 8 hr.

TABLE 6.6   IR Tentative Band Assignments for Germanium Phthalocyanine Polymers and Monomers[a]

Band	Band Assignment		$GePcCl_2$	$GePc(OH)_2$	$[GePcO]_n$ Sonic	$[GePcO]_n$ Thermal
	Ge—Cl	A Str.	312 (s)	—	—	—
1	C—C	Def. re	437 (w)	—	435 (w)	435 (w)
2	C—C	Def. re	513 (s)	—	507 (w)	509 (w)
3	C—C	Def. re	572 (s)	573 (s)	573 (m)	572 (w)
4	C—C	Def. re	644 (m)	—	642 (w)	642 (w)
	Ge—OH	A Str.	—	648 (vs)	650 (w)	—
5	C—H	Def. e	727, 723 (s)	725 (vs)	727 (vs)	729 (vs)
6	C—H	Def. e	756 (vs)	758 (vs)	756 (s)	756 (s)
7	N—H	Def. e	—	—	—	—
8	C—H	Def. e	768 (s)	770 (s)	—	764 (w)
9	C—H	Def. e	779 (s)	777 (s)	775 (sh)	777 (s)
10	C—H	Def. e	806 (m)	802 (w)	804 (w)	806 (w)
11	C—H	Def. e	868 (m)	874 (m)	871 (m)	—
	Ge—O—Ge	Str.	—	—	885–850 (bd)	890–840 (bd)
12	C—H	Def. e	885 (m)	—	883 (m)	883 (sh)
13	C—H	Def. e	905 (vs)	903 (vs)	901 (s)	902 (s)
14	C—H	Def. e	947 (w)	953 (m)	951 (w)	947 (w)
15	—		—	—	—	—
16	N—H	Def.	—	—	—	—
17	C—H	Def. i	—	1069 (s)	1072 (s)	1070 (m)
18	C—N	Str.	1082 (vs)	1082 (s)	1090 (vs)	1088 (s)
19	C—H	Def. i	—	1096 (s)	—	—
20	C—H	Def. i	1124 (vs)	1121, 1134 (s)	1124 (vs)	1124 (vs)
21	C—H	Def. i	1165 (s)	1169 (s)	1167 (s)	1168 (m)
22	C—H	Def. i	—	1186 (s)	—	—
23	C—C	Str.	1288 (s)	1288 (s)	1288 (s)	1288 (m)
24	N—H	Def.	—	—	—	—
25	N—H	Def.	—	—	—	—
26	C—N	Str.	—	1337 (s)	1337 (vs)	1333 (vs)
	C—N	Str.	1344 (vs)	1346 (s)	1346 (vs)	1344 (s)
27	C—C	Str.	1425 (s)	1423 (s)	1423 (s)	1423 (s)
28	C=N/C=C Str.		1470 (s)	1468 (m)	1468	1467 (w)
29	C=N/C=C Str.		1487 (m)	1476 (w)	1477 (m)	1477 (w)
30	C=N	Str.	1512 (s)	1502 (s)	1502 (s)	1499 (m)
31	C=C	Str.	1589 (w)	1583 (w)	1589 (w)	1589 (w)
32	C=C	Str.	1609 (s)	1612 (m)	1612 (m)	1612 (m)
	C—H	Str.	3013 (w)	3013 (w)	3011 (w)	3013 (w)
	C—H	Str.	3057 (m)	3043 (m)	3045 (m)	3048 (m)
	C—H	Str.	3074 (w)	3078 (w)	3078 (w)	3080 (w)
	O—H	Str.	—	3497 (bd)	3464 (bd)	—

[a] Photoacoustic FT-IR spectral data of $GePcCl_2$, $GePc(OH)_2$, and both thermally and sonically polymerized $[GePcO]_n$. Band numbering scheme (1–32) is in accordance with ref. A in Table 6.5A. *Key:* vs = very strong, s = strong, m = medium, w = weak, bd = broad, sh = shoulder; Def.e = out-of-plane deformation (e.g., wagging and twisting); Def.i = in-plane deformation (e.g., rocking and scissoring); Def.re = out-of-plane ring deformation. Far-infrared bands are in accordance with the following references: J. B. Davison and K. J. Wynne, *Macromolecules*, 1978, **11**, 186; R. D. Joyner and M. E. Kenney, *Inorg. Chem.*, 1962, **1**, 717.

Source: Courtesy M. D. Mohol.

temperature (440°C) conditions over extensive periods of time (12 hr).[17] As a consequence, much work on the development of $[Si(Pc)O]_n$ has been impeded. Subsequently, the apparent need for a more convenient synthetic technique was addressed in the studies of Exsted and Urban, who illustrated a one-step, room-temperature method for the synthetic preparation of $[Si(Pc)O]_n$ utilizing ultrasonic energy.[18] This is the first utilization of ultrasonic waves in polymer synthesis. Table 6.5$A-F$ shows the band assignments of various phthalocyanine polymers and monomers. Along the same lines, Hohol and Urban[19] used ultrasonic energy to polymerize germanium phthalocyanine dichloride ($Ge(Pc)Cl_2$) to poly(phthalocyanato) germanium oxide ($[Ge(Pc)]O_n$) at room temperature in the presence of sodium chalcogenide. These studies also indicated the usefulness of isotopic exchange H_2O/D_2O in revealing the role of traces of water present during the polymerization reaction. Table 6.6 provides band assignments for selected Ge-containing phthalocyanines, and reactions of metal phthalocyanines with polymeric surfaces will be discussed in Chapter 8.

6.4 SULFUR-CONTAINING COMPOUNDS

The most common functional groups in which sulfur participates are listed in Table 6.7. As was indicated earlier, such functional groups as OH and NH have a strong tendency to form hydrogen bonding. In contrast, the SH link is not capable of hydrogen bonding. Hence, there is a marginal frequency shift on passing from the liquid to dilute solutions. Although the S—H stretching is listed to cover a rather broad spectral region, the presence of various substituents will affect this frequency such that in the majority of S—H-containing species, it will be detected in the $2600-2400$-cm^{-1} region with varying intensities. Table 6.8 lists the S—H stretching frequencies for the most common molecular arrangements. In Raman spectra, the S—H is invariably strong, but IR intensity varies, depending on the neighboring substituents. While for a number of simple mercaptans such as propyl, butyl, and isoamyls, a well-defined although weak band is observed in the $2650-2550$-cm^{-1} region,[20] a comparison of mercaptanes

TABLE 6.7 Spectral Regions for Sulfur-Containing Molecules

Functional Group	Spectral Region (cm^{-1})
S—H (mercaptan)	2950–2550
S—S	500–400
S—C	715–620
=S=O (sulfoxide)	1070–1010
—SO$_2$—	1360–1290; 1170–1120
—SO$_2$—X	1450–1405; 1260–1225
—SO$_2$—N= (sulfamide)	1360–1335; 1170–1150

TABLE 6.8 Band Assignments of Various S-Containing Molecules

Functional Group	Region (cm^{-1})
S—H stretch	
Aliphatic thiols	2600–2550
Aryl S—H	2600–2450
—CS—SH (free)	2600–2500
—CS—SH (H-bonded)	2500–2400
S—S stretch	
Aliphatic S—S	520–500
Aryl S—S	500–430
Diaryl S—S	540–520
Polysulfides	500–470
C—S stretch	
CH_3—S—	710–685
RCH_2—S—	660–630 (primary)
R'R" CH—S—	630–600 (secondary)
R'R"R''' C—S—	600–570 (tertiary)
Aliphatic disulfides	710–570
$(ArS)_2$ C=O	580–560
Phenyl sulfides	715–670
=S=O	
Sulfoxides	1070–1035
Aryl sulfoxides	1060
Methyl aryl sulfoxides	535–495
—SO_2—	
Diakyl sulfones	1330–1305 (asymmetric)
	1150–1135 (symmetric)
Alkyl, aryl sulfones	1335–1325 (asymmetric)
	1160–1150 (symmetric)
Diaryl sulfones	1360–1335 (asymmetric)
	1170–1160 (symmetric)
Sulfones	610–545 (scissor)
Saturated aliphatic sulfones	525–495 (wagging)
—SO_2—X	
—SO_2—F	1415–1395 (SO_2 asymmetric stretching)
	1215–1200 (SO_2 symmetric stretching)
RO—SO_2—Cl	1455–1405 (SO_2 asymmetric stretching)
	1225–1205 (SO_2 symmetric stretching)
RO—SO_2—F	1510–1445 (SO_2 asymmetric stretching)
	1260–1230 (SO_2 symmetric stretching)
—SO_2—N=	
—SO_2—NH_2	3390–3245 (NH stretching)
—SO_2—N=	1380–1325 (SO_2 asymmetric stretching)
	1170–1150 (SO_2 symmetric stretching)
=N—SO_2—N=	1340–1320 (SO_2 asymmetric stretching)
	1145–1140 (SO_2 symmetric stretching)
S—O stretch	
~O—SO_3^-···Na^+	1040–1060 (S—O symmetric stretching)

and the corresponding sulfides was also made.[21,22] In all cases the disappearance of the SH bands was one of the major spectral changes. It should also be kept in mind that hydrogen-bonding effects are much smaller for the S—H groups as compared to the N—H or OH groups. If, however, dimers and monomers coexist, two S—H bands may be detected.[23]

Another spectral region of interest is at lower frequency due to S—S stretching vibrations. As one may expect, this band is not of as much value in IR spectrum because of relatively small dipole moment changes, which are, of course, related to the symmetric nature of the bond. Table 6.8 lists those regions where S—S may occur in IR, but by far more useful information provides Raman spectroscopy with the bands in 400–550-cm^{-1} region.

The first studies initiating the spectroscopic importance of the C—S linkages were performed by Trotter and Thompson,[21] who, on the basis of simple methyl through *tert*-butyl mercaptans, determined considerable frequency shifts and sensitivity of the C—S stretching frequency resulting from small changes of electron-donating or -withdrawing power of the neighboring groups. Progressive lowering of the C—S stretching, such as that listed in Table 6.8 for primary, secondary, and tertiary C—S compounds was also established. Both aliphatic and aromatic sulfides exhibit relatively weak to medium vibrations due to the C—S stretching vibrations in the 710–570-cm^{-1} range, with the primary sulfides at the high end and the tertiary at the lower end of the 710–570-cm^{-1} scale. An additional complexity arises when a double-bond conjugation with the =C—S— bonding occurs. This lowers even further the C—S frequency to 590 cm^{-1}, but increases intensity quite substantially.

The species containing sulfonates exhibit strong absorption band in the 1070–1035-cm^{-1} region due to the S=O stretching vibrations listed in Table 6.8. This particular vibration is also affected by the presence of H-bonding species and usually results in the shift to the 1055–1010-cm^{-1} region. Moreover, it is heavily influenced by the inductive effects of the substituents with high electronegativity. These are the substituents which tend to stabilize S=O rather than form S^{+}—O^{-}. Interestingly the S=O stretching normal vibrations exhibit remarkably constant frequency in dilute solutions because the Π electrons of any adjucent to S=O double bonds do not share the same plane as those of the S=O link and, therefore, conjugation does not occur. Thus, the vibrational energy of the S=O stretch is highly affected by the electronegativity of substituents.

One of the early works on IR characteristics of sulphones was carried out by Schrieber.[24] Apparently, these pioneering studies have established that there are two bands characteristic of sulfones in the 1360–1290- and 1170–1120-cm^{-1} regions that are absent in species containing sulfides. The two bands are due to a symmetric and symmetric stretching vibrations of the —SO$_2$— groups, and the presence of conjugation appears to affect this frequency; specifically, alkyl sulphones absorb in the range 1330–1305 and 1160–1130 cm^{-1}, while aromaticity introduced near —SO$_2$— moiety gives rise to the asymmetric and symmetric bands in the 1380–1355- and 1770–1155-cm^{-1} regions, respectively. The region that can also be helpful in identifying the presence of —SO$_2$— groups occurs in the 610–545- and 525–495-cm^{-1} regions as a result of scissoring and wagging vibrations, respectively.

A comparison of S=O stretching frequencies in —SO$_2$— and —SO$_2$—X (X = Cl, F, Br) indicates that sulfonyl chlorides, fluorides, and bromides exhibit higher S=O vibrational frequencies and are detected in the 1385–1360- and 1190–1160-cm^{-1} regions when aliphatic groups are present. For aryl sulfonyl chlorides, these regions are extended to higher frequencies, but the major difference is the presence of a doublet band attributed to symmetric S=O stretching vibrations. When fluorine is substituted instead of chlorine, and alkyl is substituted by aryl groups, the frequency increases even further owing to the presence of aromatic rings and highly electronegative substituents that increase the S=O frequency to 1500 and 1270 cm^{-1}.

The presence of a nitrogen atom attached to —SO$_2$— adds another dimension to the S=O vibrational energy. In ordinary amides, the presence of a tertiary nitrogen atom with a pair of electrons attached to the carbonyl group results in the changes of C=O frequency by lowering it. This effect, however, does not occur with sulfoneamides because the orientation of π electrons prevent that from happening. As a result, the trend that is observed is influenced mainly by substituents attached to nitrogen. This effect is similar to that described for P=O stretching vibrations. It should be kept in mind that the presence of an electronegative nitrogen atom shifts the SO$_2$ asymmetric and symmetric stretching vibrations to higer frequencies for sulfoneamides than for sulfones.

6.5 SOLID SUPERACIDS

A *solid superacid* is commonly defined as a solid material that exhibits an acid strength greater than that observed for concentrated sulfuric acid. A superacid's importance comes from the fact that the surface and catalytic properties, surface structures, and their sites are extremely suitable for alkane reactions such as skeletal isomerization of butane and pentane. Although there have been many groups and kinds of superacids proposed, there are essentially three major groups: (1) antimony pentafluoride mounted on mixed oxides of SiO$_2$–Al$_2$O$_3$, SiO$_2$–TiO$_2$, and TiO$_2$–ZrO$_2$; (2) TiO$_2$, ZrO$_2$, and Fe$_2$O$_3$ containing a small amount of sulfate ions, and (3) fluorinated sulfonic acid resins containing mixed salts such as AlCl$_3$–CuCl$_2$, AlCl$_3$–CuSO$_4$, and others. Although there have been numerous studies on various aspects of superacid chemistry, the mechanisms of generation of the surface superactive sites resulted in extensive use of vibrational spectroscopy in an effort to elucidate mechanisms and surface reactions responsible for catalytic activity.

Because of the immense amount of literature on superacids generated in recent years, the reader is referred to specific monographs treating aspects other than spectroscopy.[25,26] As an illustrative example, here we will focus on the surface superacid properties of the SbF$_5$/SiO$_2$–Al$_2$O$_3$ system, which are due to unusual activity for alkane reactions. Furthermore, the additional attractiveness of this system comes from the fact that when pyridine is adsorbed on the surface, Lewis or Bronsted acid sites are present. It is believed that the alkane reactions proceed via carbenium ion intermediates that result from the following mechanisms: (1) hydride ion abstraction from the alkane chain by Lewis acid and (2) protonation of the alkane by a Bronsted acid with an intermediate carbenium ion. This is illustrated in Fig. 6.1, but it should be kept in mind that the studies that utilized deuterium suggest that the intramolecular hydrogen (or deuterium) transfer may involve rearrangements of the carbon skeletons and the carbenium ion mechanisms may be initiated by abstraction of an H^-. It is not just antimony pentafluoride but a combination of the above and metal oxides that make catalytic activity. Therefore, the activity depends on metal oxides employed. Moreover, these observations, combined with the studies of pyridine and OH surface reactions, indicated that SbF_5 is not surface-bonded but is modified in such a manner to form superacidity.

At low temperatures, SbF_5 coordinates to the OH groups inherently present on the SiO_2/Al_2O_3 surfaces. This process increases the acid strength of the OH surface groups, and coordination to an oxygen atom enhances the acid strength of the Lewis site. When temperatures are higher, SbF_5 reacts with OH groups, giving $-OSbF_4$ and HF, followed by the HF reaction with nonoccupied OH groups giving off H_2O and leaving one F atom attached to the surface. Figure 6.2 illustrates the proposed mechanisms.

Infrared spectra of pyridine absorbed on the SiO_2/Al_2O_3 surfaces treated with SbF_5 below 373 K showed two bands at 1540 and 1460 cm^{-1} indicating the presence of Bronsted and Lewis acid sites. However, when the treatment temperature was increased to 573 K followed by outgassing, only the band at 1460 cm^{-1} was present, indicating that only Lewis acid sites exist on the surface.

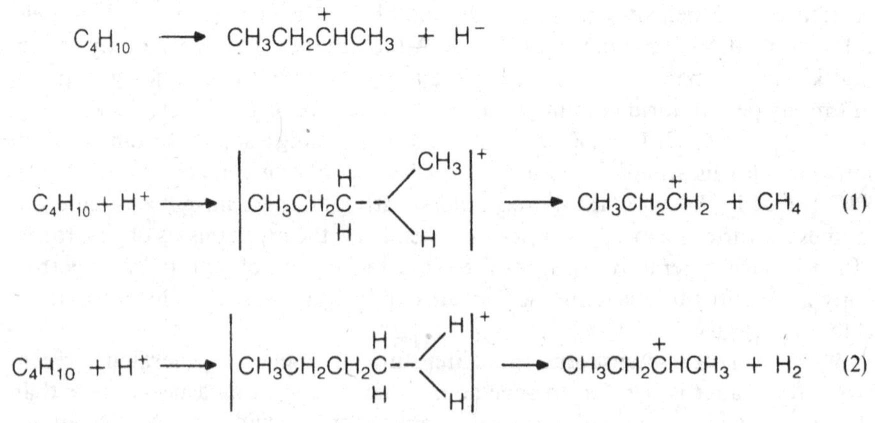

$$C_4H_{10} \longrightarrow CH_3CH_2\overset{+}{C}HCH_3 + H^-$$

$$C_4H_{10} + H^+ \longrightarrow \left| CH_3CH_2\underset{H}{\overset{H}{C}}--\overset{CH_3}{\underset{H}{\diagdown}} \right|^+ \longrightarrow CH_3CH_2\overset{+}{C}H_2 + CH_4 \quad (1)$$

$$C_4H_{10} + H^+ \longrightarrow \left| CH_3CH_2CH_2\underset{H}{\overset{H}{C}}--\overset{H}{\underset{H}{\diagdown}} \right|^+ \longrightarrow CH_3CH_2\overset{+}{C}HCH_3 + H_2 \quad (2)$$

Figure 6.1 Two possible mechanisms showing the carbenium ion formation from butane.

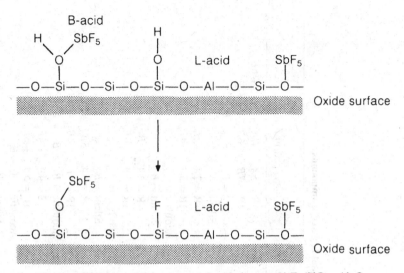

Figure 6.2 Proposed structures of acid sites on $SbF_5/SiO_2-Al_2O_3$.

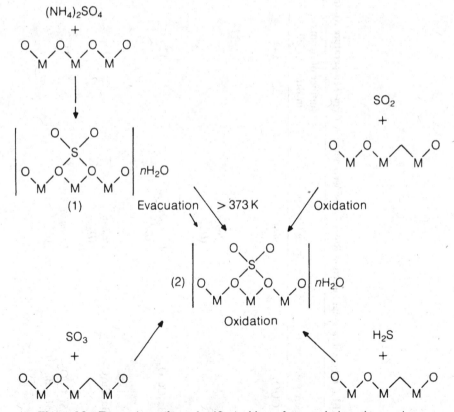

Figure 6.3 Formations of metal sulfite/oxide surface catalytic active species.

TABLE 6.9 Vibrational Selection Rules for Superconducted High-Temperature Materials and Related Compounds

Superconductor	Space Group	Number of Atoms in Bravais Cell	Selection Rules
La_2CuO_4	$Cmca$ (D_{2h}^{18})	14	$R = 5A_g + 4B_{1g} + 3B_{2g} + 6B_{3g}$ $IR = 7B_{1u} + 6B_{2u} + 4B_{3u}$ Acoustic $= B_{1u} + B_{2u} + B_{3u}$ Silent $= 4A_u$
$La_{2-x}M_xCuO_4$	I_4/mmm (D_{4h}^{17})	7	$R = 2A_{1g} + 2E_g$ $IR = 3A_{2u} + 4E_u$ Acoustic $= A_{2u} + E_u$ Silent $= B_{2u}$
$MBa_2Cu_3O_4$	P_4/mmm (D_{4h}^{1})	12	$R = 4A_{1g} + B_{1g} + 5E_g$ $IR = 5A_{2u} + 6E_u$ Acoustic $= A_{2u} + E_u$ Silent $= B_{2u}$
$MBa_2Cu_3O_7$	$Pmmm$ (D_{2h}^{1})	13	$R = 5A_g + 5B_{2g} + 5B_{3g}$ $IR = 7B_{1u} + 7B_{2u} + 7B_{3u}$ Acoustic $= B_{1u} + B_{2u} + B_{3u}$ Silent $= B_{2u}$
$M_2Sr_2CuO_6$ $(M = Bi, Tl)$	I_4/mmm (D_{4h}^{17})	11	$R = 4A_{1g} + 4E_g$ $IR = 5A_{2u} + 6E_u$ Acoustic $= A_u + E_u$ Silent $= B_{2u}$

$MCaBa_2Cu_2O_7$ (M = Bi, Tl)	$P4/mmm$ (D_{4h}^1)	13	$R = 4A_{1g} + B_{1g} + 5E_g$ $IR = 6A_{2u} + 7E_u$ Acoustic $= A_{2u} + E_u$ Silent $= B_{2u}$
$MCa_2Ba_2Cu_3O_9$ (M = Bi, Tl)	$P4/mmm$ (D_{4h}^1)	17	$R = 5A_{1g} + B_{1g} + 6E_g$ $IR = 9A_{2u} + 10E_u$ Acoustic $= A_{2u} + E_u$ Silent $= B_{2u}$
$M_2CaSr_2Cu_2O_3$ (M = Bi, Tl)	$I4/mmm$ (D_{4h}^{17})	15	$R = 6A_{1g} + B_{1g} + 7E_g$ $IR = 6A_{2u} + 7E_u$ Acoustic $= A_{2u} + E_u$ Silent $= B_{2u}$
$M_2Ca_2Ba_2Cu_3O_{10}$ (M = Bi, Tl)	$I4/mmm$ (D_{4h}^{17})	19	$R = 7A_{1g} + B_{1g} + 8E_g$ $IR = 9A_{2u} + 10E_u$ Acoustic $= A_{2u} + E_u$ Silent $= B_{2u}$
$BaBiO_3$	$Pm3m$ (O_h^1)	5	$R = $ (none) $IR = 3F_{1u}$ Acoustic $= F_{1u}$ Silent $= F_{2u}$
$Ba_2BiBi'O_4$	$Fm3m$ (O_h^5)	10	$R = A_{1g} + E_g + 2F_{2g}$ $IR = 4F_{1u}$ Acoustic $= F_{1u}$ Silent $= F_{1g} + F_{2u}$

Source: Adopted from refs. 30 and 31.

Although numerous oxides such as TiO_2 and ZrO_2 exhibit relatively weak acidic strength, the presence of small amounts of ammonium sulfate ions ranging from 0.5 to 8 w/w% places many non-active oxides in the category of superacids. As with SO_4^{2-}, addition of SO_3 to almost completely inactive Fe_2O_3 generates superacidity.[27] Infrared studies revealed that SO_4^{2-}/TiO_2, SO_4^{2-}/ZrO_2, and SO_4^{2-}/Fe_2O_3 species give a strong band due to S—O stretching vibrations, and its energy varies depending on the strength of the bonding. For zirconium, titanium, iron and aluminium oxides the following frequencies were detected: 1390, 1375, 1375, and 1398 cm^{-1}, respectively. Although these energies do not reflect the same trends as that determined for catalytic activity, the magnitude of the shift of these bands after pyridine adsorption correlates well with the catalytic activity; the greater the shift, the higher the catalytic activity. For SO_4^{2-}/TiO_2, the S—O stretching band was detected at 1410 cm^{-1}, whereas for SO_4^{2-}/Bi_2O_3, two bands at 1370 and 1320 cm^{-1} were observed. Based on these spectroscopic data and other experimental evidence, the surface structures, such as shown in Fig. 6.3, were proposed along with an alternative sulfite-like species. The latter would be characterized in IR by the presence of the S=O stretching band around 1180–1250 cm^{-1}.

6.6 SUPERCONDUCTORS

Although this section is directly related to surface vibrational spectroscopy, the topic deserves recognition as the 1987 Noble prize in physics was given to Bednorz and Muller for the discovery of seperconductivity at ~40 K based on La—Ba—Cu—O ceramic compounds.[28] Shortly threafter, Wu et al.[29] synthesized a related copper oxide-based ceramic, $YBa_2Cu_3O_{7-d}$, with a superconducting temperature of 77 K. Because vibrational spectroscopy plays a key role in characterization of bulk and surface structures, the following sections provide only fundamental spectral features of superconducting materials. A detailed spectroscopic analysis can be found in two excellent review articles by Ferraro et al.[30,31] and the references cited therein. Because the field is relatively new, there may be inaccuracies in the band assignments. Here, we provide only selection rules listed in Table 6.9 and vibrational features of selected ceramic and selected organic superconductors. Following this classification, the first section will provide vibrational features of ceramic superconductors.

6.6.1 Ceramic Superconductors

Infrared and Raman spectra of $La_{2-x}(Sr, Ba)_xCuO_4$, (LMCO), $MBa_2Cu_3O_{z\delta}$ (MBCO), Bi—(Sr, CA)—Cu—O, Tl—(Ba, Sr)—Ca—Cu—O, Pb—Sr—Ln—Ca—Cu—O, and $Ba_{1-x}K_xBiO_3$ families have been reported. Because reliable assignment of Raman bands requires polarization studies of oriented crystals, model studies and theoretical factor analysis, Table 6.10 tabulates the Raman data for Tl-based superconductors. (McCarthy, Table 6.10). Following Thomson

TABLE 6.10 Raman Bands and Their Assignments for Tl-Based Superconductor.[a]

Symmetry Species	TlCaBa$_2$Cu$_2$O$_7$ (Tl-1122) $4A_{1g} + B_{1g} + 5E_g$	TlCa$_2$Ba$_2$Cu$_3$O$_9$ (Tl-1223) $5A_{1g} + B_{1g} + 6E_g$	Tl$_2$CaBa$_2$Cu$_2$O$_8$ (Tl-2122) $6A_{1g} + B_{1g} + 7E_g$	Tl$_2$Ca$_2$Ba$_2$Cu$_3$O$_{10}$ (Tl-2223) $7A_{1g} + B_{1g} + 8E_g$
Raman →				
A_{1g}	—	—	599 [03]	601 [04]
	525 [02]	526 [03]	494 [02]	498 [03]
	475 [01]	—	407 [01]	407 [02]
	—	260 [Ca]	—	270 [Ca]
	148 [Cu]	152 [Cu2]	158 [Cu]	159 [Cu2]
	—	N. O.	130 [Tl, 02]	133 [Tl, 03]
	120 [Ba]	104 [Ba/Ca]	108 [Ba, Cu]	99 [Ba/Ca]
B_{1g}	278 [01]	238 [02]	282 [01]	245 [02]

[a] From McCarty et al. [(a) K. F. McCarty, B. Morosin, D. S. Ginley, and D. R. Boehme, *Physica C*, 1989 **157**, 135; (b) K. F. McCarty, E. L. Venturini, D. S. Ginley, B. Morosin, and J. F. Kwak, *Physica C*, 1989, **159**, 603; (c) personnel communication from K. F. McCarty] N. O. indicates that a corresponding band is expected but not observed. [02], motion of oxygen atom bridging the Tl/O sheet to the Cu/O sheets along the c axis. [01], in-phase oxygen atom motion in the Cu/O sheets [O3, O4]. motion of oxygen atom bridging the Cu/O sheets to the Tl/O sheet and motion of the oxygen atom between the Ba and Tl atoms along the c axis.

TABLE 6.11 IR bands for $Pb_2Sr_2PrCu_3O_{8+\delta}$ and $MBa_2Cu_3O_6$ (M = Sm, Eu)

Pb$_2$ Sr$_2$ PrCu$_3$ O$_{8+\delta}$			
10 K	295 K	SmBa$_2$Cu$_3$O$_6$	EuBa$_2$Cu$_3$O$_6$
606			557 (x)
567	579	571	
523	519		
(493)^a			
(475)^a			380 (x)
(394)^a		(394)^a	344 (x)
369	358	349	—
313	307	—	—
285	280	—	230 (z)
231	228	241	
215	—		
209	209		185 (z)
187	189	192	168 (x)
173	173		
170	171		
161	158	168	
145	142		
137	(134)^a		
124	—		
119	116		
114	108	112	
99	98		
96	93		
(83)^a	(81)^a		
61	63		
47	45		

^{a} Parentheses indicate polarized bands.

Source: C. Thomsen, M. Cardona, R. Liu, Hj. Mattausch, W. König, F. Garcia-Alvarado, B. Suarez, E. Moran, and M. Alario-Franco, *Solid State Commun.*, 1989, 69, 857.

et al.,[21] Table 6.11 provides selected IR data for another class of superconductors Pb—Sr—Ln—Ca—Cu—O. (Where M = Y, Sm, En, Gd, Ho, etc.).

6.6.2 Synthetic Organic Superconductors

Whereas ceramic conductivity (or superconductivity) is based on a flow of electrons with no resistance, organic superconductvity involves charge transfer in molecular materials of low densities. This requires the presence of donor/acceptor molecules

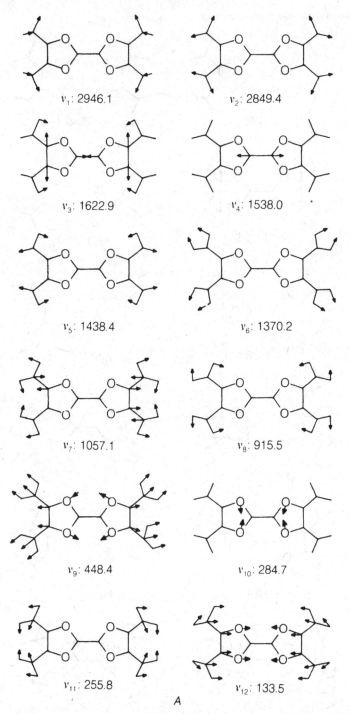

ν_1: 2946.1

ν_2: 2849.4

ν_3: 1622.9

ν_4: 1538.0

ν_5: 1438.4

ν_6: 1370.2

ν_7: 1057.1

ν_8: 915.5

ν_9: 448.4

ν_{10}: 284.7

ν_{11}: 255.8

ν_{12}: 133.5

A

Figure 6.4 Atomic displacements and normal modes in tetramethyltetraselenafulvene (TMTSF) (*A*) and tetramethyltetrathiafulvalene (TMTTF) (*B*). (From M. Meneghetti, R. Bozio, I. Zanon, C. Pecile, C. Ricotta, and M. Zanetti, *J. Chem. Phys.*, 1984, **80**, 6210.)

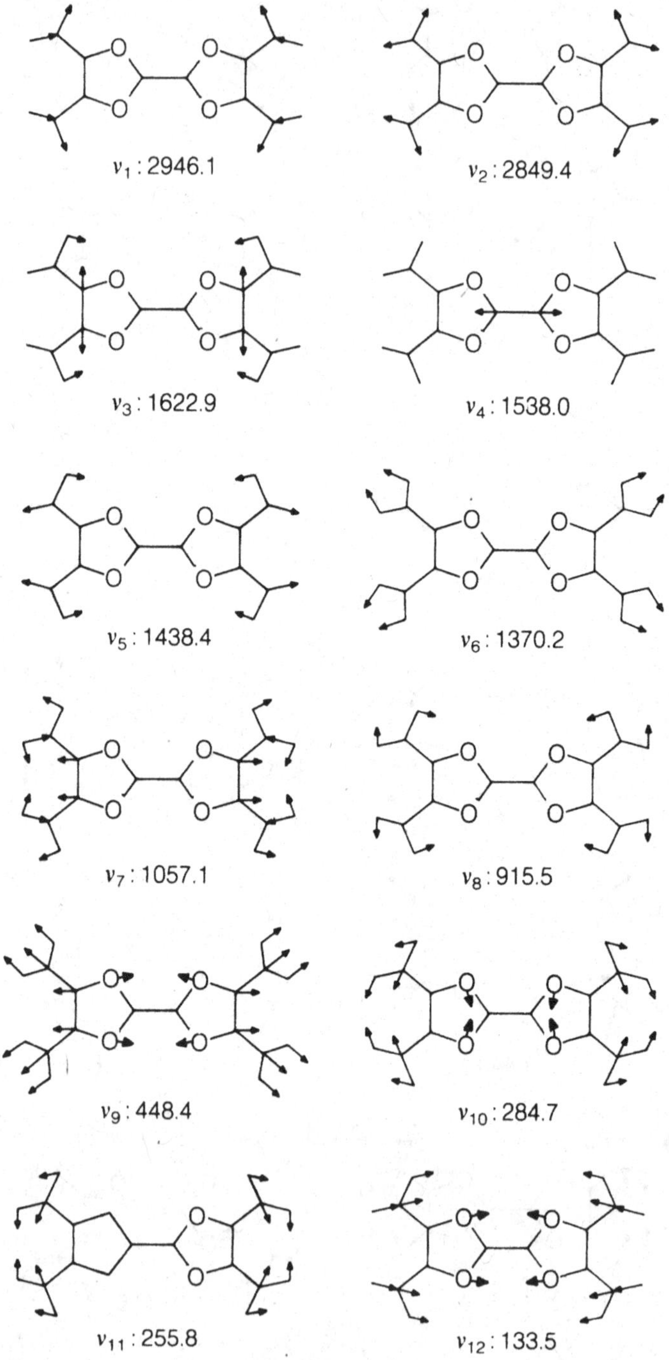

v_1 : 2946.1

v_2 : 2849.4

v_3 : 1622.9

v_4 : 1538.0

v_5 : 1438.4

v_6 : 1370.2

v_7 : 1057.1

v_8 : 915.5

v_9 : 448.4

v_{10} : 284.7

v_{11} : 255.8

v_{12} : 133.5

B

Figure 6.4 (*Continued*)

TABLE 6.12 (*A*) IR Bands $(TMTSF)_2PF_6$ and $(TMTSF)_2ReO_4$ Powers; (*B*) Raman Bands for TMTSF

A

$(TMTSF)_2 PF_6$ $T = 8$ K	$(TMTSF)_2 ReO_4$ $T = 15$ K	Assignments	
		Cation–Radical Modes	Anion Modes
254 (w)		B_{2u}, ν_{53}	
		B_{1u}, ν_{35}	
	261 (m)		
	267 (m)	A_g, ν_{10}	
	319 (w, br)		$ReO_4\text{-}(\nu_4, F_2)$
	329 (w, br)		
	436 (sh)	A_g, ν_9	
443 (w)			
	441 (m)	B_{1u}, ν_{34}	
449 (w)	465 (vw)	B_{2u}, ν_{52}	
559 (s)			$PF_6\text{-}(\nu_4, F_{1u})$
833 (sh)			
844 (vs)			$PF_6\text{-}(\nu_3, F_{1u})$
872 (vw)			
876 (vw)			
882 (m)			$PF_6\text{-}(\nu_2, E_g)$
911 (w)		βCH_3	
918 (w)			$PF_6\text{-}(2x\nu_5)$
	907 (vs)		
			$ReO_4\text{-}(\nu_3, F_2)$
	911 (vs)		
1064 (w)	1068 (vw)	B_{1u}, ν_{31}	
1081 (w)			$PF_6\text{-}(2x\nu_4, F_{1u})$
1160 (w)	1150 (w)		
	1158 (vw)	B_{2u}, ν_{47}	
1166 (m)	1365		
1395 (m)	1384 (w)	αCH_3 symmetric	
	1415 (s, br)	A_g, ν_4	
1434 (w)	1440 (m)		
1446 (m)		αCH_3 asymmetric	
1457 (w)			
1567 (w)	1545 (m)	B_{1u}, ν_{28}	
1582 (m)	1560 (w)		
	1598 (w, br)	A_g, ν_3	
	1611 (w)	B_{1u}, ν_{28}	
1639 (vw)			
1873 (vw)	1846 (w)		

[a] Nomenclature (taken from ref. 7):

ν_9, A_g, mixed, predominantly C—Se stretch, external and internal.

ν_{10}, A_g, mixed, predominantly C—Se stretch, internal and external.

ν_{28}, B_{1u}, ring C=C stretch and central C=C stretch.

ν_{31}, B_{1u}, mixed, internal and external C—Se stretch, C—CH$_3$ stretch, C—C—H bend.

Source: Refs. 30 and 31.

TABLE 6.12 *(Continued)*

v_{34}, B_{1u}, mixed, internal and external C—Se stretch, C—CH$_3$ stretch.
v_{35}, B_{1u}, mixed, external and internal C—Se stretch, ring deformation.
v_{48}, B_{2u}, mixed vibration, predominately C—CH$_3$ stretch.
v_{53}, B_{2u}, primarily Se—C—CH$_3$ bend.
See also Fig. 6.2 for vibrations $v_1 \rightarrow v_{12}$. *Key:* w = weak; m = medium; br = broad; sh = shoulder; vw = very weak; vs = verystrong.
Source: R. Bozio, C. Pecile, K. Bechgaard, F. Wudl, and D. Nalewajek, "Infrared Study of the Formation of Charge Density Waves in (TMTSF)$_2$ X(X = ReO$_4$- and PF$_6$-) at Atomospheric Pressure," *Solid State Commun.*, 1982, **41** Pergamon Press, Elmsford, New York. Reproduced by permission.

B	
Frequencies (cm^{-1})^a	Assignment^b
2906 (w)	vCH$_3$ asymmetric
1625 (s)	$A_g v_3$ ($v_{c=c}$ center)
1589 (w)	
1539 (vs)	$A_g v_4$($v_{c=c}$ ring)
1530 (w)	
1503 (vs)	
1444 (w)	αCH$_3$ asymmetric
1420 (w)	
1384 (w)	
1225 (w)	
1167 (w)	
1018 (w)	βCH$_3$
916 (m) → 752 in TMTSF-d_{12}	βCH$_3$
682 (m)	
603 (w)	
472 (m) → 439 in TMTSF-d_{12}	$A_g v_9$
453 (vs)	
328 (m) → 241 in TMTSF-d_{12}	
276 (s)	$A_g v_{10}$
263 (s) → 236 in TMTSF-d_{12}	$A_g v_{11}$
180 (w)	
173 (w)	

^{a} *Key:* w = weak; s = strong; vs = very strong; m = medium.
^{b} See Fig. 6.2 for atom displacements related to above assignments.
Source: Taken in part from Meneghetti et al., *J. Chem. Phys.*, 1984, **80**, 6210.

and anions that provide conductive or superconductive properties. These are usually the Beckgaard salts of the (TMTSF)$_2$ X type (TMTSF is tetramethyltetraselenafulvalene, and X stands for an inorganic anoin). Figure 6.4 illustrates atomic displacements of tetramethyltetrathiafulvalene (TMTSF) A_g totally symmetric mode displacements and their vibrational frequency. Table 6.12A, B lists IR and Raman bands and their tentative assignments. Another group are the salts with bis(ethyleneedithio)tetrathiafulvalene (BEDT–TTF or ET), and their virbational bands, along with the deuterated d_8 forms, are tabulated in Table 6.13.

TABLE 6.13 IR and Raman Bands in BEDT−TTF and BEDT−TTF-d_8[a]

BEDT−TTF Infrared ν_i	I	BEDT−TTF Raman ν_i	I	Assignment	BEDT−TTF-d_8 Infrared ν_i	I	BEDT−TTF-d_8 Raman ν_i	I	Assignment
2958 (w)	66				2237 (w)	66			νCD_2
2958 (w)	26			νCH_2	2225 (w)				
2916 (w)	44				2169 (vw)	26			
		1552 (m)	2		2141 (vw)		1552 (m)	2	
1505 (w)	27	1511 (m)	27		1506 (w)	27	1511 (m)	27	
		1494 (s)	3				1494 (s).	3	
1420 (w)	28				1011 (m)	29			
1406 (m)	45	1409 (vw)	56	δCH_2	1002 (vw)	5	1044 (w)	46	δCD_2
		1285 (vw)	5		1041 (m)	46	1029 (w)	57	
1282 (m)	29				1030 (w)	57	794	7	
1259 (w)	46			ωCH_2	793 (w)	31			$\nu SCD_2 + \omega CD_2$
1253 (vw)	57	1256 (vw)	57		990 (vw)	47			
ωCD_2					984 (vw)	59	984 (w)	59	
1173 (w)	21	1175 (vw)	21	$t CH_2$	930 (w)	21	935 (vw)	21	$t CD_2$
1132 (sh)	38	1132 (vw)	38		806 (w)	67	805 (w)	38	
1125 (w)	67	1126 (vw)	67				1018 (w)	58	
		1016 (vw)	59	νCC	1110 (m)	28			$\nu CC + \omega CD_2$
996 (w)	30	1002 (w)	30						
987 (w)	6	990 (w)	6	ρCH_2	741 (w)	22	741 (w)	22	
938 (vw)	22								
νCD_2									

TABLE 6.13 (Continued)

BEDT–TTF					BEDT–TTF-d_8				
Infrared		Raman			Infrared		Raman		
ν_i	I	ν_i	I	Assignment	ν_i	I	ν_i	I	Assignment
917 (s)	48	919 (vw)	60		827 (w)	50			
905 (m)	31	911 (vw)	7		905 (m)	30			
890 (m)	49	888 (vw)	49		879 (m)	49			
875 (w)	50	875 (vw)	50		879 (m)	48			
860 (vw)	61	860 (vw)	61						
772 (s)	32	765 (w)	32		772 (s)	32	784	32	
687 (w)	68				693 (m)	68			
		687 (w)	39	ρCH_2	678 (w)	39	693 (w)	39	ρCD_2
653 (w)	33	653 (m)	8		634 (w)	33	635 (m)	8	
624 (w)	51	625 (w)	62		609 (w)	51	610 (m)	62	
499 (m)	34	486 (m)	9		500 (m)	34	486 (m)	9	
450	10	440 (m)	10		452 (w)	10	439 (w)	10	
							351 (w)		
390 (m)	35	348 (w)	63		388 (m)	35	339 (m)	63	
335 (m)	52	334 (w)	64		326 (w)	52	323 (w)	64	
		308 (w)	11				296 (w)	11	
278 (m)	36	272 (vw)	36		296 (vw)	36			
257 (m)	53	260 (w)	53		257 (m)	53	255 (w)	53	
		159 (s)	65				155 (s)	65	
		151 (s)	12				147 (s)	12	
96 (m)	72	127 (vw)	43		88 (w)	72	125 (w)	43	
		52 (vs)			52 (vs)				
30 (w)	54	31 (vw)	54		30 (w)	54	30 (vs)	54	

a Relative intensities: vs, very strong; s, strong; m, medium; w, weak; vw, very weak; sh, shoulder.

Source: From M. E. Kozlov, K. L. Pokhadnia, and A. A. Yurchenko, "The Assignment of Fundamental Vibrations of BEDT–TTF and BEDT–TTF-d_8," *Spectrochimica Acta*, 1987, **43A**. Pergamon Press Elmsford, New York. Reproduced with permission.

REFERENCES

1. A. L. Smith and D. R. Anderson, *Appl. Spectrosc.*, (1984), **38**, 822.
2. L. J. Bellamy, *The Infrared Spectra of Complex Molecules*, 3rd ed., Chapman and Hall, London, 1975, Chapter 20.
3. D. R. Anderson, in *Analysis of Silicones*, D. Lee Smith, Ed., Wiley-Interscience, New York, 1974, Chapter 10.
4. G. Socrates, *Infrared Characteristic Group Frequencies*, Wiley-Interscience, New York, 1980, Chapter 18.
5. M. L. Hair, *Infrared Spectroscopy in Surface Chemistry*, Marcel Dekker, New York, 1967.
6. E. P. Pluddemann, *Silane Coupling Agents*, Plenum Press, New York, 1982.
7. A. Thomas and B. Chittenden, *Spectrochim. Acta*, 1964, **20**, 467.
8. P. K. Daasch and A. L. Smith, *Anal. Chem.*, 1951, **23**, 853.
9. C. J. Bellamy and A. C. Beecher, *J. Chem. Soc.*, **1953**, 728.
10. J. A. Elvidge and A. B. P. Lever, *Proc. Chem. Soc.*, 1959, **195**, 482.
11. R. D. Joyner and M. E. Kenney, *J. Am. Chem. Soc.*, 1960, **82**, 5790.
12. R. D. Joyner and M. E. Kenney, *Inorg. Chem.*, 1962, **1**, 717.
13. B. N. Diel, T. I. Inabe, C. R. Kannewurf, J. W. Lyding, T. J. Marks, and K. F. Schoch, Jr., *J. Am. Chem. Soc.*, 1983, **105**, 1551.
14. A. B. Anderson, T. L. Gordon, and M. E. Kenney, *J. Am. Chem. Soc.*, 1985, **107**, 192.
15. B. R. Kundalkar, T. J. Marks, and K. F. Schoch, Jr., *J. Am. Chem. Soc.*, 1979, **101**, 7071.
16. C. W. Dirk, T. I. Inabe, J. W. Lyding, C. R. Kannewurf, and T. J. Marks, *J. Polym. Symp.*, 1983, **70**, 3.
17. C. W. Dirk, T. I. Inabe, T. J. Marks, and K. F. Schoch, Jr., *J. Am. Chem. Soc.*, 1983, **105**, 1539.
18. B. Exsted and M. W. Urban, *Macromolecules*, American Chemical Society, Advances in Chemistry Series, No. 23b, Washington, DC, 1993.
19. M. Hohol and M. W. Urban, *Polym. Commun.*, in press.
20. A. Bell and L. Berlin, 1927, **B60**, 1749; 1928, **B61**, 1918.
21. S. Trotter and M. Thompson, *J. Chem. Soc.*, **1946**, 181.
22. N. Sheppard, *Trans. Faraday Soc.*, 1950, **46**, 429.
23. J. G. David and H. E. Hallmann, *Spectrochem. Acta*, 1965, **21**, 841.
24. Schrieber, *Anal. Chem.*, 1949, **21**, 1168.
25. G. A. Olah, S. Prakash, and J. Sommer, *Superacids*, Wiley, New York, 1985.
26. G. A. Olah, S. Prakash, G. K. Williams, R. E. Field, and K. Wade, *Hydrocarbon Chemistry*, Wiley, New York, 1985.
27. T. Yamaguchi, T. Jin, and K. Tanabe, *J. Phys. Chem.*, 1986, **90**, 3148.
28. J. G. Bednorz and K. A. Muller, *Z. Phys. B*, 1986, **64**, 189.
29. M. K. Wu, J. R. Ashburn, C. J. Torng, P. H. Hor, R. L. Meng, L. Gao, Z. J. Huang, Y. Q. Wang, and C. W. Chu, *Phys. Rev. Lett.*, 1987, **58**, 908.
30. J. R. Ferraro and V. A. Marroni, *Appl. Spectrosc.*, 1990, **44**(3), 351.
31. J. R. Ferraro and J. M. Williams, *Appl. Spectrosc.*, 1990, **44**(2), 200.

CHAPTER 7

BONDING TO POLYMERIC SURFACES

7.1 INTRODUCTION

The physical properties of many polymers depend on many structural factors, such as percent of crystallinity, crystalline dimension, and orientation.[1] Whereas these structural factors can be modified by various treatments such as annealing or drawing, the bulk properties will also influence surface properties. Vibrational spectroscopy has played and still plays a key role in the analysis of polymers and their surfaces. Although the spectra of linear polymers may be complicated by anisotropic effects, such as degree of crystallinity, molecular weight, and a nonuniform distribution of other properties, methods for calculating frequencies in model linear systems have been proposed. In the earlier studies,[2] a linear polymer was divided into N harmonically coupled oscillators of mass M, with bending and stretching constants. In an effort to account for crystallinity effects, each mass was coupled to a Debey lattice through a lattice constant k_L corresponding to a cutoff frequency $v_L = (2k_L/M)^{1/2}$. Using this approach, vibrational frequencies were calculated for linear alkanes[3] and agreed fairly well with the GF matrix calculations for polyethyelene.[4] However, these and other studies recognized that these methods cannot account for the modes at the center of the Brillouin zone, when $k = 0$. After all, polymers are not infinite, straight chains. They can bend, twist, or fold and as a result, various defects are present, specially at the surface, and have been classified into four types: (1) chemical, (2) conformational, (3) steric, and (4) mass defects. While the chemical changes will be reflected in the appearance of new vibrational modes, other defects may be reflected in vibrational energy and band intensity changes. For reference purposes, Fig. 9.1 illustrates most often used normal modes in the analysis of polymers.

Normal mode   Atomic displacement

Symmetric CH$_2$ stretching

Antisymmetric CH$_2$ stretching

CH$_2$ bending

CH$_2$ twisting

CH$_2$ wagging

CH$_2$ rocking

CCC bending

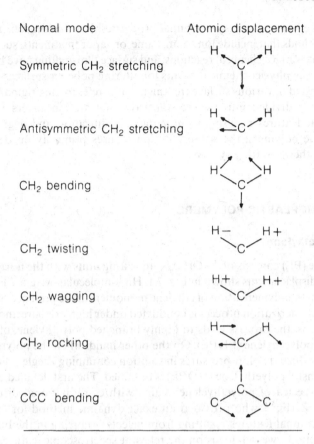

Figure 7.1   Atomic displacements in the CH$_2$ and COH groups.

It is also interesting to note that the focus of the majority of degradation studies was and still is on the carbonyl region. Of course, formation of the carbonyl groups results from polymer oxidation and is usually the final product, just like rust on an old car. Here, we will try to omit these obvious findings and rather focus on processes prior to carbonyl formation or other surface modifications.

If one adopts a concept of surface as being a boundary at which drastic property changes between the bulk and surroundings take place, surfaces of polymers are of particular interest because of mobility and flexibility of polymer chains. Of course, these properties will depend on not only the nature of polymer morphology, molecular weight, and glass tension temperature T_g but also on environmental conditions. Whereas the former properties will affect not only macroscopic surface properties, such as wetting, adhesion, or friction, overall bulk properties, such as permeability or oxidation, will also be altered. Hence, surface modifications have a significant effect on the polymer's final properties. For that reason, among less common methods that include ion beam, flame, or vapor treatments, such methods as gas/plasma, surface chemical reactions, and surface film grafting have been recognized as the key physicochemical means for altering polymer surfaces. While the reader interested in various surface treatments may refer to the original literature, this chapter is divided into two sections: the first section covers vibrational spectroscopic features resulting from surface modification and degradation in thermoplastic polymers; the second section focuses primarily on degradation processes in thermosetting systems.

7.2 THERMOPLASTIC POLYMERS

7.2.1 Polyethylene

Polyethylene (PE) consists of —$(CH_2)_n$— repeating units with the normal modes and atomic displacements shown in Fig. 7.1. High-molecular-weight PE is formed by chain growth polymerization of ethylene monomers in the presence of catalysts. When the polymerization process is conducted under high pressure and temperature conditions, the process leads to highly branched polyethylene, often called low-density polyethylene (LDPE). On the other hand, when the polymerization process is conducted at low pressures in solution containing Ziegler–Natta catalyst, high-density polyethylene (HDPE) is obtained. The first detailed analysis of vibrational spectra of polyethyelene chains with confirmational disorder was provided by Zerbi,[5,6] who provided an exact dynamic method for calculating various vibrational features resulting from defects occurring in the bulk. In the following sections, we will focus on the relevant spectroscopic features on those polymers that are of particular importance.

Approaching the problem of structure–property relationships strictly from a chemical point of view, this is the simplest polymer that exists. However, this supposedly simple polymer has generated numerous controversial articles in polymer science, although its vibrational spectrum is very simple. One rule of thumb of vibrational spectroscopy is that the fewer the bands, the more ordered the structures are expected. Let us consider the DRIFT (Chapter 3) spectrum of polyethylene shown in Fig. 7.2, tracing A. In essence, the spectrum consists of three regions of fundamental vibrations: C—H stretching ($3000 \, cm^{-1}$), (not shown) C—H bending ($1400–1500 \, cm^{-1}$), and C—H wagging ($700–800 \, cm^{-1}$) normal modes. If one were to follow the $3n-6$ (or $3n-5$ for linear molecules, where n is the number of atoms in one molecule), and calculate a number of vibrational degrees of freedom for a polyethylene molecule with a molecular weight of 100,000 that consists of over 7000 CH_2 units containing approximately $n = 21{,}000$ atoms, over 63,000 vibrational modes would be predicted. This is, of course, not observed because a high concentration of the long-range ordering of the CH_2 groups and the fewer terminal CH_3 groups are present. However, as the molecular weight decreases, the concentration of the CH_3 groups increases, and

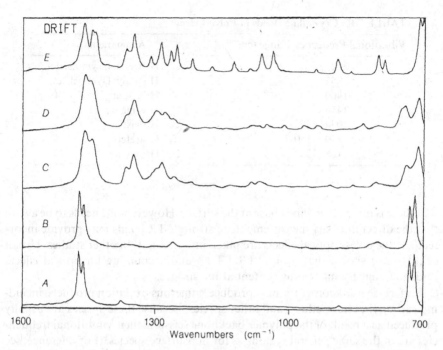

Figure 7.2 Diffuse reflectance IR-FT (DRIFT) spectra of (*A*) low-density polyethylene; (*B*) $C_{72}H_{144}$; (*C*) $C_{16}H_{32}$; (*D*) $C_{15}H_{30}$; (*E*) $C_{14}H_{28}$. (Reprinted with permission from ref. 7, copyright 1993, American Chemical Society.)

as a result of local environment changes, new symmetries impose new selection rules leading to more complex spectra. The above situation may also apply to surfaces, and for polyethylene it will result in new bands in the 1300–1400-cm^{-1} region. This is evidenced by several increasing bands in tracings *B* through *E* of Fig. 7.2,[7] which illustrates DRIFT spectra of the model compounds, $C_{72}H_{144}$, $C_{16}H_{32}$, $C_{15}H_{30}$, and $C_{14}H_{28}$, respectively.

Because most forms of polyethylene contain approximately 50% of a crystalline phase and that amount may be as high as 85%, a significant contribution to its vibrational spectra comes from crystallites. Several excellent articles on vibrational spectroscopy of polyethylene have been published, and the interested reader is referred to those monographs.[8][14] Table 7.1 lists the vibrational energies of the crystalline form.[15]

A simple cooling experiment teaches us that when the sample is quickly quenched from a melting state, the surface "freezes" first. Therefore, its amorphous "liquid state" disorder is maintained, whereas the bulk, because it cools more slowly, has enough time for crystalization. As a result, a polyethylene surface would exhibit a more amorphous character. However, Zerbi[16] have proposed that this may not necessarily be the case. Using ATR FT-IR spectroscopy, they analyzed bands sensitive to the crystallinity content at the surface and concluded

TABLE 7.1 Crystalline Bands in Polyethylene

Vibrational Frequency Range (cm^{-1})	Assignment
2925	C—H stretch (asymmetric)
2855	C—H stretch (symmetric)
1460	C—H bending
1415	C—H wagging
1050–1295	C—H twist
870–1150	C—C stretch
1168–720	C—H rocking

that there is more crystalline phase at the surface. However, one needs to be aware that the direct intensity measurements of strong ATR bands may provide incorrect results, unless optical effects are taken into account. In other studies, Urban et al.[17] have shown that using ATR FT-IR and accounting for optical effects leads to a higher amorphous content at the surface.

Surface degradation of PE may produce numerous oxidation products, including carbonyls, esters, ethers, and other species. Because these species are usually produced as a result of the polymer backbone scission, their vibrational frequencies are in the same general regions as for nonsurface species. (For reference, see vibrational band assignments for nonsurface species.) The formation of vinyl groups (relatively weak band in the 1580–1640-cm^{-1} region) usually occurs prior to the carbonyl group formation and depends on oxidizing environments.[18,19] This reinforces the introductory statement, that it is far more important to understand the chemical changes prior to carbonyl formation, in this case the presence of the C=C bonds, which may have detrimental effects on dielectric and other polymer properties.

Diffuse reflectance spectroscopy described in Chapter 3 can be used in the analysis of polymer surfaces but, as Ishida[20] pointed out, quantitative analysis must be carried out with caution. Futhermore, Ishida et al.[21] for the first time demonstrated that it is possible to obtain diffusively transmitted spectra from ultra-high-modulus polyethylene fibers coated with 12-nitrododecanoic acid. A linear relationship between the coating thickness ranging from 0 to 50 nm and the normalized intensity of the C=O stretching band in COOH was obtained. Orientation of fibers does not seem to affect quantitative analysis.

Hydrophobicity, or phobia toward water, is one of the properties that may or may not be desired. Often, however, it is important to convert the surface from hydrophobic to hydrophylic in order to improve adhesion strength, biocompatibility and other properties. The approaches used to achieve proper surface modifications of polyethylene are surface oxidations, which can be accomplished chemically or thermally, by priming, graft polymerization corona discharge, and plasma treatments. As a result of the NO and O$_2$ plasma treatments,[22] the new bands listed in Table 7.2 are generated on the surface. When oriented polyethylene was exposed to nitric acid fumes,[23] besides oxidation of the surface,

TABLE 7.2   Surface Bands Resulting from NO/plasma and O_2/plasma Treatments

New Bands (cm^{-1})		
NO/plasma	O_2/plasma	Assignment
1860, 1820	1860, 1820	$-HC\!=\!CH_2$
1790, 1760, 1730, 1710	1790, 1760, 1730, 1710	$C\!=\!O$
1660, 1590	1660, 1590	$C\!=\!C$ in diene and triene
1470		$C-N$ in amino groups
1440, 1370	1440, 1370	$C-O$ in CH_3-O-C
		and $-CH_2-O-C$
1300		
1082		$C-O$ in alcohols
890, 835		$C-H$ bend in $-CH\!=\!CH_2$

Source: Adopted from ref. 22.

TABLE 7.3   Surface Bands Resulting from Surface Treatment of Polyethylene in Nitric Acid Fumes

Band (cm^{-1})	Assignment
1710	$C\!=\!O$ in COOH
1640	$O-N$ in $-ONO_2$
1550	$N-O$ in $-NO_2$
1650	$N\!=\!O$ in $-ON\!=\!O$

Source: Adopted from ref. 23.

nitration reactions occurred. Spectroscopic features resulting from such treatments are provided in Table 7.3.

7.2.2 Polypropylene

Similar to polyethyelene, there have been numerous studies on the vibrational features of polypropylene ($-[CHCH_3]-_n$).[24][27] The surface orientation of polypropylene was also examined by ATR FT-IR.[28,29] Although a proper quantitative analysis of strong ATR bands requires that optical effects are accounted for, these studies indicated that it is possible to assess the surface crystallinity content and other surface properties of films drawn at various temperatures.[30]

7.2.3 Polystyrene

Polystyrenes are polymeric materials containing the repeating units shown below:

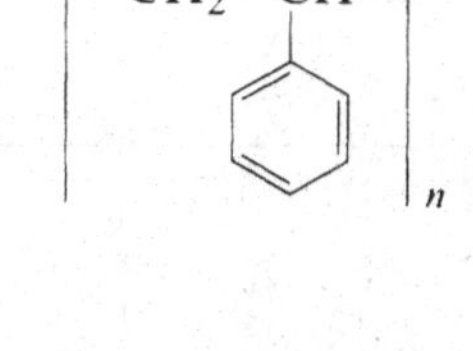

$$-\!\!\left[\!\!-CH_2-CH-\!\!\right]\!\!-_n$$

with a phenyl group attached to the CH.

The surface of polystyrene can be modified by various energy sources; one of the most popular sources is γ radiation. This surface selective modification leads to $>C=O$ (1688 cm^{-1}) and $C-O$ (1264 cm^{-1}) containing functional groups.[31] The $C-O$ band intensity at 1264 cm^{-1} was found to be proportional to a dose of γ radiation. Furthermore, weaker bands due to the formation of acid/ketone (1717 cm^{-1}), ester/aldehyde (1730 cm^{-1}), and lactone/4-memeber ring ketone (1780 cm^{-1}) groups can be found. At higher surface depths, α or β unsaturated carbonyl/acid groups can be produced. Two intense bands at 3549 and 3485 cm^{-1} can also be detected due to the $C-OH$ normal vibrations. As a matter of fact, γ radiation was used to polymerize styrene[32,33] and further crosslink[34] the polymerized polystyrene.

Exposure to ultraviolet or visible electromagnetic radiation leads to photo-degradation with a carbon–carbon bond scission.[35] When the polymer is exposed to a 365-nm wavelength in air, hydroperoxides can be formed ($-OOH$ at $\sim$ 3500–3550 cm^{-1}). On reaching a steady state, carbonyl groups due to carboxylic acids are formed as a result of chain scission. In contrast, exposures to 254-nm wave-length, with no lag between carbonyl and peroxide formation, was observed. Instead, terminal vinyl groups (995 cm^{-1}) and $C-C$ (1340 cm^{-1}) bonds were detected. No evidence of ring oxidation was detected.

Halogenation of styrene–butadiene elastomers may lead to improvement of adhesion to polyurethanes. Although various Cl-containing species including hypochloride, carbon tetrachloride, or hydrochloric acid, may be used, in recent studies[36] 1,3,5-trichloro-1,3,5-triazine-2,4,6-trione (trichloroisocyanuric acid; TIC) was used instead and such treatments appeared to provide enhanced adhe-

TABLE 7.4   IR Bands Resulting from Surface Treatment with Trichloroisocyanuric Acid (TIC)

Wavenumber (cm^{-1})		
Decreasing Bands	Increasing Bands	Assignment
3010		$C-H$ stretch
2929, 2850		$C-H$ stretch in CH_2
	1702	$C=O$ in carboxylic
1653		$C=C$ stretch aliph.
1506, 733		$C=C$ stretch aromatic
	1420, 1402	$C-H$ bend in CCl_2-CH_2
1271, 968		$C-H$ bend and $C-H$ wag
1106		$C-O$ of alkyl ether/ secondary alcohol
	1067	$C-C$ stretch of $CH_2-CCl_2-CH_2-$
918		$CH_2=CH-$ bend

Source: Adopted from ref. 36.

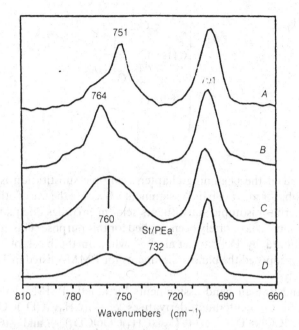

Figure 7.3 FT-IR spectra of (*A*) isotactic polystyrene; (*B*) syndiatactic polystyrene; (*C*) atactic polystyrene; (*D*) styrene/polyester crosslinked film. (Reproduced by permission from ref. 37, copyright 1991, Butterworth-Heinemann.)

sion. Infrared changes resulting from the treatments have led to the appearence of new bands and diminished intensities of the bands due to the elastomer. Table 7.4 lists these bands with tentative band assignments.

Although the focus of this section is on thermoplastic polymers, it seems appropriate to provide an example of a thermosetting system that is composed of polymerized polyester/styrene monomers because, as recent studies have shown,[37] a fraction of styrene monomer may homopolymerize. The homo-polymerization process results in atactic polystyrene, which forms physical crosslinking along with the chemical reactions between $C=C$ bonds of polyester and styrene. This is illustrated in Fig. 7.3, which compares FT-IR transmission spectra of isotactic, (tracing *A*), syndiatactic (tracing *B*), and atactic (tracing *C*) stereoisomers with the styrene/polyester crosslinked film spectrum (tracing *D*). It is quite apparent that the 760-cm^{-1} band due to the C—H bending out-of-plane normal vibrations of the phenyl groups in spectra *C* and *D* is the same in both spectra.

7.2.4 Poly(methyl methacrylate) (PMMA)

Poly(methyl methacrylate) is usually prepared by a chain growth polymerization with the repeating unit such as

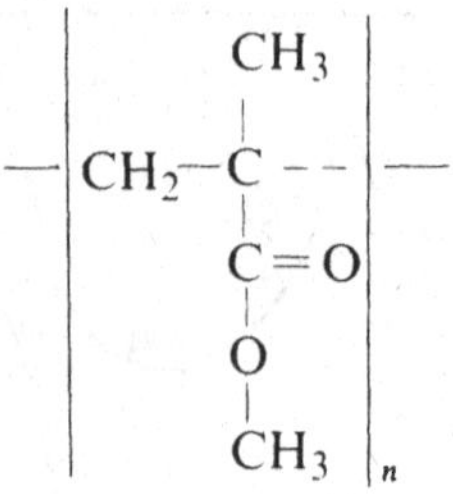

$$
-\left|\begin{matrix} & CH_3 \\ CH_2-C & -- \\ & C=O \\ & O \\ & CH_3 \end{matrix}\right|_n
$$

As indicated in the beginning chapters, isotopic substitution is a powerful molecular probe for making band assignments. Because the use of other than ^{18}O and ^2D (deuterium) isotopic substituents seldom provides adequate frequency shift, other elements have rarely been applied for this purpose. This approach was sucessfully utilized by Drilikov et al.,[38,39] who, on the basis of ^{18}O and ^2D substitutions, proposed the band assignments for PMMA listed in Table 7.5 and 7.6, respectively.

Infrared and Raman band assignments for atactic PMMA have been also assigned based on deuterated derivatives of $-(CH_2C(CD_3)COOCD_3)_n-$, $-(CD_2C(CH_3)COOCD_3)_n-$, $-(CD_2C(CH_3)COOCD_3)_n-$, and $-(CD_2C(CD_3)COOCH_3)_n-$[40]. Table 7.7 provides band assignments for specific functional groups.

Reactive ion etching of polymer surfaces often plays an important role in producing high-resolution patterns. When such an approach was applied to PMMA deposited on silver surfaces in order to achieve SERS effect,[41] etched PMMA SERS spectrum exhibited two bands at 1350 and 1580 cm^{-1}. Following the literature data, [42,43] these bands are attributed to the fused-carbon-layer ring structures resulting from oxygen ion etching. Normal, in-plane ring microcrystalline graphite vibrations shown below are believed to be responsible for these bands and correspond to E_{2g} longitudinal and transverse optical (LO, TO) modes (1375 cm^{-1}) and A_{1g} longitudinal acoustic (LA) mode (1580 cm^{-1}).

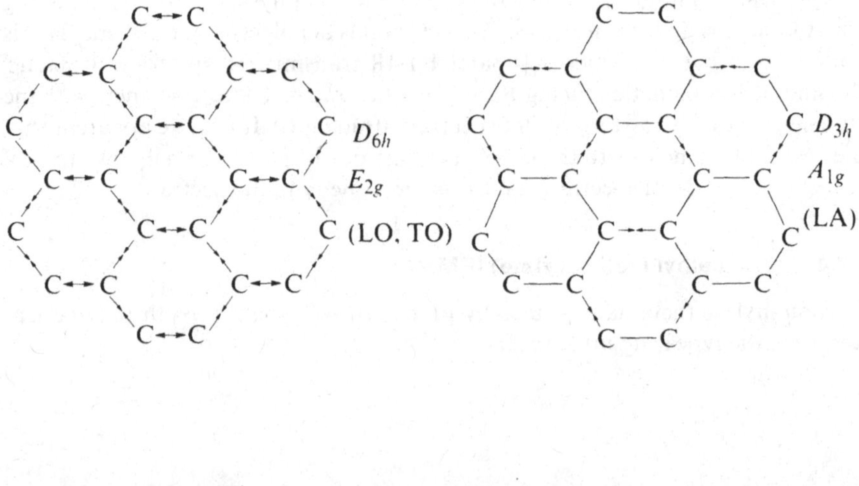

TABLE 7.5 Infrared Bands and Their Assignments in PMMA and PMMA–^{18}O^a

PMMA	PMMA-^{18}O (prepared)	PMMA-^{18}O (pure)	Assignment
cm^{-1}	cm^{-1}	cm^{-1}	
		(3015)b· c	(O)CH$_3$ stretching
2996	2996	2996	(O)CH$_3$ stretching
2952	2950	2950	(O)CH$_3$ stretching
	2930	2930	
2920		(2915)b· c	(O)CH$_3$ stretching
		(2890)b· c	(O)CH$_3$ stretching
	2852	2855	
2844	2840	2839	(O)CH$_3$ stretching
1732	1732		C$=$^{16}O stretching
	1699	1699	C$=$^{18}O stretching
	1506	1506	?
1483	1483	1483	
1465	1465	1465	(O)CH$_3$ bending
1450	1450	1450	(O)CH$_3$ bending
1438	1434	1432	(O)CH$_3$ bending
1388	1387	1387	(O)CH$_3$ bending
1368	1368	1368	
1270	1268	1268	
1242	1238	1235	Bands associated with vibra-
1194	1194		tions of the ester groups of
	1180	1180	PMMA or PMMA-^{18}O
1150	1143	1141	
1063	1060	1059	Skeletal C—C stretching
	1025	1025	
990	978	077	(O)CH$_3$ rocking coupled with
			C—O—C stretching
967	967	967	(C)CH$_3$ rocking
915	912	912	
842	842		CH$_2$ rocking (in PMMA)
	832	832	CH$_2$ rocking (in PMMA-^{18}O)
827	827		C—^{16}O—C stretching
810	810	810	
	790	790	C—^{18}O—C stretching
752	749	748	Skeletal C—C stretching
	711	711	
484	468	468	C—CO—O o.o.p. deformation

Source: Adopted from refs. 38, 39, and 40.

Oxidation demonstrated by the presence of C$=$O at 1700–1800 cm^{-1} was also detected. In other studies,[44] DRIFT was utilized to study degradation in PMMA deposited on polyethylene. A novel approach to FT-IR external reflection technique at Brewster's angle was demonstrated for PMMA.[45]

TABLE 7.6 Infrared Bands and Their Assignments in PMMA and Its Deuterated Counterpart: (A) C—H Stretching Region; (B) C—H Bending Region[a]

Type of PMMA (I, cm^{-1})					Assignment
CD_3—OCD_3	CD_2—OCD_3	CD_2—CD_3	Digital[b]	Experimental	
A. C—H Bending Vibrations Region (1350–1500 cm^{-1})[a]					
			(1380)	1370 vw	CH$_2$ twisting or wagging
1380 vw	1385 m		1385 m	1388 m	$\delta[(C)CH_3]$
		1435 s	1438 s	1438 s	$\delta[(O)CH_3]$
1446 w	1448 w	1450 sh	1447 s	1450 s	$\delta(CH_2)$, $\delta[(C)CH_3]$, and $\delta[(O)CH_3]$
		1460 sh	1460 wsh	1460 wsh	$\delta[(O)CH_3]$
1480 m	1480 m		1478 m	1438 m	$\delta[(C)CH_3]$
B. C—H Stretching Vibrations Region (2800–3100 cm^{-1})[b]					
		2845 w	2847 w	2847 w	$\nu[(O)CH_3]$
2855 vw			(2855)	(2855)	$\nu(CH_2)$
	2857 vw		(2857)	(2857)	$\nu[(C)CH_3]$
		2885 vw	(2885)	(2885)	$\nu[(O)CH_3]$
	2892 vw		(2892)	(2892)	$\nu[(C)CH_3]$
		2907 vw	(2907)	(2907)	$\nu(O)CH_3]$
2933 m	2933 w		(2933)	(2933)	$\nu_s(CH_2)$ and $\nu[(C)CH_3]$
2958 w	2958 w	2955 s	2957 vs	2957 vs	$\nu_a(CH_2)$, $\nu[(C)CH_3]$, and $\nu[(O)CH_3]$
	2998 m	2998 m	2998 s	2998 s	$\nu[(C)CH_3]$ and $\nu[(O)CH_3]$
		3025 sh	(3025)	(3025)	$\nu[(O)CH_3]$

[a] Abbreviations used are: w, weak; m, medium; s, strong; sh, shoulder; v, very.

[b] The bands given in parentheses are not observed in the spectra as a result of overlapping.

Source: Adopted from ref. 40.

TABLE 7.7   C—H Stretching and Deformation Vibrations in Deuterated PMMA

Raman Band (cm^{-1}), Atactic	Infrared Band (cm^{-1})			Region
	Atactic	Syndiatactic	Iso	
C—H Deformation and Stretching Vibrations of CH_2 Group of $PMMA-CCD_3-OCD_3$				
—	2860^c	2860^c	2860^c	Stretching vibrations
2939	2932	2934	2930	
2960	2958	2958	2958	
1450	1448	1448	1448	Bending vibrations
—	460^b	1460^b	1460^b	
1330	—	—	—	Wagging or
—	1385^a	1385^a	1385^a	twisting vibrations
—	732	732	753	Rocking vibrations
C—H Deformation and Stretching Vibrations of $C—CH_3$ Group of $PMMA-CD_2-OCD_3$				
2857^e	2855^e	2855^e	—	Stretching vibrations
2890^e	2890^e	2890^e	2890^e	
2928	2930	2928	2928	
2954	2958	2956	2958	
3004	3000	2996	3002	
1385	1385	1385	1385	Bending vibrations
1393	1395	1395	1393	
1460	1450	1450	1468^d	
1480	1483	1483	1485^d	
953	957	957	957	Rocking vibrations
967	968	968	968	
C—H Deformation and Stretching Vibrations of $—OCH_3$ Group of $PMMA-CD_2-CCD_3$				
2847	2844	2842	2845	Stretching vibrations
2880^e	2890^e	2890^e	2890^e	
2915^e	2910^e	2910^e	2910^e	
2955	2952	2952	2956	
3002	2997	2995	3004	
3031^e	3025^e	3025^e	3025^e	
1440	1436	1436	1436	Bending vibrations
1456	1456	1456	1458	
1465	1472	1472	1475	
972	972	972	973	Rocking vibration
C—H Deformation and Stretching Vibrations of $PMMA$				
1327	—^{e}	—^{e}	—^{e}	Wagging or twisting CH_2
—	1073^e	1073^e	1073^e	Wagging or twisting CH_2
1390	1388	1390	1390	CCH_3 bending
1436	1438	1438	1438	OCH_3 bending
1451	1452	1452	1452	CH_2 bending, CCH_3 bending, OCH_3 bending
1460	1465	1465	1465	CH_2 bending, CCH_3 bending, OCH_3 bending

TABLE 7.7 *(Continued)*

Raman Band (cm^{-1}), Atactic	Infrared Band (cm^{-1})			Region
	Atactic	Syndiatactic	Iso	
1487	1483	1483	1483	CCH$_3$ bending
2845	2845	2845	2842	OCH$_3$,^{e}
2893	2885	2885	2890	OCH$_3$,^{e}, CCH$_3$,^{e}
2938	2930	2930	2925	CH$_2$, CCH$_3$ stretching vibrations
2950	2953	2952	2953	CCH$_3$, OCH$_3$, CH$_2$ stretching vibrations
3002	2998	2998	3000	CCH$_3$, OCH$_3$ stretching vibrations
3031	—	—	3030	OCH$_3$,^{e}

^{a} Possible overlapping with the very strong band at 1280 cm^{-1}.
^{b} Harmonic vibration.
^{c} Probably a result of Fermi resonance.
^{d} The strong overlapping between the bands at 1468 and 1483 cm^{-1} does not allow establishment of the exact wavenumbers of their maxima.
^{e} Harmonic or combination vibrations probably due to the Fermi resonance.
Source: Adopted from refs. 39 and 40.

7.2.5 Poly(ethylene terephthalate) (PET)

Infrared and Raman spectroscopy have been used to determine orientation, percent crystallinity, and chain folding at crystalline-amorphous interfaces in PET.[46][52] The vibrational spectra of PET have been investigated frequently, and the bands at 1470, 1337, 973, and 845 cm^{-1} in IR spectra increase, whereas the bands at 1453, 1370, 1040, and 895 cm^{-1} decrease with crystallization.[53] Similarly, Raman active bands at 1096, 1000, 857, and 278 cm^{-1} increase.[54] Detailed vibra-tional analysis of PET[55] and its deuterated derivatives has also been proposed,[56] and Table 7.8 provides the observed and calculated frequencies in PET.[57] Under many circumstances, DRIFT as well as PAS (Chapter 3) can be beneficial because of a nondestructive nature of measurements. Eichinger et al.[58] utilized near-IR (1100–2400 nm) DRIFT to the analysis of PET surfaces and determined that an error involved in the crystallinity determination oscillates from 2 to 3.5%.

TABLE 7.8 Symmetry and Band Assignments for PET

Symmetry Species	v_{obs}	v_{cal}	Assignment
A_g	3085	3074	Ring C—H stretching
	3085	3072	Ring C—H stretching
	2912	2897	Methylene C—H stretching
	1730	1729	C—O stretching
	1615	1610	Ring CC stretching, ring ester CCC bending
		1574	Ring CC stretching
	1462	1450	ψ Methylene HCH bending, OCH bending
	1418	1404	ψ Methylene HCH bending, OCH bending
	1310	1310	Ring HC in plane bending
	1295	1276	Ring ester CC stretching, CO stretching
	1192	1163	Ring HC in-plane bending
	1119	1117	Ring CC stretching
	1096	1093	Ring CC stretching, CO stretching
	(1000)	997	OC stretching O—CH_2, ring torsion
	857	844	Ring CC stretching
	701	690	Ring CCC bending
	626	631	Ω ring CCC bending, ring ester CCO bending
		567	Ring ester CCO bending, ethylene glycol CCO bending
		371	Ethylene glycol CCO bending
	278	276	C=O in-plane bending, ring ester CCC bending
		152	COC bending
B_g	2968	2975	Methylene C—H stretching
		1312	CCH bending, OCH bending
		1161	OCH bending, CCH bending
	(1000)	995	Out-of-plane bend of ring C—H, ring torsion
		835	Out-of-plane bend of ring C—H
	800	801	Out-of-plane bend of C=O, ring torsion
		688	Ring torsion, out-of-plane bend of C=O
		235	Out-of-plane bend of ring ester, C—C
		76	CO torsion, ethylene glycol CC torsion
		60	CO torsion, ethylene glycol CC torsion
		35	CO torsion, ethylene glycol CC torsion
A_u (i)	2962	2965	Methylene C—H stretching
		1280	OCH bending, CCH bending
		989	Out-of-plane bend of ring C—H
	875	868	Out-of-plane bend of ring C—H
	(845)	836	CCH bending
	727	717	Out-of-plane bend of C=O, out-of-plane bend of ring C—H
		473	Ring torsion, out-of-plane bend of ring ester C—C
	430	425	Ring torsion, out-of-plane bend of ring C—H
		138	Ethylene glycol CC torsion, CO torsion
		78	Out-of-plane bend of ring ester, ring torsion, C—C
		43	Ethylene glycol CC torsion
B_u	3081	3074	Ring C—H stretching

TABLE 7.8   (*Continued*)

Symmetry Species	v_{obs}	v_{cal}	Assignment
(H.i)	3067	3072	Ring C—H stretching
	2889	2890	Methylene C—H stretching
	1727	1727	C=O stretching
	1504	1512	Ring HC in plane bending, ring CC stretching
	1475	1462	Methylene HCH bending, CCH bending
	1410	1412	Ring–ester CCC bending, ring CC stretching
	1337	1343	CCH bending, OCH bending
	1312	1295	Ring CC stretching
	1263	1257	Ring ester CC stretching, CO stretching
	1126	1133	CO stretching C—O, ring CCC bending
	1109	1114	Ring HC in-plane bending, ring CC stretching
	1018	1018	Ring CCC bending, ring HC in-plane bending
	973	968	OC stretching O—CH_2
	$(845)^b$	868	COC bending, v CO stretching
	502	506	Ring ester CC stretching, C=O in plane bending
	438	454	Ring ester CCO bending, v COC bending
	382	374	Ring ester CCO bending, v C=O in-plane bending
	145	134	Ring ester CCC bending, ethylene glycol CCO

Source: Adopted from ref. 57.

7.2.6   Poly(butylene terephthalate) (PBT)

The majority of the earlier studies focused on conformational analysis of the bulk. An infrared study of the poly(tetramethylene terephthalate) model compounds initiated by Gillette et al.[59] showed that the methylene units in the β-crystalline form are all in trans position. In contrast, detailed FT-IR and Raman studies[60] conducted on tetramethylene glycol dibenzoate, tetramethylene glycol di-*para*-chlorobenzene, and tetramethylene glycol di-*para*-nitrobenzoate in order to eluciadate possible conformations in the methyelene sequence in PBT suggested that they are in a cis configuration.

DRIFT and PA FT-IR were utilized to characterize PBT fibers and transitions on elongation.[61] The intensity changes of the 812- and 845-cm^{-1} bands on stretching reflect $\alpha-\beta$ transitions.[62]

7.2.7 Poly(aryl ether ketone) (PAEK)

PAEK contains amorphous and crystalline components, and it has been suggested that $1-2$-μm surface layers may have an access of crystallinity. As one may expect from the structure, there are essentially only a few bands of interest:[63]

Band (cm^{-1})	Assignment
810	Out-of-plane C—H deformation in phenyl rings
1146, 12,011	C—O—C stretching
1608, 1597	Aromatic C—C
1650	C=O stretch
3068	C—H stretch

Because these bands will change intensities with the changes of the crystallinity content, the use of IR and Raman spectroscopy to determine crystallinity in PAEK fibers has been reported.[64,65]

7.2.8 Poly(propylene oxide)/Poly(ethylene oxide)

Macromolecules may exist in a variety of conformational forms that range from randomly coiled chains to more spatially ordered structures. Of particular interest are the polymers that adopt helical symmetry. These symmetries result from an orderly repeated unit with internal rotational angles along the polymer backbone. In the crystalline state, polyoxyethylene exists in a helical conformation that contains seven chemical units ($-CH_2CH_2-O-$) and two turns in a backbone identity period of 19.3 Å.[66] Although this structure is well established, in the early 1970s it was a subject of debate until Koenig et al.[67] identified dihedral symmetry of this polymer on the basis that no Raman bands were detected at the same wavenumbers as the parallel IR bands. For reference, Table 7.9 provides Raman bands and their assignments for poly(ethyelene oxide).

7.2.9 Chlorinated Polymers

7.2.9.1 Poly(vinyl chloride) (PVC)

$$-\left|CH_2-\underset{\underset{Cl}{|}}{CH}\right|_n-$$

Numerous experimental and theoretical studies have been employed various force fields and force constants.[68,69]

TABLE 7.9 Bands Observed in the Raman Spectrum
of Poly(ethyelene oxide)

Observed (cm^{-1})	Assignment	Mode[b]
1484	A_1	$\delta(CH_2)_i$
1471	E_1	$\delta(CH_2)_o$
1448	E_1	$\delta(CH_2)_o$
1396	A_1	$w(CH_2)_i$, $v(C-C)$
1376	E_1	$w(CH_2)_i$, $v(C-C)$,
1286		amorphous
1283	E_1	$t(CH_2)_o$, $t(CH_2)_i$
1237	E_1	$t(CH_2)_i$, $t(CH_2)_o$
1234	A_1	$t(CH_2)_o$
1142	A_1	$v(CH_2)$
1125	A_1	$v(C-C)$, $w(CH_2)$
1073	A_1	$v(C-O)_i$, $r(CH_2)$
1065	E_1	$v(CO)_i$, $r(CH_2)$
947	E_1	$r(CH_2)_i$, $v(CO)_i$,
936		amorphous
860	A_1	$r(CH_2)_i$, $v(CO)_i$
846	E_1	$r(CH_2)_o$,
832		end-group OH,
811		amorphous
583		
537	E_1	$\delta(OCC)_o$
364	E_1	$\delta(COC)_o$, $\delta(OCC)_i$
281	A_1	$\delta(COC)_i$, $\delta(COC)$
231	A_1	$\delta(OCC)_i$, CO_i

[a] From H. Matusura, T. Miyazawa, *Butt. Chem. Soc. Jpn.*, 1968, **41**, 1798.[65]

[b] Key: v, stretching; δ, bending; w, wagging; t, torsional; r, rocking; o, out-of-plane; i, in-plane.

Source: Reproduced with permission from J. L. Koenig, *Appl. Spectrosc. Rev.*, 1971, **4**(2), 233.

The structures that develop during thermal and radiation-induced degradation of poly (vinyl chloride) have been of interest for a long time. Although several mechanisms[70,71] and structural features, including polyenes or crosslinkable species have been proposed as a result of degradation,[72][74] key feature of the PVC studies involves a correlation between the C=C frequency in the resonance Raman spectrum and the conjugated polyene length *n*. Maddams et al.[75] was the first investigator to show that the adopted C=C frequencies from polyacetylene spectrum at 1474 cm^{-1} should be 1461 cm^{-1} for the longest C=C sequence.

7.2.9.2 Poly(vinylidene chloride) (PVDC)

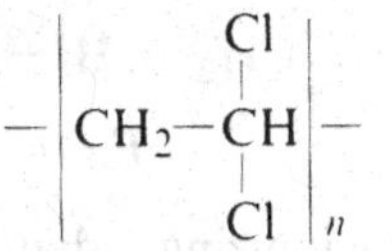

$$-\left[CH_2-\underset{\underset{Cl}{|}}{\overset{\overset{Cl}{|}}{CH}} \right]_n -$$

Detailed normal-mode analysis based on the group frequency approach was shown in the past.[76] As a result of degradation, the primary reaction involves a cleavage of the C—Cl bond; however, there are still unresolved issues related to the C—Cl band assignments. On degradation, C=O formation at around $1700\,cm^{-1}$ is usually detected, prior to which C=C should form.

7.2.10 Fluorinated Polymers

Fluorinated polymers are known to be inert, and therefore, their surface modifications cannot be achieved by commonly used methods. In fact, fairly sophisticated chemistry may be required in order to make normally unreactive surfaces functionalized for further reactions while maintaining inert interiors of the polymer. This approach has been used in numerous studies and, in essence, represents a key concept in polymer surface modifications.

7.2.10.1 Poly(chlorotrifluoroethylene) (PCTFE) Reactions of poly-(chlorotrifluoroethylene) surfaces with polystyryllithuium, (polystyrylbutadienyl)lithium, and (polystyrylthio) lithium solutions result in the formation polystyrene covalently grafted to PCTFE.[77] Besides ESCA data supporting polystyrene formation, IR bands at $3100-3000\,cm^{-1}$, (aromatic C—H stretch), the conjugated C=C bands at $1650-1500\,cm^{-1}$, C≡C at $2200-2000\,cm^{-1}$, OH at $3420\,cm^{-1}$, and the carbonyl band at $1719\,cm^{-1}$ were detected. On the basis of spectroscopic results and their analysis, the authors offered the mechanism illustrated in Fig. 7.4.

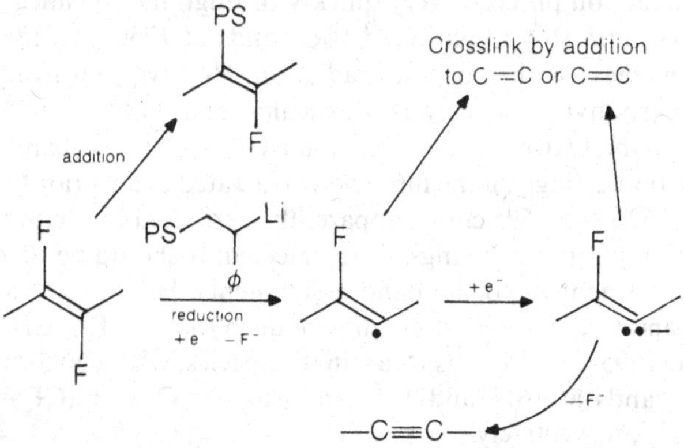

Figure 7.4 An abbreviated mechanism of C=C and C≡C formation. (From ref. 77. B. U. Kolb, P. A. Patton, and T. J. McCarthy, *Macromolecules*, 1990, **23**, 2901.)

7.2.10.2 Poly(vinyl fluoride) (PVF)

$$-\left|CH_2-\underset{\underset{F}{|}}{CH}\right|_n-$$

When PVF was irradiated using ionizing radiation in air,[78] the appearance of a strong band at 1725 cm^{-1} due to oxidation is usually detected. The same process conducted under vacuum conditions may iniciate dehydrofluorination (see section 7.2.10.3) demonstrated by the bands on the 1600-cm^{-1} region due to C=C bonds.

7.2.10.3 Poly(vinylidene fluoride) (PVDF)

$$-\left|CH_2-\overset{\overset{F}{|}}{\underset{\underset{F}{|}}{CH}}\right|_n-$$

One of the most studied surface treatment is dehydrofluorination, which is a removal of HF groups from the surface, resulting in a formation of conjugated C=C bonds. The reaction can be conducted under various conditions, but the most common one is that of phase-transfer catalysis. It appears that one of the most efficient catalysts is tetrabutylammonium bromide (TBAB) in aqueous NaOH.[79] This reaction results in the 1590-cm^{-1} band due to C=C bonds, and may shift depending on conditions. For the surface treatments at higher temperatures, often two bands at 1613 and 1717 cm^{-1} are detected.[80, 81] Both are due to —CH=CF— bonds. The 1717-cm^{-1} band is stronger when head-to-head or tail-to-tail defects are present. Often, a weaker band at 2100 cm^{-1} was found and attributed to the C—C triple bonds. Originally, dehydrofluorination was conducted using thermal degradation[82] and indicated that there are surface centers of crystallinity where bonds in allylic hydrogens are diminished. After one H is removed the reaction proceeds very quickly through the so-called zipper mechanism. On exposure to γ radiation,[83] two bands at 1750 and 1850 cm^{-1} were observed to increase with a dose of radiation. However, on irradiation under vaccum, the carbonyl presence was drastically reduced and heat treatments at 636 K produced the 1710 cm^{-1} band, attributed to —C=C— bands with various F substitutes. Interestingly, if the films were irradiated in air prior to degradation, two bands at 1620 and 1595 cm^{-1}, apparently attributed to cyclization resulting in the formation of aromatic rings, were detected. It should be pointed out that there is no agreement as to the band assignments: Hagiwara[84] attributed the 1710-cm^{-1} band to the C=C stretching mode in the —(CF=CH)— units and the 1625- and 1595-cm^{-1} bands to aromatic species, whereas Urban et al.[85, 86] assigned the bands at 1613 and 1717 cm^{-1} to C=C in —(CF=CH)— and —(CF=CF)—, respectively.

Polyvinylidene fluoride (PVDF) has been extensively studied over the last decade for its inherent piezoelectric and pyroelectric properties.[87, 88] Elimination of hydrogen fluoride from the PVDF polymeric backbone has lead to the formation of conjugated polyene structures that can also serve as synthetic, semiconductive alternatives to polyacetylene, as described previously in independent investigations by Kise and McCarthy et al.[89,90] Traditionally, dehydrofluorination of PVDF has been carried out by thermal means. More recently, however, attention has been drawn toward the dehydrofluorination of PVDF by chemical means, that is, by either alkaline or heterogeneous phase-transfer-catalyzed dehydrohalogenation.[91–93]

Although numerous papers have been published with respect to the dehydrofluorination of PVDF film, only a limited number of studies have actually been performed that involve the surface modification of the resulting dehydrofluorinated PVDF film substrate. Percec et al.[94] reported an electrophilic addition reaction of both bromine and chlorine across the conjugated double bonds of a dehydrofluorinated PVDF film, followed by the displacement of the respective halide with a fluoride anion of potassium hydrogen fluoride (KHF_2) to form a vinylidene fluoride–trifluoroethylene copolymer. In an effort to study other potential, addition reactions across the conjugated $-CH=CF-$ double bonds of a dehydrofluorinated PVDF film surface, this paper reports on an ultrasonic grafting reaction of a macrocyclic silicon phthalocyanine dichloride complex across the conjugated surface of a dehydrofluorinated PVDF film using a platinum catalyst.

As we recall from Chapter 6, with the inability of the macrocyclic phthalocyanine moiety to undergo a melting transition, coupled with its inherent insolubility in virtually all known organic solvents, the use of high-frequency sound (ultrasonic waves > 20 kHz) may be beneficial. Although the first reported use of ultrasonic energy in heterogeneous chemical reactions actually dates back to the mid-1920s,[95] succeeding publications of ultrasonic research on the surface modification of organic substrates have been sparse, especially in the field of polymeric materials.[96] Besides a two-step, dehydrofluorination–addition grafting procedure, an alternative one-step, Wurtz coupling grafting procedure for the ultrasonic surface modification of a PVDF film with $Si(Pc)Cl_2$ has been reported.[97] This axially functional phthalocyanine moiety was the primary subject of investigation because of its extreme chemical and thermal stability and its potential semiconductive nature.[98]

Attenuated total reflectance FT-IR spectral analysis revealed that the grafting of a phthalocyanine moiety across the conjugated surface of the dehydrofluorinated PVDF film required the use of a platinic acid catalyst and an axially functional phthalocyanine, such as $Si(Pc)Cl_2$. Although grafting of the silicon phthalocyanine moiety to the dehydrofluorinated PVDF film surface could readily be achieved by this method, XPS spectral analysis indicated that the grafting nature appeared to be through a C—Pt—Si linkage. Consequently, an alternative, one-step ultrasonic grafting procedure of an axially functional phthalocyanine was performed with the aid of a lithium coupling agent. Comparative UV–VIS spectra of PVDF film sonicated with and without lithium and an axially functional phthalocyanine revealed that the macrocyclic moieties were not physically adsorbed or embedded into the PVDF film substrate, but rather chemically reacted off the surface of the phthalocyanine moiety. These finding were supported by Si2p XPS spectral data, which revealed the formation of a Si—C bond between the silicon phthalocyanine moiety and the PVDF film substrate. On the basis of resulting Cls XPS spectral data, the mechanism of this Wurtz coupling procedure was suggested to occur by a nucleophilic substitution pathway.

Similarly, room-temperature ultrasonic surface modifications of PVDF with germanium phthalocyanine dichloride $(Ge(Pc)Cl_2$ in the presence of lithium lead to a formation of Ge—C bonds on the surface.[99] This is demonstrated by the appearance of the 613-cm^{-1} band due to the Ge—C stretching mode, and a parallel disappearance of the C—F and C—H modes at 764, 1182, and 1209 cm^{-1}.

7.2.10.4 Poly(tetrafluoroethylene) (Teflon)

$$—|CF_2—CF_2|_n—$$

TABLE 7.10 IR and Raman bands in PTFE

Frequency (cm^{-1})	Assignment	Phase[c]
IR Band Assignments for PTFE		
1241[c]	ν_{as} (CF$_2$)	
1213[b]	ν_s (CF$_2$)	C, A
1154[b]	ν_{as} (CF$_2$); ν (CC)	C, A
746[b]	ν (CC)	A
735[b]	ν (CC)	A
642[b]	r (CF$_2$)	C, A
552[b,c]	r (CF$_2$)	C
322[c]	ω (CF$_2$)	A
387[c]	δ (CF$_2$)	
203[c]	τ (CF$_2$)	
Raman Band Assignments for PTFE[d]		
1379 s	ν_s (CC)	
1295 w	ν (CC)	
1215 w	ν_c (CF$_2$)	
741 w	ν_s (CF$_2$)	
729 vvs	ν_s (CF$_2$)	
676 w	ω (CF$_2$)	
389 m	δ (CF$_2$)	
383 vs	δ (CF$_2$)	
308 w	τ (CF$_2$)	
291 vs	τ (CF$_2$)	
198 w	r (CF$_2$)	

[a] Symmetry: ν_s, symmetric stretching; ν_a, assymmetric stretching; r, rock; ω, wag; δ, band; τ, twist. Phase: C, crystalline; A, amorphous.
[b] Strakweather et al., *Macromolecules*, 1985, **18**, 1684.
[c] D. J. Cutler et al., *Polymer*, 1981, **22**, 726.
[d] J. L. Koenig and F. J. Boerio, *J. Chem. Phys.*, 1969, **50**, 2823.

Table 7.10 provides IR and Raman bands for PTFE. In the studies of the effect of γ radiation of PTFE[100] the formation of free radicals was followed by the formation of carbonyl groups. The bands at 3085 (OH) and 1779 (C=O) cm^{-1} were attributed to these species because the increased doses of radiation increased their intensities. Furthermore, the intensity of the doublet at 1212 and 1242cm^{-1}, attributed to symmetric and asymmetric CF$_2$ bands, indicated increased intensities for larger doses of radiation, indicating that shorter chains are produced. Other studies focused on the dichroism and dipolarization of other oxides[101] and their blends.[102] [104] The effect of Co and Zn isopropylxanthates(I) on the photodegradation of poly (2, 6-dimethyl-1, 4-phenylene oxide) (PDMPO) was also investigated,[105] and it was concluded that the photodegradation is greatly dependent on the presence of metal isopropylxanthates (Co enhances whereas Zn stabelizes), concentration, and wavelength of the light.

7.2.11   Polyurethanes

Polyurethanes are considered to be elastomers usually prepared as a result of the reaction between diisocyanate and polyol or amine functional groups. These systems typically have a segmented architecture having "hard" and "soft" portions of the polymer backbone. The hard segments are usually rigid urethanes, urea, allophanate, biuret, and aromatic groups; the soft segments consist of the flexible polyether, polyester, and polyalkyl groups from the polyols. Because of the inherent incompatibility of the hard and soft segments, many urethanes are phase-separated into hard and soft segments. As an example, let us consider the reaction between a diphenylmethane diisocyanate (MDI) and a diol of polybutadiene (PBD) such as

$$
HO-\left(\begin{array}{c} -CH_2 \\ \quad C=C \\ H \quad\quad CH_2 \end{array}\begin{array}{c} H \end{array}\right)-OH \;+\; O=C=N-\bigcirc-\overset{\overset{\displaystyle O}{\|}}{C}-\bigcirc-N=C=O
$$

(PBD)   (MDI)

$$
\cdots\cdots PBD-O-\overset{\overset{\displaystyle O}{\|}}{C}-\underset{\underset{\displaystyle H}{|}}{N}-\bigcirc
$$

As a result of the hard and soft segments, two glass transitions are observed, usually about 50–100 °C apart. The presence of the preceding structures gives certain characteristic frequencies in the IR spectra which are listed in Table 7.11. ATR FT-IR spectroscopic studies of PEU solvent-treated surfaces have shown that the following functional groups can be affected by chloroform treatment: C—N, C—O—C, free C=O, and H-bonded C=O.[106]

TABLE 7.11 Band Assignment and Characteristic Vibrational Frequencies of Polyurethanes

Band (cm^{-1})	Band Strength[a]	Phase[b]	Assignment[c]	Ref.[d]
3423	sh	UR	ν(N—H) free, 1	A
3370	sh	UR	ν(N—H) free, 2	—
3327	s	UR	ν(N—H) bonded	A
3075	w	1,2-PBD	ν(=CH$_2$) (vinyl)	B
3006	w	cis-1,4-PBD	ν(CH=CH)	B
2980	w	1,2-PBD	ν(CH=CH) (vinyl)	B
2921	vs	Any PBD	ν_a(CH$_2$)	A–C
2845	vs	Any PBD	ν_s(CH$_2$)	A–C
2266	vw	ISO	ν(—N=C=O) aged	C
2137	vw	ISO	ν(—N=C=O) fresh	C
2212	vw	ISO	ν(—N=C=O) fresh	C
1735	s	UR	ν(C=O) free	A
1712	vs	UR	ν(C=O) bonded	A
1640	w	1,2-PBD	ν(C=C) (vinyl)	B
1596	m	UR	ν(C=C) (phenyl)	A
1522	vs	UR	γ_b(N—H) + γ(C—N)	A
1448	m	(?)-PBD	δ(CH$_2$)	A, B
1438	m	(?)-PBD	δ(CH$_2$)	A, B
1413	m	1,2-PBD	δ(=CH)$_{ip}$	B
1313	m	trans-1,4-PBD	γ_b(IICC)	B
1220	vs	UR	γ_b(N—H) + ν(C—N)	A
1062	m	trans-1,4-PBD	ν_s(C—C)	A, B
1018	w	trans-1,4-PBD	γ_b(CCC)	B
994	w	(?)-PBD	No agreement	A–C
966	vs	trans-1,4-PBD	γ_w(=CH)$_{op}$	A, B
911	s	1,2-PBD	γ_w(=CH$_2$)$_{op}$	B
820	w	(?)-PBD	No agreement	A, C
770	w	UR	γ_b(COO)$_{op}$	A
735	w	cis-1,4-PBD	δ(=CH)$_{op}$	C
721	w	Any PBD	γ_r(CH$_2$)	A

[a] *Key:* vs, s, m, w,vw, sh = very strong, strong, medium, weak, very weak, and shoulder, respectively; ip, in-plane; op, out-of-plane.

[b] UR, ISO, 1,2-PBD, trans-1,4-PBD, cis-1,3-PBD = urethane, isocyanate, and the three isomers of polybutadiene: 1,2; trans-1,4 and cis-1,4.

[c] ν, a, s, γ_b, δ, ip, op, γ_w, γ_r = stretch, antisymmetric, symmetric, bend, deformation, in-plane, out-of-plane, wag, and rock, respectively.

[d] REFERENCES:

A. S. L. Zacharius and G. A. Cagle, *Proceedings of the International Electronics Packaging Society* (1983).

B. (a) R. S. Bretzlaff, S. L. Zacharius, S. L. Sandlin, and C. A. Cagle, *Conference on Electrical Insulation and Dielectric Phenomena*, Buffalo, NY, IEEE Annual Report (1985); (b) R. S. Bretzlaff and S. L. Sandlin, *International Conference on Conduction and Breakdown in Solid Dielectrics*, Erlangen, W. Germany (1986), (c) *Sadtler Infrared Spectra Atlas of Monomers and Polymers*, 1980.

C. (a) B. L. Goodlet, F. S. Edwards, and F. R. Perry, *J. Instn. Electr. Engrs.* **69**, 695 (1931); (b) S. W. Cornell and J. L. Koenig, *Macromolecules*, 1969, **2**, 540.

7.2.12 Polyimides

Liquid crystal (LC) is a special state of matter between isotropic liquid and crystalline solid. Therefore, it exhibits a combination of liquid properties, such as flow, and crystalline properties, such as rigidity. The presence of at least these two connected components is reflected in a combination of flexibility and thoughness. Perhaps the first liquid crystalline polymer that received a lot of practical attention was poly (p-phenylene terephthalamide) (PPTA), commonly known by the trade name Kevlar. The high crystallinity ranging from 65 to 80 w/w%, leads to an inactive surface, and therefore poor reactivity that is responsible for many undesirable properties such as poor adhesion. Several surface modifications of the PPTA surfaces have been attempted, including grafting,[107] substitution reactions of hydrogen of amide groups,[108] and oxygen/plasma.[109] Furthermore, a combination of plasma and priming with a coupling agent was also employed[110] and showed that the formation of $Si - CH_3$ ($1265 cm^{-1}$), $Si - O - Si$ and $Si - O - C$ (1080 and $1020 cm^{-1}$), and $Si - CH_3$ ($793 cm^{-1}$) enhances adhesion.

As indicated in the earlier chapters, a combination of vibrational spectroscopy and isotope exchange can provide valuable surface information. In the case of PPTA, such an approach was used to determine accessibility of surface NH and OH groups.[111] This can be accomplished by direct exchange of H with D on exposure of the PPTA surface to D_2O. The fraction of the N-H surface accessible to exchange by deuterium groups can be calculated from the integrated intensities of the N—H and N—D stretching bands, I_{NH} and I_{ND}, of the spectra of the deuterated surfaces using the following relationship:[112]

$$\frac{I_{ND}}{I_{NH}} = \frac{\beta_{ND}}{\beta_{NH}} \times \frac{C_{ND}}{C_{NH}} = \frac{\beta_{ND}}{\beta_{NH}} \times \frac{1 - C_{NH}}{C_{NH}}$$

TABLE 7.12   Infrared Bands in PPTA

IR Band (cm^{-1})		
Natural Abundance	Deuterated	Assignment
3431, 3326		N—H stretch
3150	3150	C—H stretch
3090	3090	C—H stretch
	2471, 2417	N—D stretch
1660	1660	C=O (amide I)
1545	1545	C=O (amide II)
1530	1530	C=O (amide II)
1436		C—N stretch
1253	1253	Amide III
1237	1237	Amide III
	991	N—D deformation

Source: Adopted from ref. 111.

where β_{ND} and β_{NH} are the absorption coefficients of the ND and NH, respectively, and C_{ND} and C_{NH} are concentrations. The latter can be obtained from deuteration of model compounds, such as dibenzanilide (Ph—NH—CO—Ph). Table 7.12 lists selected IR band assignments for deuterated PPTA modes. Surface analysis of fibers can be also accomplished using round ATR crystal on which the fibers can be wound.[113]

7.3 THERMOSETTING POLYMERS

Combined with various surface techniques, FT-IR spectroscopy shows unique promise as a nondestructive tool for characterization of chemical changes of polymer and coatings, in the bulk as well as surface, and coating–substrate interface.[114,115] Since degradation of polymers usually begins at the surface and results in the formation of characteristic compounds that can be identified in a IR spectrum, FT-IR provides a sensitive tool for monitoring degradation processes in polymers. Specific techniques including attenuated total reflectance (ATR),[116-119] photoacoustic (PA) techniques,[120-124] reflection–absorption (RA),[125-131] diffuse reflectance FT-IR (DRIFT),[132-135] photoemission,[136-139] photothermal beam deflection,[140,141] and their underlying principles were shown in Chapter 3.

The application of IR spectroscopy in oxidative degradation of polymers was reviewed by Hamid and Prichard.[142] Recent degradation studies of coatings by FT-IR techniques include the degradation of acrylated epoxy,[143,24] urethane acrylates,[144,145] poly(vinyl acetate),[146] and automotive finishes,[147,148] and the photostabilization and photodegradation of UV curable coatings.[149-153]

7.3.1 Epoxies

Because of their good adhesion and high strength, epoxy polymers are widely used in coatings and high-performance composites. Most commonly used epoxy systems are based on the diglycidyl ether of bisphenol A (DGEBA). Their applications, however, are limited by their poor exterior durability. The aromatic backbones, especially ether bonds, are not stable on exposure to UV radiation and cause the scission of polymer chains.

The degradation studies of three crosslinked epoxy resins, diglycidyl ethers of bisphenol A (DGEBA), phenolphthalein (DGEPP), and 9,9-bis (4-hydroxyphenyl) fluorene (DGEBF) by FT-IR were conducted.[153] The coatings were exposed to thermal, oxidative thermal, and photooxidative degradation conditions, and degradation processes were followed by reflection – absorption (RA) FT-IR spectroscopy. The resins were cured with trimethoxyboroxine in acetone, and casted on aluminum plates. To obtain a relative thermal stability of the functional groups in the resin, the substracted spectra were obtained by changing the subtracting parameter with cancellation of the bands at $830 \, cm^{-1}$ (p-phenylene), $2970 \, cm^{-1}$ (—CH$_3$) and $2935 \, cm^{-1}$ (—CH$_2$), respectively. In each individual spectrum, the relative stability of other functional groups to the group with band cancellation was identified by whether their corresponding bands were negative or positive. Using this approach, investigators proposed the following order of thermal stability:

Methyl, benzene ring > methylene > *p*-phenylene > ether linkage > isopropylidene

These studies also showed that the isopropylidene group can easily release the first $—CH_3$ group, while retaining the second methyl group until the latter stages of degradation. A possible reaction of the paraphenylene group related to the Claisen rearrangement was proposed and is shown below.

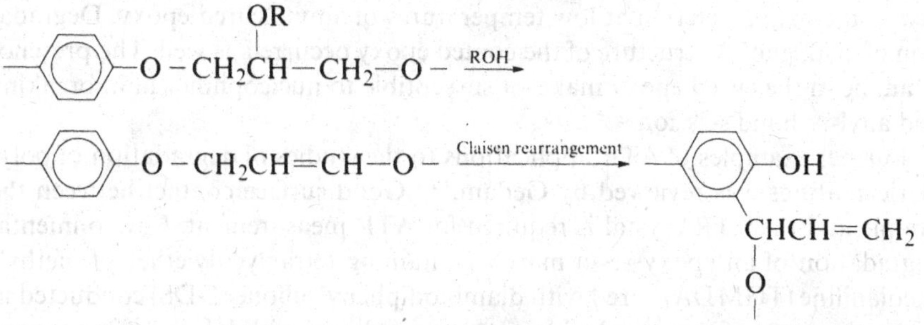

Thermooxidative degradation gives a faster deterioration rate than does thermal degradation. A formation of mixed compounds with carbonyl groups, including acid anhydride (1808, 1745 cm^{-1}), perester (1765 cm^{-1}), ester (1745 cm^{-1}), and carboxylic acid (1715 cm^{-1}), was suggested, which supports the classical free-radical autocatalytic mechanisms.[154] Ultraviolet (UV) exposure studies on the above three epoxy resins indicate that photooxidative degradation is also an autocatalyzed oxidative process. Compared with that of thermal degradation, the formation of hydroxyl groups is much faster and the acid anhydride formation is slower during the photooxidative degradation. Norrish type I and II reactions of ketone, as shown below, were considered as the primary photochemical reaction in the oxidative photodegradation mechanism, causing the formation of carboxylic acid, peracid, and perester.

Norrish type I:

$$R—\overset{\overset{\displaystyle O}{\|}}{C}—CH_2R \xrightarrow{h\nu} R—\overset{\overset{\displaystyle O}{\|}}{C}\cdot + \cdot CH_2R$$

Norrish type II:

$$RO—CH_2—\underset{\underset{\displaystyle O}{\|}}{C}—CH_2—O—CH_2—\underset{\underset{\displaystyle O—}{|}}{CH}—CH_2—OR \xrightarrow{h\nu}$$

$$RO—CH_2—\overset{\overset{\displaystyle O}{\|}}{C}—CH_3 + O{=}CH—\overset{\overset{\displaystyle O—}{|}}{CH}—CH_2—OR$$

The influence of substrate (aluminum or gold) and curing conditions (in N_2 or in air) on the degradation of epoxy coatings using the RA mode also can be examined.[154] The system under examination was the DGEBA-based epoxy cured with trimethoxyboroxine (TMB). When cured under nitrogen atmosphere, DGEBA casted on either aluminum or gold did not exhibit aldehyde or perester formation. When it was cured in air, alumina did appear to accelerate the oxidation, whereas gold did not.

As indicated in Chapter 3, RA FT-IR can be utilized for studying degradation of amine-cured epoxy on cold-rolled steel after exposure to warm, humid environments.[155] These particular studies indicated that dehydration was identified as the major reaction at low temperatures of amine-cured epoxy. Degradation of bisphenol-A structure of the coated epoxy occurred as well. The presence of amine in the cured epoxy makes it susceptible to nucleophilic chain breaking and allyl-N bond scission.

Further examples of ATR applications to the studies of degradation of polymeric coatings was reviewed by Gedam.[118] Good surface contact between the sample and the ATR crystal is required for ATR measurement. Environmental degradation of an epoxy resin matrix containing tetraglycidylether of methylenedianiline (TGMDA) cured with diaminodiphenylsulfone (DDS) conducted in a "weatherometer" saturated with water and radiated with UV light[45] indicated that the weakest linkages are the tertiary amide bonds and the secondary hydroxyl bonds. Using Ge and KRS-5 crystals as the internal reflectance elements at various angles of incidence, the surface depth profiling was performed. As expected, the degradation begins at the top surface and continues to develop toward greater depths. Parallel exposure to ambient room conditions resulted in the similar spectral changes, indicating that the same degradation reactions occur in the absence of high humidity and UV radiation, but the degradation process is much slower.

The studies of photooxidation rates of DGEBA systems crosslinked by aliphatic (diethylene triamine), cycloaliphatic (isophorone diamine), or heterocyclic (amino ethyl piperazine) amines were compared[155-158] and indicated that the amide yields are directly related to the initial α-amino methylene concentration. For the epoxy system of DGEBA crosslinked by diaminodiphenyl methane (DDM) with different stoichiometries,[81] the incompletely crosslinked sample (75% DDM) was least stable, whereas the sample containing excess of DDM was most stable. This observation indicated that photooxidative reactions on the sites of epoxy backbones are dominant over those on amine backbone, as shown below:

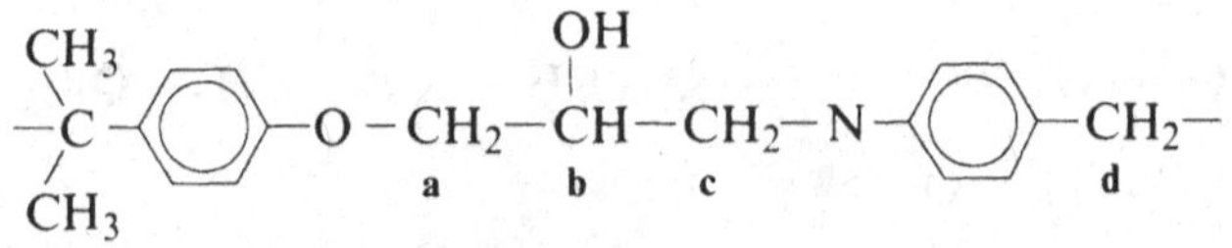

The authors suggested that the presence of residual epoxide groups is not completely crosslinked coatings (such as DDM 75%) can also be a source of the carbonyl group formation. On the other hand, enhanced molecular mobility in incompletly crosslinked systems favors intermolecular propagation of radical chains, which could explain their relatively high rates of oxidation.

The influence of structural features on the photooxidation of networks in three diepoxides containing isopropylidene or methylene bisphenol structures cross-linked by three diamines of the dianiline type containing methylene (DDM), ether (DDE), and sulfone (DDS) bridges, respectively, was estabalished.[83] The following order of stability was found: sulfone (DDS) > methylene (DDM) > ether (DDE). The amide formation resulting from degradation depends on initial amine concentration and nitrogen atom electron density. The discoloration of the sample is due to the oxidative attack of the methylene bridge in DDM systems, allowing the formation of highly conjugated structures.

Although the processes leading to exudation of small molecules in polymer matrices have been addressed for latex films,[159-162] and melamine/polyester coatings,[163] only limited studies were reported on the surface exudation in epoxy coatings. For example, Foister[164] considered the dynamic surface property changes due to amine migration during crosslinking of bisphenol-A-based epoxy/primary amine system. These studies indicated that the primary amine crosslinking agents can act as surfactants, which decrease surface tension during the initial stages of curing. The surface activity of the amine is reduced, however, when crosslinking reactions go to completion.

On the basis of these few isolated examples and the crosslink density considerations,[165] migration of small molecules will be affected by the crosslink density changes as well as other factors, including internal stresses in a polymer and its surface tension. Furthermore, if a polymer network is not completely crosslinked, it is likely that uncrosslinked molecules can migrate and, under suitable thermo-

TABLE 7.13 Tentative Infrared Band Assignments for DGEHBPA Polymer

Wavenumber (cm^{-1})	Band Assignment
3493	OH stretching
3053	CH stretch in oxirane group
2938	ν_{as} of CH_2 and CH_3
2962	ν_s of CH_2 and CH_3
1465	δ_{as} of CH_3 and δ of CH_2
1449	δ of CH_2 in cyclohexane
1386, 1369	CH_3 bending in $-C(CH_3)_2$
1310	CH_2 nonplanar wagging
1254, 910, 847	Vibrations of oxirane group
1159	ν_{C-O} in CH_2-O-
1099	ν_{C-O} in ether of cyclohexane
974	Cyclohexane ring vibration
910	Symmetric $C-O-C$ stretching
760	CH_2 wagging in cyclohexane
702	CH_2 wagging

dynamic conditions, exude to the surface. Although this process can be inhibited by the presence of rigid nonpermeable aggreggates in the form of crystalline or semicrystalline segments[166] or inorganic particles such as pigments, it is usually unavoidable due to degradation of the network. In fact, degradation studies[167] of polyurethane topcoats over epoxy/polyamide primers, in which dynamic mechanical analysis (DMA) and photoacoustic (PA) FT-IR was utilized, indicated that inorganic pigment phase enhances rigidity of the crosslinked network, but decreases crosslinking density. The latter was attributed to the presence of pigment particles, which may inhibit crosslinking reactions. It is therefore expected that the mobility of smaller segments in pigmented coatings will be altered, and it is also expected to change due to network degradation.

In the presence of the light-absorbing aromatic groups, conventional bisphenol-A-based epoxy coatings have poor gloss retention when exposed to UV radiation. Such systems can be composed of a diglycidolether hydrogenated bisphenol-A (DGEHBPA)-based epoxy resin and crosslinked with a polyamide based on dimer fatty acid and triethylenetetramine. The structures of these components are

and the tentative IR band assignments are given in Tables 7.13 and 7.14.[168]

TABLE 7.14 Tentative Infrared Band Assignments for Polyamide

Wavenumber (cm^{-1})	Band Assignment
3287	ν_s NH
2926	ν_{as} CH$_2$
2855	ν_s CH$_2$
1663	Amide I band of secondary amide
1612	Amide I band of tertiary amide and N—C=N of stretching of imidazoline
1558	Amide II band (N—H bending and C—N stretching)
1458, 1366	CH$_2$, CH$_3$ bending.
1258	Amide III band (C—N stretching and N—H bending)
1182, 1138	C—N bending.
1011	Cyclohexene ring
948	δ_{CH} in CH=CH of aliphatic chain.
916	δ_{CH} in CH=CH of cyclohexene.
723	CH$_2$ wagging in (CH$_2$)$_n$ ($n \geqslant 4$)

7.3.2 Polyamides

Although the behavior of aliphatic polyamides on exposure to UV radiation has received much attention, most of the studies were concerned with vacuum photolysis,[169] photooxidation at wavelengths shorter than 290nm, and auto-oxidation and photooxidation of low-molecular-weight model compounds.[170,171] In recent years, some of the work on the photooxidation of aliphatic polyamides at long and short UV wavelengths has been published.[172 174] It is generally accepted that under longer-wavelength radiation (>290nm), degradation of polyamides involves abstraction of a hydrogen atom in α position to nitrogen, leading to formation of free radicals, which combine with O_2 and form carbonyl groups. When shorter-wavenumber (<290nm) radiation is used, a direct photoscission of the C—N bond on amide groups occurs. A study of polyundecanamide films using FT-IR and UV spectroscopy[89] showed that at longer wavelengths, intermediate photoproducts, such as hydroperoxides, can accumulate until a fairly high photostationary concentration is reached. Photochemical decomposition of hydroperoxides into imide groups and N-1-hydroxypolyundecanamide was observed. The imide groups can be either hydrolyzed or photolyzed into acidic and amide groups.

The mechanisms of photooxidation of polyamides 6, 11, and 12 at short[175] and long wavelengths[176] indicated that in the case of short wavelength (254 nm), the formation of amines ($3400\,\text{cm}^{-1}$) was independent of the nature of the polyamide, whereas the rate of formation of the acidic groups ($1710\,\text{cm}^{-1}$) was higher in polyamides 11 and 12 than in polyamide 6. Aldehyde groups, detected as a 290-nm band in UV spectra, appear as the key products of photooxidation. However, the IR spectra of the photooxidized samples did identify the presence of imide groups, which are the key products of long-wavelength oxidation of polyamides. In the case of long-wavelength radiation, $N-1$-hydroxylated groups ($3400\,\text{cm}^{-1}$) were minor photoproducts, and all hydroperoxides converted into imide groups ($1735\,\text{cm}^{-1}$).

When nylon is exposed to light, photo- and photooxidative degradation may occur.[177] The gradual changes of films under UV radiation, and the gases evolved from the photolysis of samples were detected by FT-IR spectroscopy. On UV radiation in N_2 atmosphere, the formation of primary amide groups was demonstrated by increases in the bands at 3400, 3220, 1690, and $1647\,\text{cm}^{-1}$, and the formation of double bonds ($1645\,\text{cm}^{-1}$) was accompanied by the consumption of methylene groups. Carbon monoxide was evolved during degradation, which suggested that the amide bonds break through Norrish type I reactions. In the case of photooxidative degradation, as expected, the degradation products included hydroperoxide ($3400\,\text{cm}^{-1}$), double-bond ($1620\,\text{cm}^{-1}$), and mixed carbonyl groups ($1800-1600\,\text{cm}^{-1}$). It was found that the amorphous region ($1148\,\text{cm}^{-1}$) degraded faster than did the crystalline region ($935\,\text{cm}^{-1}$). The photodegradation of nylons containing purposefully inserted carbonyl groups was conducted, and acetone was found among the evolved gases. It was concluded that the Norrish type II mechanisms, in addition to the type I, were mainly involved in degradation processes.

7.3.3 Melamines

Melamine crosslinked coatings have been widely used, especially in the auto-mobile industry. When melamine crosslinked coatings are exposed to acid and aqueous environments, the crosslinks between the polyol and the melamine can be hydrolyzed and melamine methylol self-condensation reaction can occur.[178] Since high-solids coatings usually contain low-molecular-weight binders, more crosslinks are present in the networks. Consequently, they are more sensitive to chemical degradation.

Owing to the optical properties of pigments, pigmented coatings cannot be studied by conventional transmission IR techniques. Furthermore, ATR measurements may cause difficulty due to poor surface contact. English et al.[135] took advantage of diffuse reflection FT-IR technique (DRIFT) to characterize curing and degradation of pigmented melamine formaldehyde crosslinked acrylic copolymer coatings as a function of depth. The coating panels were exposed in Florida for 24 months. Depth profiling of coatings was accomplished by abrading the panel with a cloth containing silicon carbide in a water medium. After a short period of abrasion, the removed material was collected, filtered, and dried in a vacuum to remove residual water, and the diffuse reflectance IR was recorded. This procedure was repeated 10–30 times on each panel until the substrate was barely visible. By monitoring of IR bands at $816 \, cm^{-1}$ (melamine triazine ring deformation), 813 and $870 \, cm^{-1}$ (methoxymethyl deformation), $3570 \, cm^{-1}$ (OH stretching), and $3350 \, cm^{-1}$ (NH stretching), the degradation and chemical components at different surface depth were followed. The possible crosslinking and hydrolytic degradation reactions are shown in Scheme 7.1.

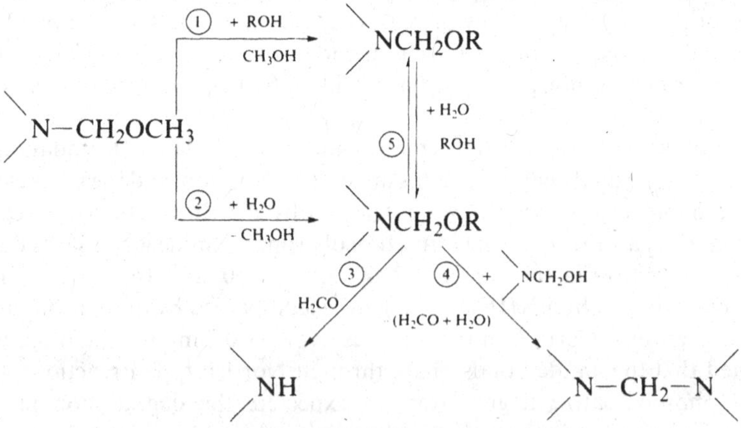

Scheme 7.1

The primary crosslinking reaction (reaction 1, above) goes to completion and no other reaction takes place to a significant extent when the cure is completed at $120 \, °C$ for 30 min. It was also suggested that degradation under realistic exposure conditions is facilitated near the surface of $2–5 \, \mu m$, and there is no evidence for any significant amount of melamine–melamine self-condensation at any time during cure or exposure. A plausible pathway for degradation in this melamine formaldehyde crosslinked acrylic coating system was suggested. At the early stages of degradation, initially the hydrophobic surface of the coating is oxidized either photolytically or thermally, thus increasing the concentration and mobility of water and the mobility of the residual acid. The residual acid, in turn, catalyzes the hydrolysis of $> NCH_2OCH_3$ to form free amine. At the very surface of the pigmented coating, where there is insufficient pigment to provide an effective solar screen, ROH and $> NCH_2OH$ resulting from the hydrolysis of $> NCH_2OR$ in reaction 5 (above) are consumed quickly through photooxidation and reaction 3.

Hirayama and Urban[163] examined the distribution of melamine in melamine/polyester coatings using ATR FT-IR spectroscopy. The 1545/1725- and 815/725-cm^{-1} bands in the IR spectra of coatings were used to determine the relative ratio of melamine to polyester. The 1545- and 815-cm^{-1} bands are attributed to the in-plane and out-of-plane deformation modes of triazine ring in melamine resin, and the 1725- and 725 cm^{-1} bands are due to the $C{=}O$ stretching and out-of-phase bending modes of the aromatic $C{-}H$ groups in polyester resin, respectively. The extent of melamine reaction was indicated by the $CH_3{-}O{-}$ deformation band of melamine at 910 cm^{-1} and for a low-OH-value melamine/polyester system, an excess of melamine was present near the film–air interface. As suggested in these as well as other experiments, the main source of this non-uniform distribution was the volatility of amine, which results in increased acidity near the film–air interface and causes fast self-condensation of melamine. The distribution of melamine may also depend on other factors such as curing conditions, concentration of acid catalyst, and film thickness. When melamine was replaced by functionalized phthalocyanine as a crosslinker, the properties of polyester polyols can be further improved.[179]

7.3.4 Polyester/Polyurethanes

Photoacoustic (PA) FT-IR spectroscopy was used to monitor degradation processes in coatings by following the chemical changes that occur during exposure of a polyester/urethane clearcoat to a variety of different weathering conditions including Florida exposure, xenon arc, carbon arc, and UV-B fluorescent bulbs.[123] From the difference spectra before and after exposure, slight formation of carbonyl groups, chain scission of urethane crosslinks, and no loss of isophthalate groups from the polyester were detected after exposure in Florida. However, exposure to artificial light sources causes loss of isophthalate groups, rapid loss of amide II band intensity, and massive oxidation. These experiments indicated that none of the conventional accelerated tests based on artificial light sources can replace natural exposure, because of the excessive amounts of lights with wavelengths shorter than 290 nm in the artificial sources, which causes much faster degradation.

In another study,[124] the photodegradation of single- and two-component polyester urethane resin films was studied using PA and ATR modes. Since a deeper surface depth of penetration can be obtained from PA, the combination of PA and ATR can provide broader depth-profiling characteristics. A decrease of the urethane–amide band at $1540 \, cm^{-1}$, a broadening and decrease of bands in the $1200–1000$-cm^{-1} region, and a broadening of the carbonyl stretching band at about $1700 \, cm^{-1}$ suggest the oxidation of urethane linkages in the polymer. The disappearance of the strong sharp band at $731 \, cm^{-1}$ indicates the elimination of the isophthalate group.

7.3.5 Alkyds

Curing processes of nonpigmented and pigmented alkyd coatings was studied by Urban et al.[180 181] using PA FT-IR. Table 7.15 provides tentative IR band assignments for alkyds. It should be kept in mind, however, that considering

TABLE 7.15 Tentative Infrared Band Assignments for Alkyds

Band (cm^{-1})	Assignment
3520	O—H stretching (free)
3430	O—H stretching (bonded OH)
(3390)[a]	
3050	C—H stretching in C=C, aromatic
(3010)	C—H stretching in aliphatic —HC=CH—
2950	C—H stretching in CH_3 (solvents)
2930	C—H stretching in —CH_2 (symmetric and unsymmetric)
2885	
(1767)	C=O stretch, ketones and ester
1737	C=Ostretch in ester
(1705)	C=O stretch in carboxylic acids
(1694)	C=C stretch
	C=O stretch in a, b-unsaturated ketones and esters
1600	C=C stretching in aromatics
1585	
1465	C—H bending in CH_2, CH_3
1392	C—H deformation (long-chain acids with CH_2 adjacent to C=O)
1380	C—O stretching in esters (solvent)
	C—H deformation in CH_3
(1304)	
(1296)	C—O stretching
1290	C—O stretching and OH deformation in esters
1252	C—O stretching in esters (solvent)
1178	C—O stretching, OH in-plane deformation
1125	C—O stretching in R—OH
1074	C—O stretching on R—OH and esters
740	CH_2 rocking in —(CH_2)— sequences
710	

[a] Paretheses indicate bands not present in initial spectrum.

Source: Adopted from refs. 180 and 181.

complexity of the alkyd composition, there may be other bands present. During the curing of alkyd to form three-dimensional networks, formation of low-molecular-weight species and degradation products was also found, demonstrated by an increase in the intensities of the 1709-, 1694-, and 1737- cm^{-1} bands due to carbonyl stretching vibrations in unsaturated ketones and esters, and saturated ketones, respectively. The authors suggested the hydroperoxide resulting from the molecular oxygen attack on binder backbone may lead to the formation of both crosslinks and degradation products, as shown below.[182]

Formation of crosslinked network:

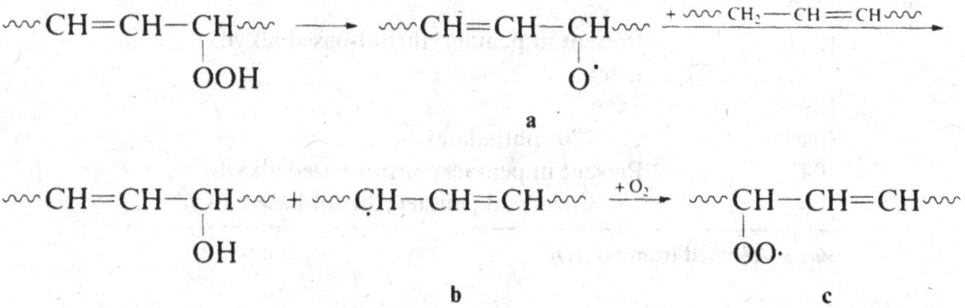

$$\text{\raise2pt{\sim\!\sim}} CH\!=\!CH\!-\!CH\text{\sim\!\sim} \quad\longrightarrow\quad \text{\sim\!\sim} CH\!=\!CH\!-\!CH\text{\sim\!\sim} \xrightarrow{+ \text{\sim\!\sim} CH_2 - CH = CH\text{\sim\!\sim}}$$

with OOH group below first, O· below second

a

$$\text{\sim\!\sim} CH\!=\!CH\!-\!CH\text{\sim\!\sim} + \text{\sim\!\sim} \overset{.}{C}H\!-\!CH\!=\!CH\text{\sim\!\sim} \xrightarrow{+O_2} \text{\sim\!\sim} CH\!-\!CH\!=\!CH\text{\sim\!\sim}$$

OH below first; OO· below product

b   **c**

a + a or b; b + b or c ———→ crosslinked network

Formation of degradation products:

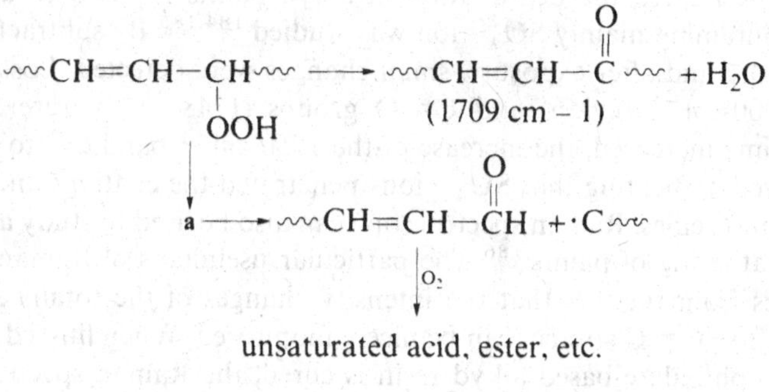

$$\text{\sim\!\sim} CH\!=\!CH\!-\!CH\text{\sim\!\sim} \longrightarrow \text{\sim\!\sim} CH\!=\!CH\!-\!\overset{\overset{\displaystyle O}{\|}}{C}\text{\sim\!\sim} + H_2O$$

OOH (below); (1709 cm – 1)

$$a \longrightarrow \text{\sim\!\sim} CH\!=\!CH\!-\!\overset{\overset{\displaystyle O}{\|}}{CH} + \cdot C\text{\sim\!\sim}$$

$\downarrow O_2$

unsaturated acid, ester, etc.

These studies indicated that crosslinking processes of alkyds are also responsible for their degradation, which begins almost at the same time when crosslinking is completed. Such degradation products as unsaturated acids, ketones, aldehydes, alcohols, and water were detected. In the case of pigmented alkyds, hydrogen bonding between pigment particles and binder molecules is responsible for a mobility of pigment particles before crosslinking is completed.

Degradation of alkyd paints pigmented with uncoated titanium dioxide using PA FT-IR[183] indicated that, on the basis of the predegradation – postdegradation difference spectrum of the sample, the loss of aliphatic species during degradation was indicated by the negative bands at 2855 and 2930 cm^{-1} as a result of the C—H stretching vibrations. The formation of species containing hydroperoxide and hydroxyl groups was found, as demonstrated by the positive

TABLE 7.16 Linseed Oil/Pentaerythritol/*o*-Phthalate-based
Alkyd Resin Raman Spectral Features

Band (cm^{-1})	Assignment
3070	Aromatic C—H stretch (*o*-phthalate)
2920, 2855	Aliphatic C—H stretch
1725	C=O stretch (ester)
1640	C=C ("linseed portion")
1450	CH$_2$ deformation ("linseed portion")
1395	Present in pentaerythritol-based alkyd
1300	Long-chain CH$_2$ twist and rock-based alkyd
1260	Ester
1217	Present in pentaerythritol-based alkyd
1155	Ester
1085	Ester
1040	C—C (*o*-phthalate)
945	Present in pentaerythritol-based alkyd
850	C—C skeletal pentaerythritol-based alkyd

Source: Adopted from ref. 186.

intensity of the 3500-cm^{-1} band. In the 500–700-cm^{-1} region, which is attributed to the titanium dioxide pigment, the negative band intensity suggested loss of pigment from the sample surface.

Using PA FT-IR, the degradation of alkyd paints exposed to a acid-rain solution containing mainly SO$_4^{2-}$ ion was studied.[184,185] By subtraction of the spectra before and after exposure, small changes were detected. Loss of C—H groups (3000-cm^{-1} region) and C=O groups (1748 cm^{-1}) were found. As exposure time increased, the increase of the 1050-cm^{-1} band due to SO$_4^{2-}$ ion was observed, indicating that SO$_4^{2-}$ ions penetrated the coatings and attacked the binder molecules. Raman spectroscopy can also be used to study artificial or natural weathering of paints.[186] The particular usefulness of Raman spectroscopy comes from the fact that the intensity changes of the totally symmetric modes, such as C=C stretch, can be nicely monitored. When linseed oil/pentaerythritol/*o*-phthalate-based alkyd resin is cured, the Raman spectral features listed in Table 7.16 are expected.

7.3.6 Polycarbonates

Webb et al.[187] have shown an experimental approach in studying photochemical, thermal, and environmental degradation of polymer films on metals using the RA FT-IR technique. Polycarbonate films were coated onto gold and aluminum substrates, and an apparatus especially designed for that purpose was employed to control the experimental parameters. Based on the subtraction spectra of polycarbonate films before and after the sample was irradiated in a solar simulator at different time in air, the chemical changes were monitored upon degradation. The increases in absorbance in the 3500–3250 cm^{-1} region and at

1690, 1620, 1590, 1490, 1340, and 1260 cm^{-1} suggest the accumulation of the ester groups coming from the photo–Fries rearrangement of carbonate groups as shown below:

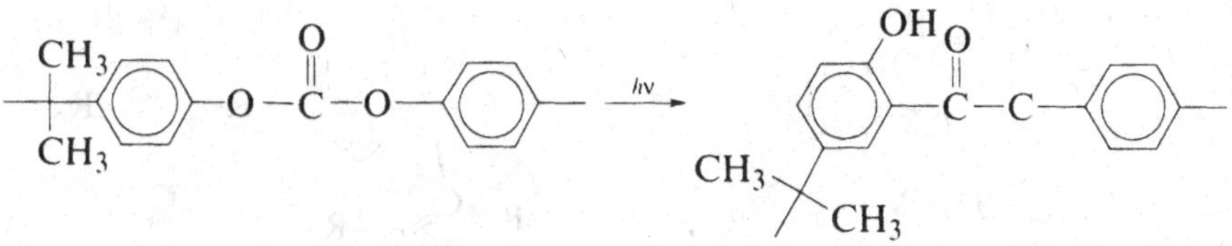

CH$_3$ — O — C(=O) — O — — $\xrightarrow{h\nu}$ — OH, O, C—C — ; CH$_3$, CH$_3$

The loss of polycarbonate functionalities is shown by bands decreases at 1790, 1775, 1510, 1270, and 1084 cm^{-1} and other frequencies characteristic of polycarbonate. During degradation, the changes of chemical structures are usually accompanied by the changes of surface morphology. Therefore, FT-IR microscopy becomes a potentially powerful tool for the study of coating defects and the degradation processes that may potentially be initiated at the defect.[188,189] The IR microscope basically consists of an optical component, which allows the sample to be viewed in a conventional manner, and optics for recording the IR spectrum. It is designed to allow visible light and IR light to be coincident so that the IR spectrum of the exact area that is viewed is recorded.

The FT-IR microscope can be used to study the waterspotting defects in coatings. The coatings exposed in Florida for 6 months may exhibit a small, discolored, and considerably softer appearance at these locations. A small amount of coating in both defective and nondefective areas were collected, and a FT-IR spectrum of each sample was recorded in, transmission mode.[190] The accumulation of ester groups on the defective spots was found.

7.4 CONJUGATED POLYMERS

Although a dramatic color transformation of diacetylenes has been known for many years, it was not until Wegner[191–194] proposed that as a result of solid-state polymerization, individual molecular segments become connected to each other and change optical properties. This process is schematically depicted in Scheme 7.2.

A quick look at the structures produced during this reaction indicates that perhaps Raman spectroscopy would provide more detailed information as the C=C and C≡C bonds exhibit strong polarizibility tensor changes. The presence of C=C (carbon–carbon double) and C≡C (carbon–carbon triple) bonds will be reflected in the two strong Raman bands at around 1400–1650 and 2000–2250 cm^{-1}, respectively. While further band intensity enhancement can be achieved in the resonance Raman effect, the carbon–carbon triple-and double-bond vibrational energies will be affected by the presence of pendant groups on a polymer backbone. Following Sandman et al.[195,196], Table 717 clearly illustrates the effect of electron-donating or -withdrawing groups on the vibrational

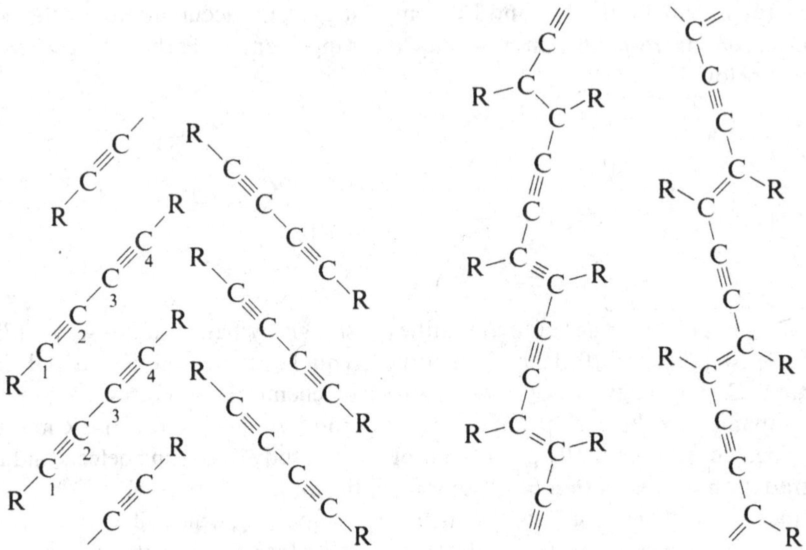

Monomer single crystal   Polymer single crystal

Scheme 7.2

TABLE 7.17   Shifts Detected in Resonance Raman Effect for Carbon–Carbon Double and Triple Bonds in the Presence of Various Substituents

PDA Structure	Band Assignment (cm^{-1})	
	v_1	v_2
CH_2R $\{C-C{\equiv}C-C\}_n$ RCH_2	2086	1485
$R = -O-SO_2-\bigcirc-CH_3$ 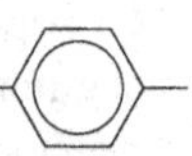	2115	1527
$R = -(CH_2)_3-O-\overset{O}{\overset{\|}{C}}-NH-C_2H_5$	2080	1456
$R = $ 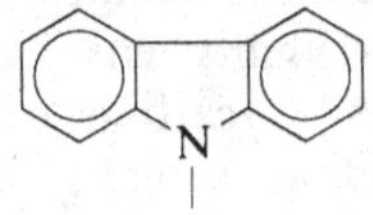 carbazole	2089	1494, 1472 1456, 1426
$R = (\bigcirc)_2 N-$ 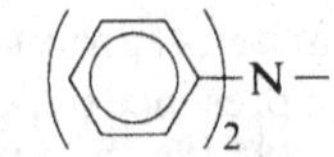	2111	1472

Source: Adopted from ref. 196.

energies of these bonds. Raman studies initiated by Butera and Lando[197] on poly (1,8-nonadiyne) ($[[(CH_2)_5C=CC=C]_x—$) have opened new insights into the sol- vatochromic and thermochromic behavior. The analysis of the resonance Raman band intensities at 1463 ($C=C$ double-bond stretch) and 2107 cm^{-1} ($C=C$ triple-bond stretch) indicated that chromic transition in these species is an order–disorder transition type, similar to that detected for PDA.

The conjugated polymers may also be prepared using other routes. The preparation, such as that shown below for Durham $(CH)_x$ (**A**) and poly (phenylenevinylene) (PPV) (**B**) involves a two-stage synthesis using a soluble precursor intermediate.[198]

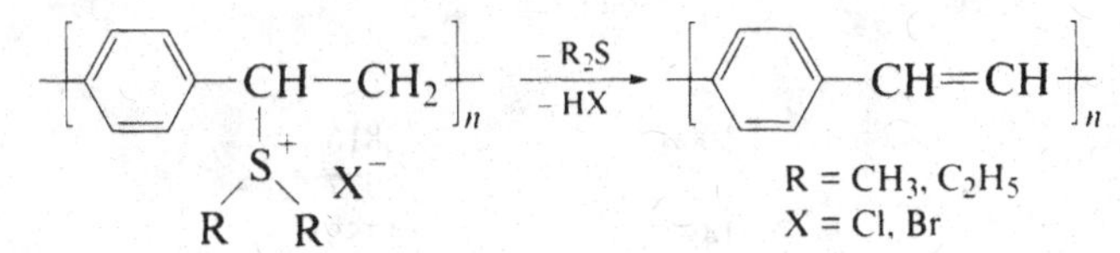

R = CH₃, C₂H₅
X = Cl, Br

A

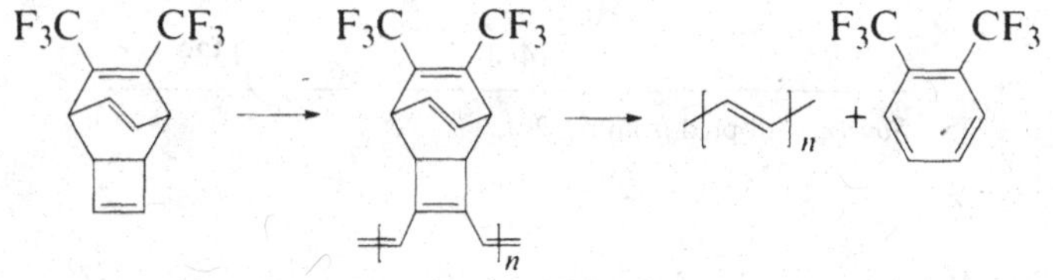

B

Table 7.18 lists Raman bands and their assignments for PPV.[199] Raman bands for hydrogenated $[(C_6H_4)_x]$ and deuterated $[(C_6D_4)_x]$ poly (*para*-phenylene) (PPP) are given in Table 7.19.[200] The IR spectra of electropolymerized poly (*para*-phenylene) films indicated that the bands at 800, 760, and 690 cm^{-1} can be used to determine the average length to approximately 14 of the number of phenyl rings in the polymer chain.[201,202]

The UV- and thermal-induced cis–trans isomerization is a well-established reaction.[203] *cis*-Polyacetylene [*cis*-$(CH)_x$—] and its ^{13}C-substituted [*cis*-$(^{13}CH)_x$—] and deuterated [*cis*-$(CD)_x$—] analogs and their trans *conterparts* were studied using IR and Raman spectroscopy.[204,205] Table 7.20 provides observed bands and their assignments for *trans*-polyacetylenes. Although an excellent theoretical treatment by Zerbi et al.[206,207] allowed us to advance many band assignments, in

TABLE 7.18   Band Assignments for Poly(phenylenevinylene)

$v(cm^{-1})$	Assignment
837	Phenylene ring CH out-of-plane bend
965	*trans*-Vinylene CH out-of-plane bend
1423, 1519, 1594	C—C ring stretch
3024	*trans*-Vinylene CH stretch

Source: Adopted from ref. 199.

TABLE 7.19 Raman Bands and Their Assignments in $(C_6H_4)_x$ and $(C_6D_4)_x$

Symmetry	Bands Assignments (cm^{-1})	
	$(C_6H_4)_x$	$(C_6D_4)_x$
A_k	—	—
	1220	894
	1276	1262
	1598	1568
B_{1k}	—	—
	—	—
	—	—
	—	—
B_{3u}	1000	816
	—	977
	1482	1355
B_{2u}	—	—
	—	—
	1401	1329

Source: Adopted from ref. 201.

TABLE 7.20 Infrared (A), Raman (B), and Calculated (C), Vibrational Energies for trans-Polyacetylenes

A Assignments of IR Absorption Bands in trans-Polyacetylenes[a]

$(CH)_x$	$(^{13}CH)_x$	$(CD)_x$	Assignment
		1994	$v_2 + v_8$ (1201 + 797 = 1998)
		1952	$v_2 + v_7$ (1201 + 746 = 1947)
		1827	$2v_4$ (2 × 916 = 1832)
1910–1720	1883–1708	1550–1348	$v_7 + v_8$
		1710	$C\!=\!O$ stretch
	1632	1654	$C\!=\!O$ stretch
1630–1468	1573–1447	(1550–1348)	v_1^b
		1458	C—H bend of CH_2 and CH_3 (catalyst)
1425	1412		CH_2 scissors ($=\!CH_2$?)
1380	(1380)	(1380)	CH_3 symmetric deformation (catalyst)
		1298	$v_7 + v_8$ (623 + 676 = 1299)
1292	1257	1262, 1201, 1080	v_2
		1048	Unassigned
1250	1231	963, 916	v_4
1170, 1082	1166, 1068	861, (916)	v_3
1012	1010	746, 623	v_7
892, 740	872, 720	797, 676	v_8

TABLE 7.20 (*Continued*)

B Assignments of Raman Bands in *trans*-Polyacetylenes[a]

(CH)$_x$	(^{13}CH)$_x$	(CD)$_x$	Assignment
		1704	$2v_3$
1457	1434	1347	v_1
		1299	$v_7 + v_8$
1294	1254	1201	v_2
1066, 1174	1045, 1169	852	v_3
1013	1011	748	v_7
884	855	816	v_8
610	590	545	v_5

C Calculated Vibrational Frequencies (in cm^{-1}) of Infinite
trans-Polyene Chains at $\delta = 0$ and π

	(C$_2$H$_2$)$_n$		(^{13}C$_2$H$_2$)$_n$		(C$_2$D$_2$)$_n$	
Branch[a]	$\delta = 0$	$\delta = \pi$	$\delta = 0$	$\delta = \pi$	$\delta = 0$	$\delta = \pi$
$v_1(A_g)$	1470	1579	1447	1521	1352	1544
$v_2(A_g)$	1302	1324	1256	1302	1191	1259
$v_3(A_g)$	1079	1175	1054	1147	854	920
$v_4(B_u)$	1235	1306	1231	1297	908	948
$v_5(B_u)$	0	524	0	505	0	496
$v_6(B_u)$	0	479	0	461	0	433
$v_7(A_u)$	1014	938	1011	937	744	604
$v_8(B_g)$	887	744	866	735	795	676
$v_9(A_u)$	0	387	0	372	0	379
$v_{10}(B_g)$	0	223	0	217	0	194

[a] Symmetry of the vibration with $\delta = 0$ is given in parentheses. A_g and B_g modes are active in the Raman spectrum; and A_u and B_u, in the IR Spectrum.
[b] Overlapped by $v_7 + v_8$ bands in (CD)$_x$.
Source: Adopted from refs. 204 and 205.

certain circumstances the resonance Raman enhanced bands shift. Apparently, certain phonon Raman modes at 1065 and 1460 cm^{-1} shift to 1124 and 1500 cm^{-1} when the excitation line is changed from 647.1 to 457.9 nm. To complicate the issue even further, sample preparation has a significant effect on the behavior of these bands.

7.5 POLYDIMETHYLSILOXANES

In Chapter 6 the band assignments for silicon-containing polymers were given. Here, surface modifications of these elastomers will be treated. It appears that inertness of these networks combined with modifiable elastic properties of surfaces make these elastomers good candidates for biomedical applications. Figure

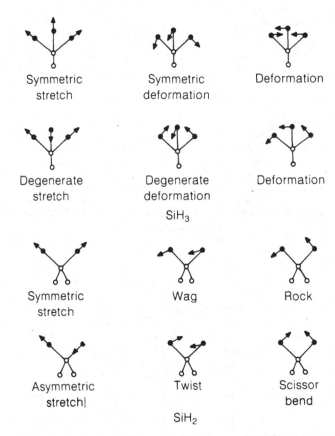

Symmetric stretch	Symmetric deformation	Deformation
Degenerate stretch	Degenerate deformation	Deformation

SiH₃

Symmetric stretch	Wag	Rock
Asymmetric stretch|	Twist	Scissor bend

SiH₂

Figure 7.5 Normal vibrations of SiH$_2$ and SiH$_3$ groups. (Adopted from C. X. Cui, M. Kertesz, *Macromolecules*, 1992, **25**, 1103.)

7.5 illustrates normal vibrational modes of the SiH$_3$ and SiH$_2$ groups. While useful network theories have provided further understanding of the network formation, Gaboury and Urban[208] developed a microwave plasma treatment in an inert-gas environment that results in Si—H formation. As shown in Fig. 7.6, using ATR FT-IR spectroscopy two distinct bands at 2158 and 912 cm^{-1} due to Si—H stretching and bending modes were detected. Furthermore, the effect of chlorine-containing species on surface modifications was addressed.[209] Using ammonia/plasma treatments, amide groups can be formed, but the presence of a chloro-functional molecule, such as an initiator or residual freon from the cleaning process, will lead to the formation of water-soluble ammonium chloride. The

Figure 7.7 ATR FT-IR spectras of microwave plasma treatments of PDMS in the presence of acrylamide, proppoinamide, maleic anhydride, and succinic anhydride.

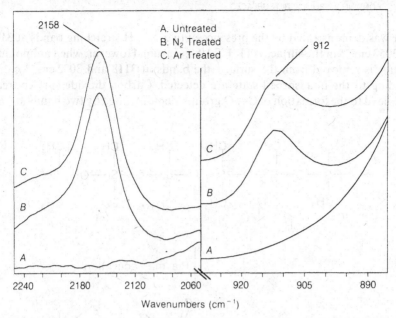

Figure 7.6 ATR FT-IR spectra of gas/microwave plasma-treated PDMS surfaces: (*A*) untreated PDMS; (*B*) N$_2$/plasma-treated; (*C*) Ar/plasma-treated. (Reproduce with permission from ref. 208, copyright 1991, Butterworth-Heinemann.)

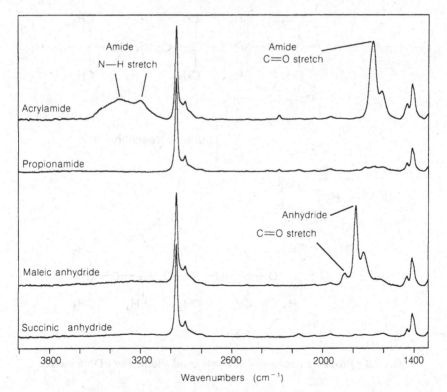

latter was demonstrated by the presence of two N—H stretching bands at 3150 and 3053 cm^{-1} in the surface ATR FT-IR spectrum. However, when ammonium chloride is removed from the surface, the bands at 3118 and 3032 cm^{-1} corresponding to the unadsorbed state are detected. Carbon dioxide surface treatments lead to the formation of C=O groups, demonstrated by two bands at 1725

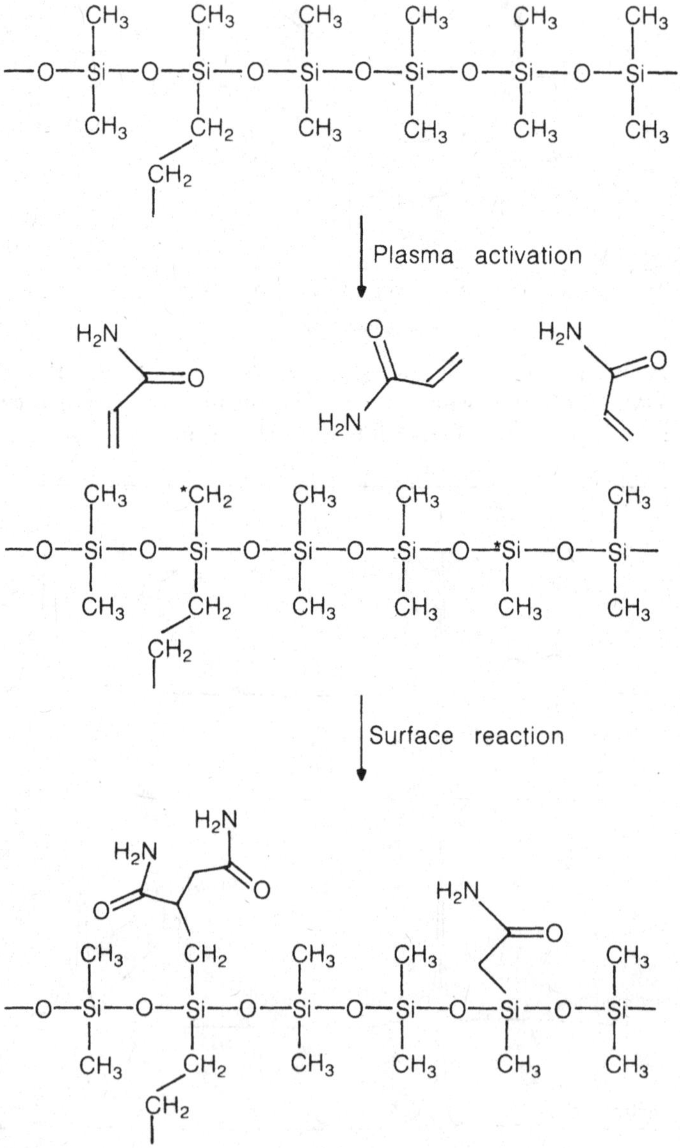

Figure 7.8 Proposed mechanism of C=C bond addition to PDMS surface.

and $1700\,\text{cm}^{-1}$.[210] In addition, two bands at 1675 and $1596\,\text{cm}^{-1}$ can be detected, all attributed to the $-O-CO-C(CH_3)=CH_2$ species attached to the silicon backbone.

Modification of the surfaces of silicone elastomers can be beneficial if the surface contains groups that can be further reacted. Again, localized microwave plasma treatments in the presence of solid-phase monomers containing $C=C$ bonds have shown to be successful.[211] Figure 7.7 illustrates ATR spectra of acrylamide- and propionamide-treated PDMS surfaces. It is quite apparent that the presence of a $C=C$ bond in acrylamide results in a stable chemical bonding to the surface. In contrast, propionamide shows no reactions. A similar scenario is observed for maleic anhydride, which reacts with the surface, and succinic anhydride, which does not. On the basis of these spectroscopic data, the surface plasma reactions shown in Fig. 7.8 were proposed.

REFERENCES

1. R. T. Samuels, *Structured Polymer Properties*, Wiley-Interscience, New York, 1974.
2. M. G. Broadhurst and F. I. Mopsik, *J. Chem. Phys.*, 1971, **55**, 3708.
3. M. G. Broadhurst and F. I. Mopsik, *J. Chem. Phys.*, 1971, **54**, 4239.
4. T. Kitagawa and T. Miyazawa, *Rept. Prog. Polym. Phys. Jpn.*, 1968, **11**, 219.
5. G. Zerbi, L. Piseri, and F. Cabassi, *Mol. Phys.*, 1971, **22**, 241.
6. G. Zerbi, *Pure Appl. Phys.*, 1971, **26**, 499.
7. M. W. Urban, in *Structure–Property Relations in Polymers; Spectroscopy and Performance*, M. W. Urban and C. D. Crower, Eds., American Chemical Society, Washington, DC 1993.
8. T. Shimanouchi, *Colloquium on NMR* (Aachen, 1971), Vol. 4, Springer-Verlag, Berlin, p. 287.
9. P. J. Hendra and M. A. Cudby, *Spectrochim. Acta*, 1972, **13**, 104.
10. J. L. Koenig, *Appl. Spectrosc. Rev.*, 1971, **4**, 233.
11. R. G. Snyder and J. H. Schachtschneider, *Spectrochim. Acta*, 1963, **19**, 85.
12. S. Krimm and J. Jakes, *Macromolecules*, 1971, **4**, 605.
13. W. Glenz and A. Peterlin, *Makromol. Chem.*, 1971, **150**, 163.
14. F. J. Boerio, *Diss. Abstr.* (*B*), 1971, **32**, 1766.
15. N. Sheppard, *Adv. Spectrosc.*, 1959, **1**, 288.
16. G. Zerbi, *Polymer*, 1990.
17. J. B. Houng, J. W. Hong, and M. W. Urban, *Polymer*, 1992, **33**, 5173.
18. G. C. Furnleaux and K. J. Ledbury, *Polym. Degrad. Stab.*, 1980, **3**, 431.
19. S. H. Hamid and W. H. Prichard, *Polym. Plast. Techn. Eng.*, 1988, **27**(3), 303.
20. H. Ishida, *Rubber Chem. Techn.*, 1987, **60**, 497.
21. A. Taboudoucht and H. Ishida, *Appl. Spectrosc.*, 1989, **43**(6), 1016.
22. N. Inagaki, S. Tasaka, H. Kawai, and Y. Kimura, *J. Adhes. Sci. Techn.*, 1990, **4**(2), 99.
23. R. Trilla, J. M. Perena, and J. G. Fatou, *Polym. J.*, 1983, **15**(11), 803.

24. G. Zerbi, *J. Chem. Phys.*, 1968, **49**, 3840.

25. T. Yasukawa, N. Kimura, N. Watanabe, and Y. Yamada, *J. Chem. Phys.*, 1971, **55** 983.

26. D. K. Roylance and K. L. DeVries, *J. Polym. Sci.*, Part B, *Polym. Lett.*, 1971, **9**, 443.

27. J. L. Koenig and P. D. Vasko, *Macromolecules*, 1970, **3**, 597.

28. F. M. Mirabella, *J. Polym. Sci., Polym. Phys. Ed.*, 1984, **22**, 1283.

29. F. M. Mirabella, *J. Polym. Sci., Polym. Phys. Ed.*, 1984, **22**, 1293.

30. A. E. Tshemel, V. I. Vettegren, and A. F. Zolotarev, *J. Macromol. Sci. Phys.*, 1982, **B21**, 243.

31. E. C. Onyiriuka, L. S. Hersh, and W. Hertl, *Appl. Spectrosc.*, 1990, **44**(5), 808.

32. J. Westlake and R. Y. Huang, *J. Polym. Sci.*, Part A-1, 1972, **10**, 2149.

33. P. W. Moore, F. W. Ayscough, and J. G. Clouston, *J. Polym. Sci., Polym. Chem. Ed.*, 1977, **15**, 1291.

34. A. Charlesby, *Radiat. Phys. Chem.*, 1977, **10**, 177.

35. E. P. Otocka, *J. Appl. Polym. Sci.*, 1983, **28**, 3227.

36. J. C. Fernandez-Garcia, A. C. Orgiles-Barcelo, and J. M. Martin-Martinez, *J. Adhes. Sci. Techn.*, 1991, **5**(12), 1065.

37. M. W. Urban, S. R. Gaboury, and T. Provder, *Polym. Commun.*, 1991, **32**(6), 171.

38. S. Drilikov and J. L. Koenig, *Appl. Spectrosc.*, 1979, **33** (6), 551.

39. S. Drilikov and J. L. Koenig, *Appl. Spectrosc.*, 1979, **33** (6), 555.

40. B. Schneider, J. Stokr, P. Schmidt, M. Mihailov, S. Drilikov, and N. Peeva, *Polymer*, 1979, **20**, 705.

41. S. Asada and K. Mori, *J. Vac. Sci. Technol.*, 1986, **B4** (1), 318.

42. C. A. Murray and D. L. Allara, *J. Chem. Phys.*, 1980, **76**, 1290.

43. H. G. Craighead, R. E. Howard, J. E. Sweeney, and D. M. Tennant, *J. Vac. Sci. Techn.*, 1982, **20**, 316.

44. J. A. J. Janson and W. E. Haas, *Polym. Commun.*, 1988, **29**, 77.

45. Y. Ishino and H. Ishida *Appl. Spectrosc.*, 1992 **46**(3), 504.

46. I. M. Ward, *Chem. Ind.* (London), 1957, 905.

47. G. Farrow and I. M. Ward, *Polymer*, 1960, **1**, 330.

48. P. G. Schmidt, *J. Polym. Sci., A*, 1963, **1**, 1271.

49. A. J. Melveger, *J. Polym. Sci., A-2*, 1972, **10**, 317.

50. L. D' Esposito and J. L. Koenig, *J. Polym. Sci., Polym. Phys. Ed.*, 1976, **14**, 1731.

51. S.-B. Lin and J. L. Koenig, *J. Polym. Sci., Polym. Phys. Ed.*, 1982, **20**, 2277.

52. L. J. Fina and J. L. Koenig, *Macromolecules*, 1984, **17**, 2572.

53. A. Miyake, *J. Polym. Sci.*, 1959, **38**, 479.

54. G. E. McGraw, *Polymer Characterization: Interdisciplinary Approaches*, C. D. Craver, Ed., Plenum Press, New York, 1971.

55. C. Y. Liang and S. Krimm, *J. Mol. Spectrosc.*, 1959, **3**, 554.

56. F. J. Boerio and S. K. Bahl, *J. Polym. Sci., Polym. Phys. Ed.*, 1976, **14**, 1029.

57. S. K. Bahl, D. D. Cornell, and F. J. Boerio, *Polym. Lett. Ed.*, 1974, **12**, 13.

58. C. E. Miller and B. E. Eichinger, *Appl. Spectrosc.*, 1990, **44**(3), 496.

59. P. C. Gillette, S. D. Drilikov, L. L. Koenig, and J. B. Lando, *Polymer*, 1982, **23**, 1759.

60. A. Palmer, S. Poulin-Dandurand, and F. Brisse, *Can. J. Chem.*, 1985, **63**, 3079.

61. M. W. Urban, E. G. Chatzi, B. C. Perry, and J. L. Koenig, *Appl. Spectrosc.*, 1986, **40**, 1103.

62. I. M. Ward and M. A. Wilding, *Polymer*, **18**, 327.

63. J. K. Agbenyega, G. Ellis, P. J. Hendra, W. F. Maddams, C. P. Assingham, and H. A. Willis, *Spectrochim. Acta*, 1990, **46A**, 197.

64. J. D. Louden, *Polymer*, 1986, **27**, 82.

65. N. J. Everall, J. Lumsdon, and J. M. Chalmers, *Spectrochim. Acta*, 1991, **47A**, 1305.

66. R. F. Schaufele, *J. Opt. Soc. Amr.*, 1967, **57**, 105.

67. J. L. Koenig, *Appl. Spectrosc. Rev.*, 1971, **4**(2), 233.

68. S. Krim, *Pure Appl. Chem.*, 1968, **16**, 369.

69. W. H. Moore and S. Krimm, *Makromol. Chem. Suppl.*, 1975, **461**, 1.

70. A. A. Miller, *J. Phys. Chem.*, 1959, **63**, 1755.

71. G. J. Atchison, *J. Polym. Sci.*, 1961, **49**, 385.

72. D. L. Gerrard and W. F. Maddams, *Macromolecules*, 1975, **8**, 54.

73. D. L. Gerrard and W. F. Maddams, *Macromolecules*, 1975, **47**, 111.

74. D. L. Gerrard and W. F. Maddams, *Macromolecules*, 1977, **10**, 1221.

75. A. Baruya, D. L. Gerrard, and W. F. Maddams, *Macromolecules*, 1983, **16**, 578.

76. M. S. Wu, P. C. Painter, and M. M. Coleman, *J. Polym. Sci., Phys. Ed.*, 1980, **18**, 95.

77. B. U. Kolb, P. A. Patton, and T. J. McCarthy, *Macromolecules*, 1990, **23**, 366.

78. M. Hagiwara, G. Ellinghorst, and D. Hummel, *Makromol. Chem.*, 1977, **178**, 2901.

79. H. Kise and H. Ogata, *J. Polym. Sci., Chem. Ed.*, 1983, **21**, 3443.

80. K. J. Kuhn, B. Hahn, V. Percec, and M. W. Urban, *Appl. Spectrosc.*, 1987, **41**, 843.

81. M. W. Urban and E. M. Salazar-Rojas, *Macromolecules*, 1988, **21**, 372.

82. A. J. Lovinger and D. J. Freed, *Macromolecules*, 1980, **13**, 989.

83. M. Hagiwara, G. Ellinghorst, and D. Hummel, *Makromol. Chem.*, 1977, **178**, 2901.

84. M. Hagiwara, G. Ellinghorst, and D. Hummel, *Makromol. Chem.*, 1977, **178**, 2913.

85. K. J. Kuhn, B. Hahn, V. Percec, and M. W. Urban, *Appl. Spectrosc.*, 1987, **41**, 843.

86. M. W. Urban and E. M. Salazar-Rojas, *Macromolecules*, 1988, **21**, 372.

87. M. G. Broadhurst and G. T. Davis, *Ferroelectronics*, 1984, **60**, 3–13.

88. D. Geiss, R. Danz, A. Janke, and W. Künstler, *IEEE Trans. Elect. Insul.*, 1987, **EI-22**(3), 347–351.

89. H. Kise and H. Ogata, *J. Polym. Sci., Polym. Chem. Ed.*, 1983, **23**, 3443–3451.

90. A. Dias and T. J. McCarthy, *J. Polym. Sci., Polym. Chem. Ed.*, 1985, **23**, 1057–1061.

91. E. Barth, *Kunststoffe*, 1982, **72**, 300.

92. K. J. Kuhn, B. Hahn, V. Percec, and M. W. Urban, *Appl. Spectrosc.*, 1987, **31**, 843–847.

93. M. W. Urban and E. M. Salazar-Rojas, *Macromolecules*, 1988, **21**, 372–378.

94. B. Hahn and V. Percec, *J. Polym. Sci.*, 1987, **25**, 783–804.

95. A. L. Loomis and W. T. Richards, *J. Am. Chem. Soc.*, 1927, **49**, 3086.

96. K. Matyjaszewski, F. Yenca and Y. L. Chen, *Polym. Prepr., Am. Chem. Soc., Div. Polym. Chem.* 1987, **28**(2), 222–223.

97. B. Exsted and M. W. Urban, *J. Organomet. Polymn. Chem.*, 1993, **3** (2), 105.

98. F. L. Moser and A. L. Thomas, *The Phthalocyanines*, Vol. 1, CRC Press, Boca Raton, Fla., 1983.

99. M. D. Hohol and M. W. Urban, *Polymer*, 1993, **34** (9), 1995.

100. H. Vanni and J. F. Rabolt, *J. Polym. Sci., Phys. Ed.*, 1980, **18**, 587.

101. J. J. Lindberg and G. Lundstrom, *Acta Chemica Scand.*, 1978, **A32**, 367.

102. D. Lefebvre, B. Jesse, and L. Monnerie, *Polymer*, 1981, **22**, 1616.

103. A. K. Mukherji, M. A. Butler, and D. L. Evans, *J. Appl. Polym. Sci.*, 1980, **25**, 1145.

104. C. Bouton, V. Arrondel, V. Rey, P. Sergot, J. L. Manguin, B. Jasse, and L. Monnerie, *Polymer*, 1989, **30**, 1414.

105. R. Chandra, B. P. Singh, S. Singh, and S. P. Hendra, *Polymer*, 1981, **22**, 523.

106. T. G. Vargo, D. J. Hook, J. A. Gardella, M. A. Eberhardt, A. E. Meyer, and R. E. Baier, *Appl. Spectrosc.*, **45**, 1991, 448.

107. M. Takayanagi, *Pure Appl. Chem.*, 1983, **55**, 819.

108. M. Takayanagi and T. Katayose, *J. Polym. Sci., Polym. Chem. Ed.*, 1981, **19**, 1133.

109. M. R. Wertheimer and H. P. Schreiber, *J. Appl. Polym. Sci.*, **26**, 2987.

110. N. Inagaki, S. Tasaka, and K. Kawai, *J. Adhes. Sci. Techn.*, 1992, **6**, 279.

111. E. G. Chatzi, M. W. Urban, H. Ishida, and J. L. Koenig, *Polymer*, 1986, **27**, 1850.

112. J. K. Smith, W. J. Kitchen, and D. B. Mutton, *J. Polym. Sci.*, 1961, **C2**, 499.

113. A. M. Tiefenthaler and M. W. Urban, *Appl. Spectrosc.*, 1988, **42**, 163.

114. M. W. Urban and J. L. Koenig, in *Application of FT-IR Spectroscopy*, Vol. 18, J. R. Durig, Ed., Elsevier Science, New York, 1990, Chapter 3.

115. M. W. Urban, *Prog. Org. Coat.*, 1989, **16**, 321.

116. N. J. Harrick, *Internal Reflection Spectroscopy*, 2nd ed., Harrick Scientific, Ossining, 1979.

117. J. Fahrenfort, *Spectrochim. Acta*, 1961, **17**, 698.

118. P. H. Gedam and P. S. Sampathkumaran, *Prog.Org. Coat.*, 1983, **11**, 313.

119. G. A. Luoma and R. D. Rowland, *J. Appl. Polym. Sci.*, 1986, **32**, 5777.

120. A. Rosencwaig and A. Gersho, *Science*, 1975, **190**, 556.

121. A. Rosencwaig and A. Gersho, *J. Appl. Phys.*, 1976, **47**, 64.

122. A. Rosencwaig, *Photoacoustics and Photoacoustic Spectroscopy*, Wiley, New York, 1980.

123. D. R. Bauer, M. C. Paputa Peck, and R. O. Carter, III, *J. Coat. Technol.*, 1987, **59** (755), 103.

124. R. O. Carter, III, M. C. Paputa Peck, and D. R. Bauer, *Polym. Deg. Stab.*, 1989, **23**, 121.

125. S. A. Francis and A. H. Ellison, *J. Opt. Soc. Am.*, 1959, **49**, 130.

126. R. G. Greenler, *J. Chem. Phys.*, 1966, **44**, 310.

127. S. C. Lin, B. J. Bulkin, and E. W. Pearce, *J. Polym. Sci., Polym. Chem., Ed.*, 1979, **17**, 3121.

128. C. S. Chen, B. J. Bulkin, and E. M. Pearce, *J. Appl. Polym. Sci.*, 1983, **8**, 1077.

129. T. Nguyen and E. Byrd, *J. Coat. Technol.*, 1987, **59**(748), 39.

130. S. Yoshida and H. Ishida, *J. Adhesion*, 1984, **16**, 217.

131. D. L. Allara and C. W. White, in *Stabilization and Degradation of Polymers*, D. L. Allara and W. L. Hawkins, Eds., American Chemical Society, Washington, DC, 1978, p. 273.

132. M. P. Fuller and P. R. Griffiths, *Anal. Chem.*, 1978, **50**, 1906.

133. M. P. Fuller and P. R. Griffiths, *Am. Lab.*, 1978, **10**, 69.

134. M. T. McKenzie, S. R. Culler, and J. L. Koenig, *Appl. Spectrosc.*, 1984, **38**, 786.

135. A. D. English, D. B. Chase, and H. J. Spinelli, *Macromolecules*, 1983, **16**, 1422.

136. M. J. D. Low and I. Coleman, *Spectrochim. Acta*, 1966, **22**, 369.

137. P. R. Griffiths, *Appl. Spectrosc.*, 1973, **26**, 73.

138. V. W. King and J. L. Laurer, *J. Lubrc. Techn.*, 1981, **103**, 65.

139. Y. Nagasawa and A. Ishitani, *Appl. Spectrosc.*, 1984, **38**, 168.

140. M. J. D. Low and C. Mortera, *Carbon*, 1983, **21**, 275.

141. M. J. Mortera and A. G. Severdia, *Appl. Surf. Sci.*, 1985, **20**, 317.

142. S. H. Hamid and W. H. Prichard, *Polym. Plast. Technl. Eng.*, 1988, **27**, 303.

143. T. Bendaikha and C. Decker, *J. Radiat. Curing*, 1984, **11** (2), 6.

144. N. S. Allen, P. J. Robinson, N. J. White, and D. W. Swales, *Polym. Deg. Stab.*, 1987, **19**, 147.

145. C. Decker, K. Moussa, and T. Bendaikha, *J. Polym. Sci., Polym. Chem. Ed.*, 1991, **29**, 739.

146. E. Y. L. Vaidergorin, M. E. R. Marcondes, and V. G. Toscano, *Polym. Deg. Stab.*, 1987, **18**, 329.

147. J. H. Hartshorn, *Polym. Mater. Sci. Eng.*, 1987, **57**, 880.

148. D. J. McEwen, M. H. Verma, and R. O. Turner, *J. Coat. Technol.*, 1987, **59**(755), 123.

149. D. R. Bauer, J. L. Herlock, and D. F. Mielewski, *Polym. Deg. Stab.*, 1990, **28**, 115.

150. N. A. R. Falla, *Polym. Paint Colour J.*, 1989, **179**, 550.

151. D. R. Bauer, *Polym. Deg. Stab.*, 1990, **28**, 39.

152. K. Ohashi, K. Nakai, S. Kitayama, and T. Kawaguchi, *Polym. Deg. Stab.*, 1988, **22**, 223.

153. N. S. Allen, P. J. Robinson, N. J. White, and R. Clancy, *Polym. Deg. Stab.*, 1989, **23**, 245.

154. B. Ranby and J. F. Rabek, *Photodegradation, Photo-oxidation and Photostabilization of Polymers*, Wiley, New York, 1975.

155. V. Bellenger and J. Verdu, *J. Appl. Polym. Sci.*, 1983, **28**, 2599.

156. V. Bellenger and J. Verdu, *J. Appl. Polym. Sci.*, 1985, **30**, 363.

157. V. Bellenger and J. Verdu, *J. Appl. Polym. Sci.*, 1983, **28**, 2677.

158. V. Bellenger, C. Bouchard, P. Claveirole, and J. Verdu, *Polym. Photochem.*, 1981, **1**, 69.

159. M. W. Urban and K. W. Evanson, *Polym. Commun.*, 1990, **31**, 279.

160. K. W. Evanson and M. W. Urban, *J. Appl. Polym. Sci.*, 1991, **42**, 2287.

161. K. W. Evanson, T. A. Thorstenson, and M. W. Urban, *J. Appl. Polym. Sci.*, 1991, **42**, 2297.

162. K. W. Evanson and M. W. Urban, *J. Appl. Polym. Sci.*, 1991, **42**, 2309.

163. T. Hirayama and M. W. Urban, *Prog. Org. Coat.*, 1992, **20**, 81.

164. R. T. Foister, *J. Colloid Interface Sci.*, 1984, **99**, 568.

165. P. J. Floury, *Principles of Polymer Chemistry*, Cornell University Press, Ithaca, N. Y., 1953.

166. B. W. Ludwig and M. W. Urban, *Polymer*, 1992, **33** (16), 3343.

167. S. G. Croll, *Prog. Org. Coat.*, 1987, **15**, 223.

168. M. He and M. W. Urban, *J. Appl. Polym. Sci.*, still in press.

169. B. F. Sagar, *J. Chem. Soc.*, 1967, B, 1047.

170. J. F. McKellar and N. S. Allen, *Macromol. Rev.*, 1978, **13**, 241.

171. J. F. McKellar and N. S. Allen, *Photochemistry of Man Made Polymers; Applied Science*, London, 1979, P. 125.

172. L. Tang, J. Lemaire, D. Saller, and J. M. Mery, *Makromol. Chem.*, 1981, **182**, 3467.

173. L. Tang, D. Saller, and J. Lemaire, *Macromolecules*, 1982, **15**, 1432.

174. L. Tang, D. Sallet, and J. Lemaire, *Macromolecules*, 1982, **15**, 1437.

175. A. Roger, D. Sallet, and J. Lemaire, *Macromolecules*, 1985, **18**, 1771.

176. A. Roger, D. Sallet, and J. Lemaire, *Macromolecules*, 1986, **19**, 579.

177. C. H. Do, E. M. Pearce, and B. J. Bulkin, *J. Polym. Sci., Polym. Chem. Ed.*, 1987, **25**, 2301.

178. D. R. Bauer, *J. Appl. Polym. Sci.*, 1982, **27**, 3651.

179. B. Exsted and M. W. Urban, *J. Appl. Polym. Sci.*, in press.

180. E. M. Salazar-Rojas and M. W. Urban, *Prog. Org. Coat.*, 1989, **16**, 371.

181. M. W. Urban and E. M. Salazar-Rojas, *J. Polym. Sci., Polym. Chem. Ed.*, 1990, **28**, 1593.

182. M. W. Urban, *Prog. Org. Coat.*, 1989, **16**, 321.

183. J. Hodson and J. A. Lander, *Polymer*, 1987, **28**, 251.

184. M. W. Urban and J. L. Koenig, *Anal. Chem.*, 1988, **60**, 2408.

185. R. S. Williams, *Am. Paint. Coat. J.*, 1988, **37**.

186. C. M. Jenden, *Polymer*, 1986, **27**, 217.

187. J. D. Webb, P. Schissel, A. W. Czanderna, A. R. Chughtai, and D. W. Smith, *Macromolecules*, 1981, **35**, 598.

188. R. G. Messerschmidt and M. A. Harthcock, *Infrared Microspectroscopy Theory and Practice*, Marcel Dekker, New York, 1988.

189. M. A. Harthcock, L. A. Lentz, B. L. Davis, and K. Krishnan, *Appl. Spectrosc.*, 1986, **40**, 210.

190. D. J. Skrovanek, *J. Coat. Technl.*, 1989, **61**, (769), 31.

191. G. Wegner, *Z. Naturforsch.*, 1969, **24B**, 824.

192. G. Wegner, *Makromol. Chem.*, 1971, **145**, 85.

193. G. Wegner, *Makromol. Chem.*, 1972, **154**, 35.

194. G. Wegner, in *Molecular Metals*, W. E. Hatfield, Ed., Plenum Press, New York, 1979.

195. D. J. Sandman, Y. J. Chen, B. S. Elman, and C. S. Velazques, *Macromolecules*, 1988, **21**, 3112.

196. D. J. Sandman and Y. J. Chen, *Polymer*, 1989, **30**, 1027.

197. R. J. Butera, J. B. Lando, and B. Simic-Glavaski, *Macromolecules*, 1987, **20**, 1722.

198. W. J. Feast, in *Handbook of Conducting Polymers*, Vol. 1, T. Skotheim, Ed., Marcel Dekker, New York, 1986.

199. D. D. C. Bradley, R. H. Friend, T. Hartman, E. A. Marseglia, M. M. Sokolowski, and P. D. Townsend, *Synth. Metals*, 1987, **17**, 473.

200. J. P. Buisson, S. Krichene, and S. Lefrant, *Synth. Metals*, 1987, **21**, 229.

201. S. Krichene, S. Lefrant, Y. Pelous, G. Froyer, M. Petit, A. Digua, and J. F. Fauvarque, *Synth. Metals*, 1987, **17**, 607.

202. S. Krichene, J. P. Buisson, and S. Lefrant, *Synth. Metals*, 1987, **17**, 589.

203. H. Shiragawa, T. Ito, and S. Ikeda, *Makromol. Chem.*, 1978, **179**, 1565.

204. E. Cernian, L. D' Ilanrio, M. Lupoli, E. Mantovani, M. Schwarz, P. Benni, and A. Zanobi, *J. Polym. Sci., Polym. Chem. Ed.*, 1984, **22**, 3393.

205. H. Takeuchi, T. Arakawa, Y. Furukawa, and I. Harada, *J. Mol. Struct.*, 1987, **158**, 179.

206. G. Zerbi and G. Dellepiane, *Gaz. Chimica Italiana*, 1987, **117**, 591.

207. C. Castiglioni, J. T. Lopez Navarrete, G. Zerbi, and M. Gussoni, *Solid State Commun.*, 1988, **65**(7), 625.

208. S. R. Gaboury and M. W. Urban, *Polym. Commun.*, 1991, **32**(13), 3990.

209. S. R. Gaboury and M. W. Urban, *J. Appl. Polym. Chem.*, 1992, **44**, 401.

210. M. W. Urban and M. T. Stewart, *J. Appl. Polym. Sci.*, 1990, **39**, 265.

211. S. R. Gaboury and M. W. Urban, *Langmuir*, 1994.

CHAPTER 8

SURFACES, INTERPHASES, AND INTERFACIAL REGIONS

8.1 INTERFACIAL BONDING AND INTERPHASE FORMATION

When two compounds form chemical bonds with each other, the process is called *chemical reaction*. When two substrates form bonds with each other, or adhere to each other, the process is called *adhesion*.

Chemists usually associate adhesion with the energy liberated when two surfaces meet to form an intimate contact called an interface. Conversely, it may be defined as the energy required to destroy the interface between two substances. Physicists or engineers customarily describe adhesion in terms of forces, with the force of adhesion being the maximum force exerted when two previously adhered substances are separated. While all definitions of adhesion describe more or less the same phenomenon, the measurements of energy or forces to separate two materials do not reveal the origin of the interfacial bonding; instead, they tell us how strong the bonding is or how much energy is required to destroy it. This is why specroscopists have many opportunities to enhance a molecule-level understanding of adhesion.

One can visualize the process of interphase or interfacial region formation as that depicted below and designated as **A** and **B**:

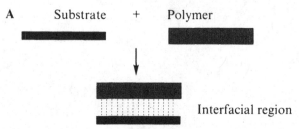

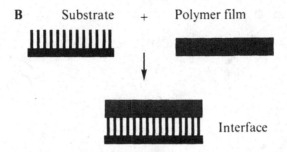

Interface

When a polymer is deposited on a substrate, formation of an interphase results from reactions when two substrates are in contact. However, for such a reaction to occur, it is necessary that each surface contain functional groups accessible for bonding. It is therefore often desirable to modify one of the surfaces to increase reactivity, which, in macroscopic terms, would imply a modification of the surface energy. This is schematically depicted above and designated as case **B**. In contrast to case **A**, where an interfacial region develops as a result of chemicophysical interactions, in case **B**, a polymeric coating is deposited on a pretreated substrate. As a result, thicker interfacial region, or interphase, is created.

Although quantitative characterization of the interfacial regions between metal surfaces and organic polymers (or coatings) has long been a goal of sciences, there are still many issues to be sorted out. The primary problems with metal–polymer adherence are usually associated with intrusion of water[1] and subsequent electrochemical processes at the interface,[2] often followed by crack formation and propagation due to internal stresses and thermal cycling. A broad spectrum of possibilities responsible for either interfacial failure or adhesion leaves us with not much choice but to identify the origin of the forces responsible to chemicophysical changes, and on that basis, to modify the chemistry to improve bonding. Hence, characterization of interfacial regions should allow designing of the interfaces that resist environmental degradation and other external changes. Polymer–metal bonding is particularly important as its macroscopic function is to transfer stresses between the two substances. Although there may be many chemical reactions involved in forming the interfacial layer, bonding is usually accomplished through formation of bridges between the organic phase and metal oxides or by metal–polymer coordination where polymer – metal complexes can be formed. The first case is commonly known as *priming*, which according to the literature[3] suggests that an inorganic–metallic substrate is treated with a silane coupling agent applied from a dilute aqueous solution. The process that leads to silane hydrolysis and subsequent polymerization on the surface is shown in Fig. 8.1. In the second category, metal–polymer bonding occurs when the organometallic group is a pendant group of a polymer backbone, the metal complex acts as a crosslinking agent, the metal complex is enmeshed in a polymer, or the metal is an integral part of a polymer backbone. In this second category, polymer–metal complexes are usually formed during synthesis through complexation of polymeric ligand to metal ion, polymerization leading to metal complexes, or cluster complex formation. Since in the system

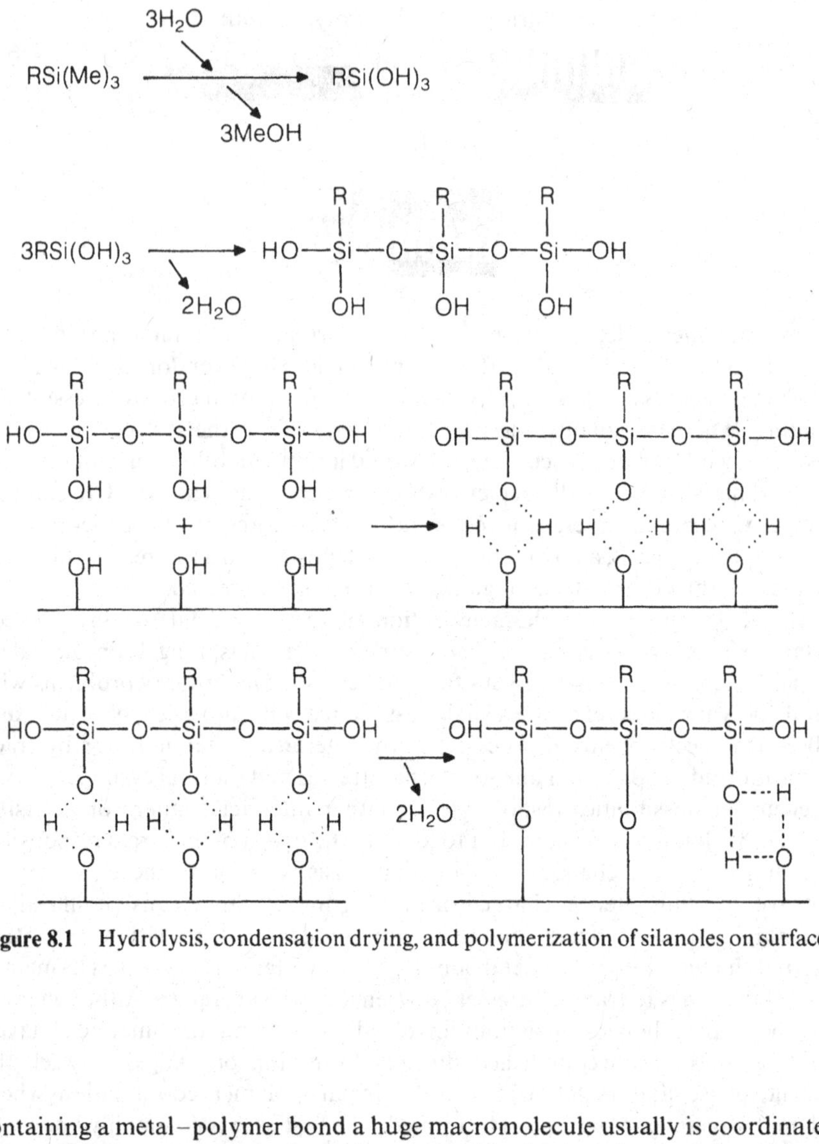

Figure 8.1   Hydrolysis, condensation drying, and polymerization of silanoles on surfaces.

containing a metal–polymer bond a huge macromolecule usually is coordinated to metal, the concentration of metal–polymer bonds per volume of the system is relatively small and often metal–polymer bonds are difficult to detect. Because of this, most of the studies have focused on model surface reactions, which will be described in the following sections. Furthermore, the fact that the electronic environment of the metal–polymer bond may be strongly affected by the presence of the macromolecules further complicates the issue. As a result, the characteristic vibrational features of polymer – metal complexes in solution will differ for the same system deposited on a metal substrate. This is simply related to the multilayer nature of the interface.

8.2 SURFACE–INTERFACIAL PRIMERS

It is common practice to use "primers" to improve strength of adhesive bonds to metals. They are commonly referred to as coupling agents because they couple two dissimilar substances. Figure 8.1 illustrates a general hydrolysis reaction leading to the formation of polymerized surface species that promotes adhesion. As soon as water evaporates, the oligomer molecules polymerize, forming a layer that may adhere to the metal oxide surface through silanol groups. If these primers have defined thickness, they are considered as separate phases or "interphases." There are numerous studies where the physical or chemical adsorption of silane coupling agents have been addressed, and further details concerning spectroscopic features of Si-containing polymers are presented in Chapter 7. Here, we focus only on the studies directly related to interfacial or interphases regions which address the issue of adhesion.

As we recall the experimental considerations in Chapter 3, the primary spectroscopic studies used are surface FT-IR and/or Raman–surface-enhanced Raman (SERS) spectroscopies. The initial model studies indicated that the reactions between aminosilane and anhydride-cured adhesives conducted at 150°C lead to formation of cyclic imides, which are unable to bond across the surface.[4] Culler et al.[5] demonstrated that drying of aminoprimer on KBr plates in air leads to the formation of bicarbonates, unless the reactions are conducted above 95°C. However, above 125°C, imine groups may be formed. Reactions of anhydride/epoxy adhesive cured at 150°C for 3 hr on aminosilane coupling agent coated Ge crystals were investigated by ATR spectroscopy.[6,7] These studies showed that the amino groups of the silane primer reacted with anhydrides, producing amides but no evidence for imide formation. Since water was unable to escape during curing, the equilibrium conditions were such that the amide–imide conversion could not occur. The effect of drying temperature on the structural changes in γ-aminopropyltriethoxysilane (γ-APS) adsorbed onto KRS-5 and sapphire from aqueous solutions at pH 10.4 was investigated by Sung et al.[8]

TABLE 8.1 Si—O Stretching Vibrational Bands at the E-Glass/γ-APS Interface

Wavenumber (cm^{-1})	Si—O Stretching
1198	SiO$_2$
1133	Multilayers of γ-APS
1122	γ-APS bulk
1111	Chemisorbed γ-APS
1113	SiO$_2$
1072	Chemisorbed γ-APS
1045	Multilayers of γ-APS

Source: Ref. 82.

TABLE 8.2 Selected IR Bands Due to Reactions of Silane Coupling Agents with Atmospheric CO_2 and Their Tentative Assignment

Compound	Formula	RNH	RNH_3^+	CH	OH	NCl_2^-	CO_2 Asymmetric	NCO_2^-	OCO_2 Asymmetric
Sodium bicarbonate	$HO-C{\overset{O}{\underset{O^-}{}}}$ Na$^+$	—	—	—	2600–2475	—	1700–1600	—	1400–1300
Ammonium bicarbonate	$HO-C{\overset{O}{\underset{O^-}{}}}$ NH$_4^+$	—	3200–2850	—	2570–2540	—	1600	—	1500–1370
Ammonium carbamate	$H_2N-C{\overset{O}{\underset{O^-}{}}}$ NH$_4^+$	3469, 3302 doublet	3200–2800	—	—	1625	—	1530	—
n-Propylammonium chloride	$CH_3(CH_2)_2NH_3^+Cl^-$	—	3200–2250	3100–2700	—	—	—	—	—
i-Propylamine	$(CH_3)_2CHNH_2$	3409, 3280 doublet	—	—	—	—	—	—	—
i-Propylamine·CO_2	$(CH_3)CHNHCO_2^-$ $NH_3CH(CH_3)_2$[d]	3370 singlet	3000–2190	2980–2120	—	1649	—	1561	—
n-Butylamine	$CH_3(CH_2)_3NH_2$	3462, 3370 doublet	—	3050–2860	—	—	—	—	—
n-Butylamine-CO_2	$CH_3(CH_2)_3NHCO_2^-$ $NH_3(CH_2)_3CH_3$	3332 singlet	3174–2140	3000–2170	—	1635	—	1561	—
1,4-Diaminobutane	$NH_2(CH_2)_4NH_2$	3361, 3284 doublet	3000–2150	2985–2790	—	1660, 1636'	—	1556	—

1,4-Diamino-butane·CO₂	$^+NH_3(CH_2)_4NHCO_2^-$	3332 singlet	—	2960–2750	—	—
n-Butylaminodimethyl methoxysilane	$NH_2(CH_2)_4Si(CH_3)_2(OCH_3)$	3371, 3292 weak doublet	—	2960–2800	—	—
n-Butylamino dimethylmethoxy-silane·CO₂	$^-O_2CNH(CH_2)_4Si(CH_3)_2(OCH_3)\cdot{}^+NH_3(CH_2)_4Si(CH_3)_2(OCH_3)$	3298 Singlet	3000–2170	3000–2850	1650	1508
n-Propylamino triethoxysilane	$NH_2(CH_2)_3Si(OCH_2CH_3)_3$	3368, 3295 doublet	—	3000–2850	—	—
n-Propylamino triethoxysilane·CO₂	$^-O_2CNH(CH_2)_3Si(OCH_2CH_3)_3\cdot{}^+NH_3(CH_2)_3Si(OCH_2CH_3)_3$	3302 singlet	3100–2180	2950–2800	1660, 1620	1580
Ethylenepropylene diaminotrimethoxysilane	$NH_2(CH_2)_2NH(CH_2)_3Si(OCH_3)_3$	3339, 3280	—	2950–2800	—	—
Ethylene propylene diaminotrimethoxy silane·CO₂	$^-O_2CNH(CH_2)_2NH(CH_2)_3Si(OCH_3)_3\cdot{}^+NH_2(CH_2)_2NH(CH_2)_3Si(OCH_3)_3$	3240, trilet 3295 singlet	2950–2180	—	1620 very weak shoulder	—
Ethyl carbamate (urethane)	$NH_2CO_2CH_2CH_3$	3413, 3345 3267, 3209 split doublet	—	—	—	—
Ethyl N-ethylcarbamate (N-ethylurethane)	$CH_3CH_2NHCO_2CH_2CH_3$	3300 singlet	—	2950–2860	1695	1517
N,N′-Methylene-bis-(ethylcarbamate)	$CH_2(NHCO_2CH_2CH_3)_2$	3344 singlet	—	3075–2780	1680	1526

Source: Adopted from ref. 13.

These studies indicated that bicarbonates can be easily decomposed under dry vacuum conditions at 110°C. Slow, parallel oxidation of amino to imino groups was also detected. In other studies,[9] Sung et al. investigated the effect of γ-APS primers on the peel strength of polyethylene. Apparently the decreased strength was attributed to a lack of interpenetration between polyethylene and highly polymerized silane primer. Table 8.1 summarizes the band assignments proposed for γ-APS deposited on E-glass fibers (55% SiO_2). In fact, γ-APS is perhaps the most studied coupling agent, and the formation of bicarbonate salt resulting from the presence of CO_2 in air has been shown in numerous studies.[10 12] Table 8.2[13] summarises selected IR bands that are due to structural features resulting from the reactions of CO_2 with amines.

Boerio was the first to realize that "the metal substrates have a significant effect on the molecular structure of interphases formed when polymers are cured against them."[14,15] For example, characteristic bands of carboxylate species are present in the anhydride/epoxy adhesives cured against aluminum and steel substrates, whereas other substrates may not exhibit such features. Table 8.3 illustrates that the bands due to SiOH and SiOSi have lower vibrational energy for aluminum than steel and titanium. This observation is attributed to a lower degree of polymerization on aluminum 2024 resulting from hydrogen bonding between silanol groups and hydroxyl groups on the oxidized surface of the aluminum. These studies also indicated that on heating at 110°C, there was no additional polymerization because the band intensity at $1100\,cm^{-1}$ did not change. But the bands due to bicarbonate disappeared at the expense of the new bands at 1600 and $1660\,cm^{-1}$, which are attributed to the imine formation in the oxidation process of amine groups:

$$2RCH_2NH_2 + O_2 = 2RCHNH + 2H_2O$$

When γ-APS is adsorbed onto mechanically polished titanium mirrors from 1% aqueous solutions at pH 10.4, there are several characteristic features in the spectrum, demonstrated by a strong band at $1070\,cm^{-1}$ along with the weaker bands at 1570, 1470, and $1330\,cm^{-1}$. Because the bands due to Si—O—C, charac-

TABLE 8.3   Vibrational Frequencies of γ-APS Polymerized on Aluminum and Steel

Steel Substrate	Assignment	Aluminum (2024)	Ref. -
γ-APS 920	SiOH	890	14, 15
1040	Si—O—Si		14, 15
1120	Si—O—Si	1100	14, 15
1330		1330	14, 15
1470	Adsorption of CO_2 and formation of amine bicarbonates	1470	14, 15
1570		1570	14, 15
1640		1640	14, 15

teristic of γ-APS monomer normally observed at 1105, 1080, and 960 cm^{-1} are not present in the spectrum, γ-APS has hydrolized to form a polymeric network. The band at 1070 cm^{-1}, attributed to Si—O—Si stretching vibrations, indicates the polymerized form of silane in a form of polysiloxane.

The bands at 1570, 1470, and 1330 cm^{-1} are attributed to the formation of amine bicarbonate molecules resulting from absorption of atmospheric CO_2. Specifically, the 1570-and 1470-cm^{-1} bands are due to the deformation modes of protonated amino groups (NH_3^+), whereas the 1330-cm^{-1} band is assigned to the symmetric stretching mode of bicarbonate. Although these assissments indicate that a stable polymerized siloxane film, should be formed, the spectral changes after the film exposure to ambient conditions indicate that this is not the case. The band at 1070 cm^{-1} splits to two bands at 1130 and 1040-cm^{-1}, indicating further polymerization of silanes. Furthermore, the 1470 cm^{-1} band decreases, whereas the 1330-cm^{-1} band increases in intensity, suggesting the disappearance of unstable bicarbonate in the presence of atmospheric moisture. Table 8.4 lists other spectral changes resulting from heat and acid treatments. Recently,[16] the analysis of structures formed by a multicomponent silane primer consisting of a 50:50 molar ratio of aminosilane and phenylsilane applied to aluminum showed that the oxidation of the primer occurred during curing process of epoxy coating. Observed IR bands and their tentative band assignments are listed in Table 8.5.

One commonly used anaerobic adhesives consists of triethylene glycol dimethacrylate (TRIEGMA) and a redox cure system composed of acetylphenylhydrazine (APH), o-benzoic sulfimide (BS), and cumene hydroperoxide (CHP).[17] This adhesive mixture has been known for rapid cure in the presence of copper or iron and the absence of oxygen. Interestingly, cure on aluminum and zinc was

TABLE 8.4 Vibrational Frequencies Due to Silane Coupling Agent γ-APS Treatment of Ti Surface

	Sample Treatment		
Untreated	Heat[a]	Acid[b]	Water[c]
—[d]		1600 NH_3^+ Cl^-	
		1500 NH_3^+ Cl^-	
1570	—	—	
1470	—	—	
1330	—	—	
1130	1145	Increased	
1040		Stronger	
850			

[a] 100°C for 20 min in air.
[b] 1% HCl aqueous solution.
[c] Prior to siloxane deposition, the titanium substrate was oxidized at 300°C to form a surface oxide 150 Å thick.
[d] — not observed (throughout).
Source: Adopted from ref. 15

TABLE 8.5 Observed IR Bands and Their Tentative Assignments for a 50 : 50 Molar Ratio of Amino and Phenyl Silane (Dow Corning Z-6020)

Z-6020	Phenyl Silane	Partially Hydrolyzed Mixture	Assignment
	656		$Si(Cl)_n'$
691			
	698	698	Phenyl $C\!=\!C$
	745	737	Phenyl
815	812	809	Si—O—C
929		948	Si—O—Et or Si—C_2H_5
1071	1088	1097, 1055	Si—O—C or Si—O—Si
1191	1188	1193, 1197	CH_3 rock of Si—O—Me
1285		1285	Si—C_2H_5
1312		1312	CH_2 wag of Si—CH_2—R
	1308, 1325		Various bands of monosubstituted
1343			phenyls
		1381, 1375	
1411		1411	CH_2 deformation of Si—CH_2
	1429	1431	Si—O—phenyl

Source: Adopted from ref. 16.

TABLE 8.6 Raman Bands of (*A*) Acrylic Adhesive, Saccharin, and Their Sodium Salts Observed in SERS and Their Assignments and (*B*) Benzoic Acid and Adhesive with Benzoic Acid, and Their Sodium Salts Observed in SERS and their Assignments

		A		
Adhesive (SERS)	Saccharin	Sodium Salt	Saccharin (SERS)	Assignment
	437			$v(18a)$
552	535	555	556	$\delta(SO_2)$
	602			
609	628	620	612	$v(6a)$
658		658	649	
720	712	722	725	$v(1)$
	785	786		$v(12)$
		806		
		1020		
1025	1025	1030	1026	$v(18b)$
	1125			
	1138			
		1153		
1155	1170	1161	1160	$v(15)$
1175	1185	1176	1178	$v_a(SO_2)$

TABLE 8.6 *(Continued)*

A

Adhesive (SERS)	Saccharin	Sodium Salt	Saccharin (SERS)	Assignment
1266	1265	1264	1270	$v(3)$
1306	1308	1308	1308	$v(14)$
	1344			
1360	1362	1350	1362	$v_a(SO_2)$
1470	1472	1470	1472	$v(19b)$
1600	1604	1606	1605	$v(8a)$
		1644		$v(C=N)$
		1658		
	1707			$v(C=O)$

B

Adhesive (SERS)	Benzoic Acid	Sodium Salt	Benzoic Acid (SERS)	Assignment
452				
		408	450	$\delta(COO^-) + v(6a)$
	426			$\delta(COOH) + v(6a)$
620	620	620	628	$v(6b)$
694		683	688	$\delta(COO^-)$
	815	827	831	
852		850	848	$v(1)$
1015	1005	1010	1010	$v(12)$
	1031	1034	1035	$v(18a)$
1151	1140	1148	1148	$v(15)$
1172	1165	1163	1163	$v(9a)$
	1188	1188		
1320				
1400			1394	$v_a(COO^-)$
1466	1450	1440	1450	$v(19b)$
1518			1498	$v(19a)$
		1594		
1612	1610	1610	1608	$v(8a)$
	1640			$v(COOH)$

Source: Adopted from ref. 18.

much slower. In other studies, Boerio et al.[18] used surface-enhanced Raman scattering (SERS) to characterize model acrylic adhesions of silver. Tables 8.6*A* and 8.6*B*, provide a comparison of the band assignments between the Raman bands observed from TRIEGMA/BS/CHP/APH adhesive, saccharin, benzoic acid, and respective salts.

8.3 COMPLEX FORMATION

Table 8.7 summarizes IR and Raman bands of polyethylene glycol (PEG; molecular weight (MW) = 3000) complexed with NaSCN in a ratio of 4:1 and indicates that there are significant differences between uncomplexed and complexed PEG molecules caused by conformational changes resulting from complexation. It is believed that these spectroscopic changes are attributed to the presence of double-heli structures in which Na^+ is incorporated within the space of two helical chains bridged by thiocyanate anions.[19]

Almost all studies dealing with the metal–polymer bonding originate with the model system that on further studies, is used in the analysis of actual metal–polymer interfaces. In the studies of poly(N-vinylimidazole) (PVI) with NO^{2+}, the spectra of neutral, fully protonated PVI and PVI – Ni complexes in a ratio of 6:1 were analyzed[20] and indicated that the most probable stoichiometry is between 4 and 6. On complexation, the PVI ring modes due to $v_{C\ C}$ and $v_{C\ N}$ at $1500\,cm^{-1}$ shifts to $1511\,cm^{-1}$. At the same time, the band at $1085\,cm^{-1}$ due to combination of $\delta_{CH} + v_{ring}$ shifts to $1096\,cm^{-1}$, and the 915-cm^{-1} ring mode decreases in intensity concurrent with the formation of a new band at $945\,cm^{-1}$.

In other studies,[21] PMMA – salts were examined, and the results excluded the possibility of the methoxy oxygen atom as being an interactive site with metal ion. The polymerization of MMA causes changes of the $C = O$ bands because the conjugation with the $C = C$ bond conjugation is removed. The bands at 1310 and $1290\,cm^{-1}$ in MMA are detected at 1275 and $1238\,cm^{-1}$ for PMMA. The studies

TABLE 8.7 IR and Raman Vibrational Energies in Polyethylene Glycol (PEG) and PEG–NaSCN Complexes

PEG		PEG–NaSCN		
IR	Raman	IR	Raman	Assignment
845	846	835	835	CH_2 rocking
		825	827	
1065	1064	1075	1079	CH_2 rock
1155	1128			
1250	1143			
1285	1239			
1345	1281			
1370	—			
1420	1397			
1480	1447			
—				
—				
2895				
2955				

Source: Adopted from ref. 19.

of McCluskey et al.[22] on other polyacrylate systems apparently have shown that the degree of complex formation can be measured using FT-IR.[23]

Infrared spectra of polybutadiene obtained at the polymer–air interface on various metallic substrates indicated that the hydrogen bonding at the polymer–metal interfaces plays an important role in the metal–polymer bonding and follows relative order of activity as earlier proposed by Cullis and Berr.[24] For the most part, internal reflection spectroscopy was utilized in the studies of polymer–metal interfaces. These studies were pioneered by Allara[25] and showed that carboxylate groups being formed after oxidation of a thin poly(1-butene) film on copper. The carboxylate ions were also responsible for bonding of adsorbed poly(acrylic) acid on an aluminum oxide surface.

When a thin film of pyromellitic dianhydride (PMDA)–oxydianiline (ODA) derived polyamic acid is baked at 120°C and exposed for 10 sec to a 0.2M sodium metasilicate (SMS), followed by curing at 160°C, the species are unimidized because the condensation pathway requiring a carboxylic acid ortho to an amide group is blocked.[26] The situation is different at 350–400°C, where imidization can be completed with virtually all amic acid groups, and Na-containing films undergo a thermal decomposition reaction that results in a cleavage of the polymer backbone into subliming free amine groups and decarboxylating dicarboxylic acid groups.

In the studies of polyimide–metal interactions,[27] polyamic acid, a precursor of polyimide, was used as a model system. While the primary tool in these studies again was internal reflection spectroscopy, digital spectral subtraction was used to reveal the difference between metal-bonded and non-metal-bonded polyamic acid. Apparently the negative bands at 1528 and 1410 cm^{-1} indicate that imidization is enhanced at the cobalt–polymer interface, whereas a positive 1377-cm^{-1} band due to C—N stretching modes in polyamide suggested a higher degree of imidization at the interface. Furthermore, the spectrum recorded at the cobalt–polyimide interface revealed the increase of the band at 1670 cm^{-1} due to carbonyl groups, along with the bands attributed to symmetric carboxylate stretching vibrations at 1584 and 1397 cm^{-1}. More general discussion about various polyimide—metal morphologies can be found in the literature.[28]

The interactions derived from pyromellitic dianhydride (PMDA)–oxydianiline (ODA) polyamic acid with metal surfaces were also analyzed using FT-IR spectroscopy. While thermal curing of such films can be completed at 350–400°C, sodium-containing films undergo thermal decomposition reactions that appear to cleave the polymer into free amine and sodium salt of dicarboxylic acid.[29] In 1992, Boerio et al.[30] examined molecular structures that can be produced on silver surfaces during curing of polyamic acid of pyromellitic diahydride (PMDA) and oxydianiline (ODA). While these studies showed that curing at the metal–polymer interfaces may be inhibited by the interactions with the substrate, an interesting spectroscopic aspect was a comparison of Raman, SERS, transmission IR, and reflection–absorption IR. Table 8.8 provides the bands and their band assignments. Reflection — absorption FT-IR spectroscopy was also used for the analysis of interactions with and orientation of PMMA on various metal

TABLE 8.8 Tentative Band Assignments for Polyamic Acid Before and After Curing Detected in Raman, SERS, IR, and RA-IR Mode of Detection

| Raman Heat Treatment | | SERS Heat Treatment | | IR Heat Treatment | | RA-IR Heat Treatment | | Assignment |
Before	After	Before	After	Before	After	Before	After	
				3255 (w)	3372 (w)			ν(N—H)
				3205 (w)				ν(20)—C_6H_2
				2926 (w)		2921 (w)		ν(CH)
	1802 (m)		1798 (w)		1777 (s)		1778 (w)	ν(C=O), in-phase imide I
				1720 (s)		1722 (w)		ν(C=O), acid
1698 (w)								ν(C=O), acid
					1725 (s)		1730 (s)	ν(C=O), out-of-phase
				1670 (s)			1660 (m)	ν(C=O), amide I
1623 (m)	1623	1620 (s)						ν(8a)—C_6H_2
	1600 (w)		1606 (s)	1610 (w)			1609 (w)	ν_{as}(COO⁻), ν(8a)—C_6H_2 and —C_6H_4
								ν(8a)
1574 (w)		1576					1570 (w)	ν_{as}(COO⁻)
	1525 (w)							ν(8a)—C_6H_4
				1545 (m)		1542 (w)		ν(C—N—H), amide II
				1499 (s)	1500 (s)	1500 (s)	1500 (m)	ν(19a)—C_6H_4
1450 (w)					1456 (w)			ν(19b)—$C\delta H_2$
	1403 (s)	1412 (w)	1408	1407 (m)		1400 (w)		ν_6(COO⁻)
						1410 (w)		β(OH), acid
					1378 (s)		1379 (w)	ν(C—N—C) axis imide II 1340 (m)
								ν(C—N)
		1345 (m)						ν(C—OH)
				1303 (m)	1291 (w)	1304 (w)		ν(7a)—C_6H_4

								Assignment
								ν(C—N—H), amide
1263 (w)		1269 (s)						
								ν(C—O—C), ODA
				1237 (s)	1239 (w)	1240 (w)	1240 (w)	
				1216 (m)		1219 (w)		ν(9a)—C₆H₄ or ν(13)—C₆H₂
1177 (w)	1180 (w)	1178 (w)	1176 (w)	1170 (w)		1170 (w)		ν(C—N—C), transverse
					1169 (m)	1169 (w)	1169 (w)	ν(9a)—C₆H₄, ν(13)—C₆H₂
	1134 (w)		1142 (w)					ν(C—N—C), transverse imide III
				1113 (m)	1116 (m)	1110 (w)	116 (w)	ν(18b)—C₆H₄
					1096 (w)			ν(18a)—C₆H₄
							1093 (w)	π(OH), acid
					1015 (w)			Isoimide
940 (w)					920 (w)		920 (w)	ν(17b)—C₆H₂
	865 (w)			872 (w)	882 (m)			ν(1)—C₆H₄
			840 (w)			833 (w)		δ(COO⁻)
						823 (m)	825 (m)	ν(17b)—C₆H₄
								β(C=C), acid
761 (m)		733 (w)		754 (w)	726 (m)	726 (w)	726 (w)	C—N—C, out-of-plane bending imide
736 (w)	732 (w)							ν(1)—C₆H₂
717 (w)								ν(12)—C₆H₄
				661 (w)				π(C=O), acid
636 (w)	625 (w)		633 (w0)					ν(6b)—C₆H₄ imide ring, in-plane bending
					605 (w)			ν(12)—C₆H₂
586 (w)								ν(3)—C₆H₂
					518 (w)			ν(16b)—C₆H₄

Source: Adopted from ref. 30.

surfaces.[31] It appears that the eryl ether rings are preferentially parallel to the metal surface, whereas other moieties are perpendicular. No chemical polymer–metal bonding was detected.

The molecular structures that develop on deposition of aminophenyltrimethoxysilanes (APTMS) on metals are of considerable interest as the presence of phenyl groups provides a thermally stable environment for high-temperature applications. The species are initially hydrolizded in solution, and on condensation on a metal substrate, are polymerized by heating it at 250°C.[32] Table 8.9 illustrates a comparison between the band assignments for pure APTMS, APTMS deposited on aluminum, and APTMS polymerized on aluminum.

TABLE 8.9   Infrared Bands Observed for Neat Aminophenyltrimethoxysilanes (APTMS) and for APTMS Adsorbed onto Aluminum Mirrors

APTMS	Film on Al, as Deposited	Film on Al, Heated 250°C for 2 hr	Assignment
3462	3460	3460	$\nu_{as}(NH_2)$
3370	3370	3357	$\nu_s(NH_2)$
3234	3236	3214	$1624 + 1599$
3022	3028	3029	$\nu(CH)$
2943			$\nu(CH_3)$
2841			$\nu(CH_3)$
1624	1624	1624	$\nu(8b)$
1599	1599	1599	$\delta(NH_2)$
1577	1577	1577	$\nu(8a)$
1510	1510	1510	$\nu(19a)^a$
1484	1484	1484	$\nu(19b)$
1439	1439	1439	$\nu(19a)$
1297	1297	1297	$\nu(14)$
1274	1274	1274	$\nu(3)$
1189			$\gamma_r(SiOCH_3$
1129			$\nu(Si\text{-}\phi)$
1078			$\nu(SiOCH_3)$
	1132	1132	$\nu(SiOSi)$
	1073	1073	$\nu(SiOSi)$
993	993	993	
		960	$\nu(AlOAl)$
	915		$\nu(SiOH)$
900			
810	819	810	$\nu(17b)^a$
	792	785	$\nu(11)$
	760	754	$\nu(11)^b$
730			
699	699	699	$\nu(4)$

[a] In the *p*-isomer.
[b] In the *o*-isomer.
Source: Adopted from ref. 32.

Some of the surface primers are already on the surface. For example, the factors controlling the mechanical behavior of polymer–coupler–metal oxide systems can be considered in terms of the weak link in the chain. Polyimides have become important for their good adhesion and easy processing. Adhesion to metals such as Cr can be be enhanced when polyimide surfaces are treated by the radiofrequency (RF) plasma. According to these studies,[33] RF sputtering of polyimide surfaces prior to metal deposition breaks the hydrogen-bonded carbonyls in polyimide, creating reactive sites on polyimide and thus enhancing adhesion. The ratio of the $C-N$ at $1120 \, cm^{-1}$ to the aromatic phenyl absorption at $1490 \, cm^{-1}$ was found to decrease with increasing RF power, indicating that the RF sputtering also breaks the $C-N$ bonds in polyimide.

8.4 ACID–BASE INTERACTIONS

One of the fastest-growing and most interesting areas of focus on interfacial bonding is intermolecular acid–base interaction. Back in 1938, Lewis[34] defined acids as electron acceptors and bases as electron donors. In 1952, Mulliken[35] characterized acid–base interactions as charge-transfer complexes having two major contributions from electrostatic and covalent interactions, leading to further advancements proposed by Pearson[36] in the form of hard–soft/acid–base (HSAB) interactions. Whereas Pearson focused on the equilibrium contents and rates of complexations, Drago's[37] approach was on the heats of acid–base complexation. These heats of acid – base interactions of polymers can be determined from the IR shifts because vibrational-energy changes of, for example, carbonyl groups will be affected by a magnitude of complexation. In the model studies,[38] a comparison of the $C=O$ stretching frequency at $1764 \, cm^{-1}$ on ethyl acetate in liquid and vapor was used to estimate the magnitude of the van der Waals dispersion forces and acid–base contributions to bonding. It appears that the magnitude of the van der Waals interactions is proportional to the van der Waals contribution to the surface tension or surface free energy. The acid–base contributions to adhesion, on the other hand, are proportional to the heats of acid–base interactions. On the basis of calorimetric analysis of the $C=O$ frequency shifts for ethyl acetate, this magnitude was estimated to be $4.24 \, cm^{-1} \, kcal^{-1} \, mol^{-1}$ (or $0.99 \, cm^{-1} \, kJ^{-1} \, mol^{-1}$).

8.5 INTERFACIAL METAL–POLYMER CHELATION

One reaction pattern that often occurs is chelation of a polymer to a metal atom inserted into the polymer backbone. Because polymeric chelates exhibit good thermal film stability and catalytic activity, such thin films have been utilized in biologically active materials. The system that has been extensively studied is the polychelate reaction of transition ions with poly[4,4'-dihydroxy-3,3'-diacetyl biphenyl] (DABP)–4,4'-diamine diphenyl methane (DDM).[39] IR spectra provided valid information about the bonding sites. As a result of chelation, the primary spectral changes is the appearance of 20–40-cm^{-1} shift of the band due to azomethine groups. The shift to higher frequency is due to the fact that the oxygen atom is involved in bonding.

A chelation of poly(O-isophthalolisophthalamide oxime) with Cu(II) ions[40] results in the disappearance of the amide bands at 3250 cm^{-1} due to N—H stretching vibrations and marked increase of new bands at 1560, 1430, and 1340 cm^{-1}, attributed to the chelation reaction of NH_2 and C=O groups with Cu(II) ion.

A series of poly-yne polymers containing σ-bonded transition metals in the backbone were reported by Takahashi et al.[41] Figure 8.2 illustrates a series of IR spectra of Pt-poly-yne polymer films with the characteristic bands due to C≡C

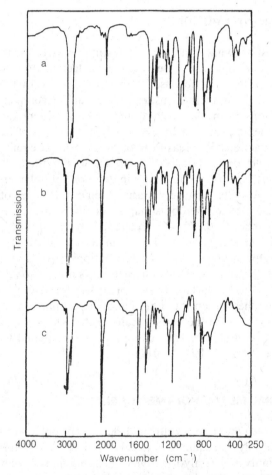

Figure 8.2 FT-IR spectra of Pt-poly-yne polymer films: (A)—(Pt (PBu$_3$)$_2$—C=C—C=C)$_n$—; (B)—(Pt (PBu$_3$)$_2$—C=C—⟨O⟩—C=C; (C) — (Pt (PBu$_3$)$_2$—C=C—C=C—C=C—⟨O⟩—C=C)$_n$—. (Reprinted with permission from ref. 41).

TABLE 8.10   The C—C Trible-Bonds Stretching Frequency for Various Pt-Containing Structures

Structure	$v_{c=c}$	λ_{max} (nm)
Monomer		
—[Pt—C≡C—C≡C]$_2$—	2147	318
cis-[Pt—C≡C—C≡C]$_n$—	2145 2153	301
—[Pt—C≡C—⟨◯⟩—C≡C]$_2$—	2098	338
cis-[Pt—C≡C—⟨◯⟩—C≡C]$_2$—	2120 2130	308
Polymer		
—[Pt—C≡C—C≡C]$_n$—	2000	384
—[Pt—C≡C—⟨◯⟩—C≡C]$_n$—	2095	380
—[Pd—C≡C—C≡C]$_n$—	2240	342
—[Pd—C≡C—⟨◯⟩—C≡C]$_n$—	1980 2095	349
—[Ni—C≡C—C≡C]$_n$—	2120 1980	414
—[Ni—C≡C—⟨◯⟩—C≡C]$_n$—	2075	402
—[Pt—C≡C—C≡C—C≡C—C≡C]$_n$—	2135 1985	501
—[Pd—C≡C—⟨◯⟩—C≡C—C≡C—⟨◯⟩—C≡C]$_n$—	2090	403
—[Pt—C≡C—⟨◯⟩—C≡C—C≡C—⟨◯⟩—C≡C]$_n$—	2090	410

Source: Adopted from ref. 41.

in the $2000 \, cm^{-1}$ region. Table 8.10 lists vibrational frequencies of the $C\equiv C$ stretching bonds for various Pt-containing poly-ynes.

Chemical bonding across the polymer–metal interface is usually considered as a short-range effect that has a tremendous influence on adhesion and adhesive bond durability. There are also substantially longer-range effects that involve at least several molecular layers adjacent to the interface. Although one could debate the significance of each, both types of interactions profoundly affect stability of the interface. For example, when polyethyelene oxidizes on copper, long-range effects dominate,[42] or oxidation of polybutadiene on metal surfaces can be catalyzed.[43,44] Infrared reflection–adsorption spectroscopy has been utilized in the studies of interfacial reactions that may occur during curing of epoxy–dicyandiamine on steel and zinc galvanized steel.[45] Table 8.11 illustrates the spectral changes and tentative band assignments for the thin diacyandiamide films cured on zinc surfaces.

To produce methacrylamide polymer with pendant tridentate phosphine ligands the following condensation reaction can be used:[46]

$$\begin{array}{c} \overset{O}{\underset{\displaystyle |}{\overset{\displaystyle \diagdown}{C}}}\overset{Cl}{\overset{\displaystyle \diagup}{}} \\ +CH_2\!-\!\underset{\displaystyle CH_3}{\overset{\displaystyle |}{C}}\!\xrightarrow{}_n + H_2NR_i \quad\xrightarrow[CH_2Cl_2]{NEt_3^+}\quad +CH_2\!-\!\underset{\displaystyle CH_3}{\overset{\displaystyle |}{C}}\!\xrightarrow{}_n + NE_3^+\!\cdot\!HCl \end{array}$$

$$i = 1, 2, 3; \quad R_1 = CH_2(CH_2)_2 \, PPh_2(P\!-\!phosphorus);$$
$$R_2 = CH_2(CH_2)_2 \, PPh(CH_2)_2 PPh;$$
$$R_3 = CH_2(CH_2)_2 P(CH_2CH_2PPh)_2$$

As one would expect, IR spectra will contain a strong carbonyl band at about $1670 \, cm^{-1}$, but in addition, acid chloride bands appear at 1710 and $1770 \, cm^{-1}$ as a result of failure of some sites of the original polymer to react with the amide functional groups. As a result, poly-P_3 may react with $Mo(N_2)_2(PPh_2Me)_4$, which was monitored using FT-IR, giving rise to the shift of the dinitrogen ligand from 1925 to $1940 \, cm^{-1}$. Similar reactions of $Mo(N_2)_2(PPh_2Me)_4$ with poly-P_2 and a 2:1 mixture of poly-P and poly-P_3 result in the formation of polymer pendant complexes with the bands at 1952, 1925, and $1943 \, cm^{-1}$, respectively.

The formation of Li complexes with polymers has been considered in the studies conducted by Eschman et al.[47] In the case of polyethylene oxide mixed with $LiClO_4$ and polyethylene oxide mixed with $LiAsO_4$, a new band due to Li^{+} ion complexation at $879 \, cm^{-1}$ was observed.

As indicated by Pleuddemann,[48] adhesion can be improved when acid groups react with acidic functionalities. In the studies of benzimidazole (BIMH) reacted with copper in the presence of oxygen it was found that the coupling mechanism relies on the formation of benzimidazolato copper ($+1$) complexes.[49] The argument is that there is an absence of the N—H stretching and bending modes at 3200 and $1587 \, cm^{-1}$, respectively, in the surface spectra recorded for adsorbed BIMH. According to Lundberg,[50] chelation of O,O'-dibenzoylisophthalamide

TABLE 8.11   Infrared Bands Observed for Dicyandiamide and Zinc Reaction Products[a]

| Thin Films on Zn | | Bulk | | |
Heated	Unheated	Zn Complex	Dicyandi Amide[b]	Vibrational Assignment[b]
	3434 (m)			
		3440 (s)		
	3382 (m)		3380 (s)	
3332 (m)	3336 (m)	3341 (s)	3330 (s)	NH_2 asymmetric stretch
	3242 (m)	3159 (sh)	3185 (s)	
3176 (w, br)	3190 (m)		3154 (s)	NH_2 symmetric stretch
	3156 (m)	2237 (s)		
2204 (sh)	2211 (s)		2208 (m)	
		2195 (s)		$N-C{\equiv}N$ asymmetric stretch $(C{\equiv}N)$
2160 (m)	2169 (s)		2165 (m)	
2018 (m)				?
	1666 (s)		1658 (s)	
1638 (m)		1642 (s)		NH_2 deformation
	1647 (m)		1639 (s)	
1568 (s)	1576 (s)	1551 (s)	1587 (m)	$N-C-N$ asymmetric stretch
	1510 (m)		1506 (m)	$N{=}C-N$ asymmetric stretch $(C{=}N)$
		1433 (w)		
1421 (m)				?
1277 (sh)		1285 (w)		
1255 (m)	1257 (m)	1245 (vw)	1254 (m)	$N{\equiv}C-N$ symmetric stretch $(C-N)$
1184 (m)				
1146 (m)				?
1118 (m)				
	1098 (m)	1091 (w)	1096 (w)	NH_2 rock
		971 (vw)		
929 (w)	931 (w)	929 (w)	929 (m)	$N-C-N$ symmetric stretch $(C-N)$
813 (w)				
720 (w)	722 (vw)	724 (vw)	720 (w)	$N{=}C-N$ wag
670 (w)	671 (w)	663 (w)	669 (m)	$N{=}C-N$ deformation
588 (m)	571 (w)		554 (m)	NH_2 wag
508 (m)			528 (m)	$N{\equiv}C-N$ twist
493 (m)			500 (m)	$N{=}C-N$ rock

[a] Abbreviations: w, weak; m, medium; s, strong; v, very; sh, shoulder; br, broad; asym., asymmetric; sym., symmetric. All frequencies in cm^{-1}.

[b] Dicyandiamide vibrational frequencies and assignment consistent with W. J. Jones, and W. J. Orville-Thomas, *Trans. Faraday Soc.*, 1959, **55**, 193.

Source: Adopted from ref. 45.

oxime with Cu^{2+} results in the IR bands at 3250 (N—H stretch), 1730 (C=O stretch), and the 1630-cm^{-1} band characteristic of C=N stretching vibrations. In addition, the chelation was demonstrated by the presence of bands at 1560, 1430, and 1350 cm^{-1}.

Derivatives of imidazole are important corrosion inhibitors for metals.[51] Specifically, benzotriazole (BTAH) and 2-mercaptobenzimidazole (MBI)[52] have been used for controlling surface oxidation on copper and copper alloys. When MBI is deposited on the copper oxide surface, the bands at 1170 and 600 cm^{-1} due to C=S vibrations are present.[53]

We have already considered the requirements for obtaining IR spectra of thin films on metal surfaces. The usable spectra of films as thin as monolayer thickness can be obtained using the reflection–absorption (RA) technique with parallel polarized IR light reflected from the metal surface at a grazing angle. Although Raman spectroscopy is a complementary technique to IR measurements, the sensitivity of Raman spectroscopy is considered to be somewhat lower, and therefore, only some applications to the problems involving adhesives and coatings on metal are found.

8.6   COMPOSITE INTERFACES

Thermally stable fibers and matrices are only prerequisites to produce stable composites because no matter how good the stability of both components is, the bonding between them will usually dictate the final properties. In an effort to establish what functional groups on silane coupling agents may provide suitable interfacial thermal stability for Nextel ceramic fibers (SiO_2, B_2O_3, and Al_2O_3), photoacoustic FT-IR measurements can be employed.[54] These studies indicated that a combination of epoxy and aromatic functionalities into the silane backbone tremendously enhances interfacial thermal properties. Figure 8.3A illustrates a series of photoacoustic FT-IR spectra of silane coupling agent deposited on the surface of Nextel fibers exposed to various temperatures. As shown in Fig. 8.3B, integrated intensities of the C—H stretching band indicate that even exposure to 500°C for one hour does not remove epoxy/aromatic ring-containing silanes from the surface.

In composite technology, hydrolytic stability of the polymer matrix and interface between the fiber and matrix is essential in order to maintain mechanical and chemical stability of a composite. In the studies of Nextel/polyimide interfaces again using the nondestructive photoacoustic FT-IR approach, it was found that the most thermally stable interface ought to contain a combination of epoxy and aromatic functionalities. It appears that this combination provides a useful chemical entity, forming a crosslinked interface.

Polyimides are a broad class of high-performance polymers that meet the needs for elevated temperatures and hydrolytically stable matrix materials.[55] Their preparation usually involves polymerization of monomeric reactants (PMR), such as 3,3′,4,4′-benzophenonetetracarboxylic diahydride, 5-norbor-

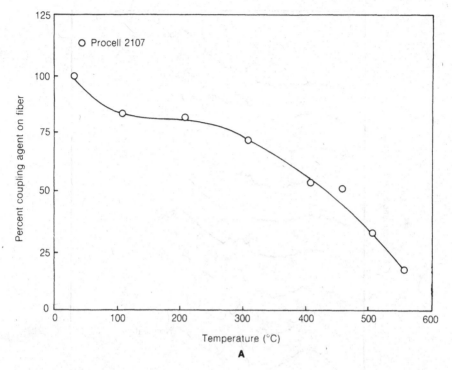

Figure 8.3 (*A*) Percent of coupling agent on the surface of Nextel fibers plotted as a function of temperature; (*B*). PA FT-IR spectra in the C—H stretching region of the silane coupling agent on the surface of Nextel fibers treated at various temperatures. (Reprinted with permission from ref. 54).

nen-2, 3-dicarboxylic anhydride, and 4, 4'-methylene dianiline. Urban et al.[56] have shown that photoacoustic FT-IR spectroscopy can be used to monitor interfacial stability of the Nextel fiber/polyimide composite interfaces fractured and exposed to aqueous environments. On the basis of spectral changes, several types of interfacial interactions have been identified. They are shown in Fig. 8.4 along with the vibrational bands attributed to these interactions.

Interfacial properties of graphite fibers are quite significant because they may affect many composite properties. While extensive research is currently being conducted by many groups, spectroscopic contributions of Ishida et al.[57] on oxidized graphitized carbon fiber interfaces are of particular importance. Besides experimental ATR advances demonstrating that the presence of surface acid groups can activate interfacial crosslinking reactions, an exact optical approach was utilized for quantitative analysis of the results. As a result of oxidation, three main bands due to carbonyl ($1720 \, cm^{-1}$), quinone ($1580 \, cm^{-1}$), and the C—O stretching and O—H bending modes ($\sim 1200 \, cm^{-1}$) were detected.

Many model studies on composite interfaces were carried out for carbon and graphite fibers. Perhaps the most convincing experimental and theoretical

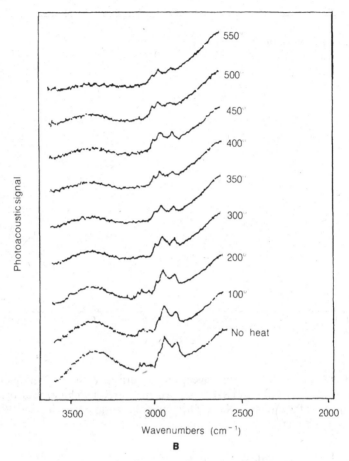

Figure 8.3 *(Continued)*

approaches came out of Ishida's group.[58,59] Because graphite exhibits semi-conducting nature and the high magnitude of its absorption coefficient, it can enhance IR bands of a thin film of poly(vinyl acetate).[60] Carbon-filled polymers have been studied using IR spectroscopy.[61] In fact, they have a broad spectrum of uses, ranging from fibers in composites to pigmentation and fillers. Claybourn et al.[62] have taken an interesting and useful approach involving the use of specular reflection at normal incidence in IR microscope experiments, followed by Kramers–Kronig analysis of the spectra. It should be realized that these and as well as other recent studies on carbon-filled polymers disprove previous believes that surface analysis of carbon black by IR transmission is made difficult by the nonreflecting, highly absorbing nature of the material. There are many surface functional groups, including hydrogen-bridged OH at $3425\,cm^{-1}$, CH_3 and CH_2 groups at 2923 and $2860\,cm^{-1}$, $C{=}O$ at 1600 cm^{-1} due to highly conjugated quinone conformation or aromatic ring stretching bands, and COOH groups with a broadband at $1335–1000\,cm^{-1}$.[63]

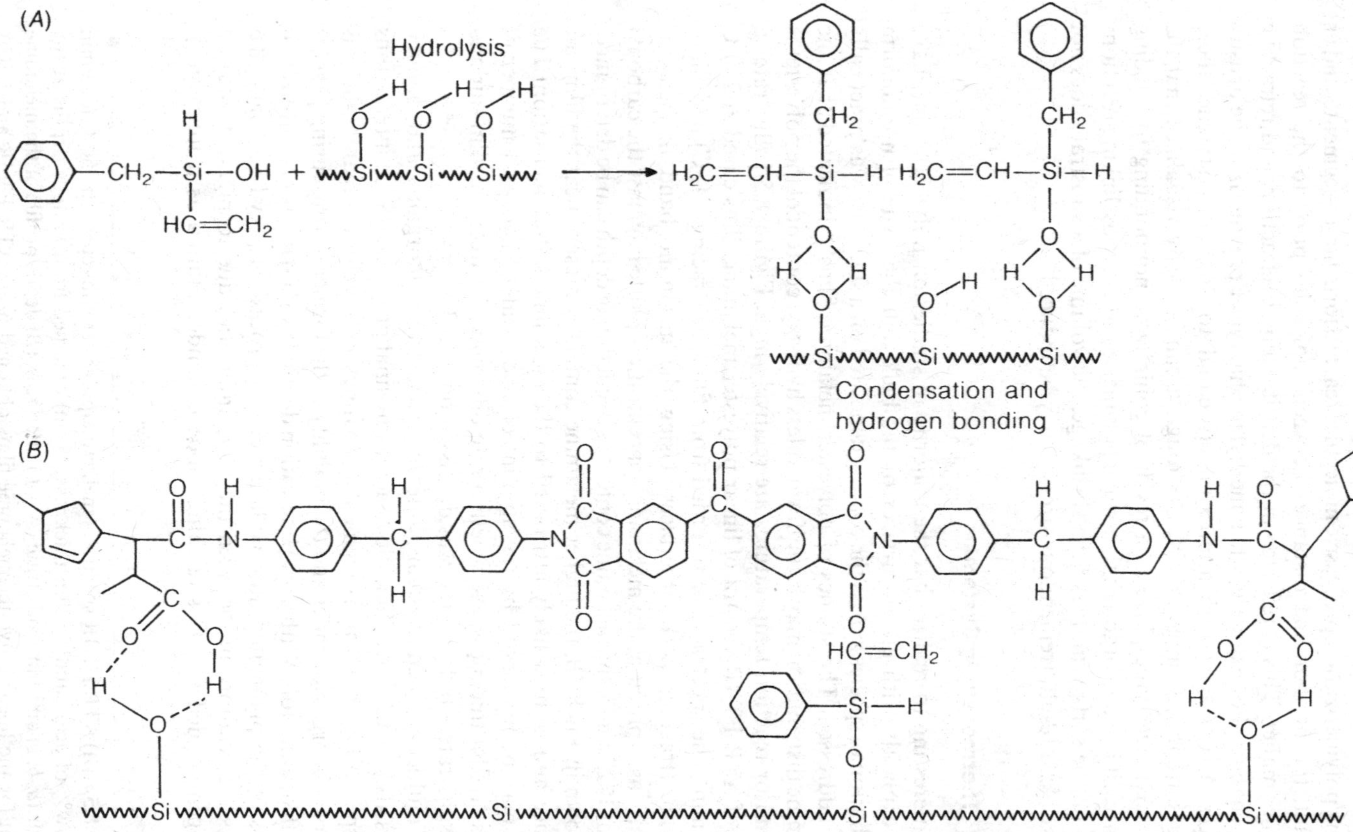

Figure 8.4 (A) Hydrolysis and condensation of vinyl terminated coupling agents; (B) formation of hydrogen bonding at the interfaces of polyimide- and surface-treated fibers. (Adopted from ref. 56).

325

8.7   FILM FORMATION AND CROSSLINKING ON SURFACES

During polymerization processes, molecules react to form larger segments, and as a result, the nonbonding distances between molecules prior to the reaction become smaller. This is reflected in the density, molar refraction, and refractive index changes. For example, when methyl methacrylate polymerizes, the density changes $\Delta\rho/\rho \times 100\%$ are approximately equal to 26%. For styrene, these changes are in the range of 17%. Although it can be easily visualized that the volume changes take place as a result of going from nonbonding to bonding distances, it is often forgotten that when polymerization or crosslinking reactions occur on the surface and result in shrinkage, the volume of a substrate does not change. As a result, tremendous stresses may be induced.

8.7.1   Thermosetting Systems

It is interesting to realize that the majority of studies on polymeric films have dealt explicitly with either surfaces or the bulk, but the issue of nonuniform distribution of species across the coating deposited on a substrate was practically never addressed. The studies of polyester/melamine films[64] indicated that the melamine distribution may be uniform unless hydroxyl content of the polyester is low and/or reaction temperatures are relatively high. Figure 8.5A illustrates a series of ATR FT-IR spectra of linear polyester/melamine films cured at 250°C for 1 min. The spectra were recorded from film–air (tracings A–C) and film–substrate (tracings D–F) interfaces. Using the melamine band intensity at 815 cm^{-1} as a probe of the melamine concentration plotted against the carbonyl band at 1725 cm^{-1}, the melamine content as a function of depth can be determined. As shown in Fig 8.5B, the higher melamine content is present near the film–air interface and is most likely attributed to the melamine self-condensation. The latter may be influenced by the amount of acid catalyst, hydroxyl number of polyester, film thickness, the reaction rate differences between the film–air and film–substrate interfaces and was discussed on p. 283.

In multicomponent systems, interactions between inorganic and organic phases also play an essential role in achieving final properties. Such interactions are especially important in pigmented coatings in which an organic phase is immersed in organic binder and the mobility of the former during curing process may effect adhesion. With a set penetration depth, PA FT-IR has been utilized in studies of the pigmented autooxidative curing process of alkyds.[65] Figure 8.6 depicts the relationship between the band intensities due to pigment and the hydrogen-bonded C=O stretching mode of binder plotted as a function of

--→

Figure 8.5   (A) ATR FT-IR spectra of melamine/polyester films cured at 250°C for 1 min (0.5 w/w% of acid catalyst; film thickness 25 μm) obtained from various depths at the film–air (a, b, c) and film–substrate (d, e, f) interfaces; (B) depth profile of the melamine content for the linear polyester/melamine films obtained using ATR FT-IR and ESCA measurements. (Reprinted with permission from ref. 64).

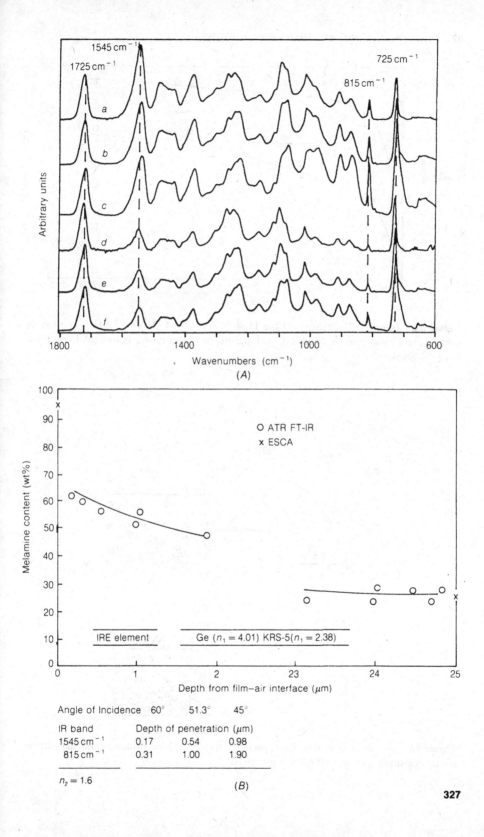

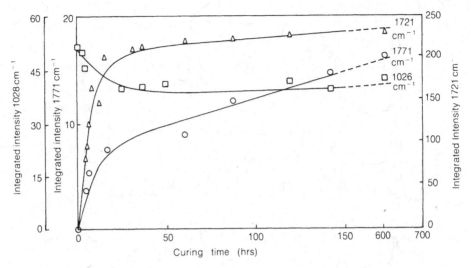

Figure 8.6 Integrated intensities of the 1028-, 1771-, and 1721-cm^{-1} bands plotted as function of curing time. (Reproduced with permission from ref. 65, copyright 1990, John Wiley & Sons).

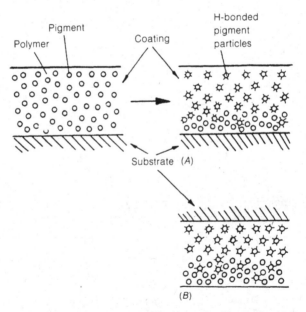

Figure 8.7 A distribution of pigment particles across the film surface: (*A*) normal curing (*B*) upside-down curing. (Adopted from ref. 65).

curing time. It is interesting to note that as the band due to the C$=$O stretching normal vibration increases, the 1026-cm^{-1} band decreases. This observation indicates that the binder particles undergo a stratification process during curing that parallels the increase of the hydrogen-bonded carbonyl band at 1721 cm^{-1}. In order to reveal the nature of forces that govern this process, the coating was cured in the upside-down configuration. Although the changes in C$=$O stretching intensity were the same, the band due to pigment increased, indicating that pigment particles moved toward the surface. These data illustrate that the gravitational forces and hydrogen bonding between pigment and binder are responsible for the distribution of inorganic phase within the binder. This phenomenon is schematically depicted in Fig. 8.7 and may inherently influence the film adhesion to a substrate since uneven distribution of pigment particles will result in the differences of pigment volume content (PVC) across the film.

8.7.2 Latex Films, Surfactants, and Copolymers

Before we identify spectroscopic features of selected latex systems, let us realize that surfactants play a vital role in latex technology, both in the synthesis of the latex and as postsynthesis additives. The latter has a major influence on the stabilization of the latex against coagulation, the modification of rheologic properties, and aids in the dispersion and stabilization of added pigments. Despite their utility, these low-molecular weight species can also give rise to a host of undesirable properties. If the surfactant is incompatible with the polymer system, it can migrate to the latex–film interface, resulting in optical defects, premature degradation, or loss of adhesion.

Although the importance of surfactants and the potential problems associated with their use are well established, little work has been done with regard to surfactant–polymer compatibility and surfactant mobility within applied latex films. A series of styrene–butadiene (SB) latices prepared with nonylphenol ethylene oxide surfactants were examined via electron microscopy. The compatibility of the surfactants with the nonpolar SB latices was found to decrease with increasing polarity of the surfactant.[66] On the other hand, in the case of more polar vinyl acetate–vinyl acrylate latices prepared with the same nonionic surfactants, surfactant adsorption increased as a result of higher polarity of the latex which resulted form its greater vinyl acetate content.[67] Apparently, the hydrolysis of vinyl acetate groups to form poly(vinyl alcohol) provides OH functionality for the polyether oxygens of the surfactant to interact with. This is why the formation of polymer–surfactant complexes may affect copolymer–surfactant compatibility, and the classification of various surfactants into penetrating and nonpenetrating types may be valid.[68] This classification was proposed[67] for a series of vinyl acetate–vinyl acrylate latices, which exhibit increasing copolymer polarity, resulting in induced penetration of anionic surfactants into the latex particles. This conclusion was based on viscosity studies, and the effect was attributed to the formation of poly-electrolyte type solubilized polymer–surfactant complexes. Apparently the polymer chains may uncoil resulting in acetyl groups being pushed into the aqueous phase where surfactant molecules may readily adsorb on the polymer chains leading to the increased solubility of these segments.[68] As one may expect, penetration of the surfactant was found to depend on a critical size, the charge density at the polymer–water interface, and a shape conducive to penetration. In the case of nonionic surfactants, the surfactant concentration was found to decrease with increasing polarity while, at the same time, inhibiting penetration by the anionic surfactants.

The presence of interfacial surface tension at the polymer–water interface may affect the degree of surfactant adsorption on various polymer surfaces.[69] The surface area per molecule of sodium dodecyl surface (SDS) surfactant on a latex polymer particle was shown to increase with the increasing polarity of the polymer–water interface. It was suggested that the driving force for the adsorption of surfactant at various polymer–water interfaces is related to the differences in the interaction energy between the surfactant molecules and the surface in question.[67,70]

On a similar note, it is reasonable to expect that the surface free energy of the substrate may influence the distribution of surfactant within the latex film. Bradford and Vanderhoff[71] prepared styrene–butadiene copolymer latex films on a variety of substrates including poly(tetrafluoroethylene) (PTFE), Mylar (polyethylene terephthalate) polyester, rubber, and mercury and examined the film–air and film–substrate interfaces via electron microscopy. Differences were observed for the film–substrate interfaces, but they were determined to be too small for further analysis and considerations. It was concluded that the surfactant exudation and film formation behavior at the film–substrate interface was closely parallel to that at the film–air interface.

When Zhao et al.[72,73] employed ATR FT-IR, x-ray photoelectron spectroscopy (XPS), and secondary-ion mass spectrometry (SIMS) to examine the film–air and film–substrate interfaces of butyl acrylate–methyl methacrylate latices prepared using the anionic surfactants sodium dodecyl sulfate (SDS) and sodium dodecyl diphenyl disulfonate (SDED), the story was quite different. The latex films exhibited enrichment at both the film–air and film–substrate interfaces. The extent of enrichment was found to be dependent of the nature of the surfactant, the coalescence time, the global concentration of the surfactant, and the interface involved, with the film–air interface showing a greater degree of enrichment. SDS was found to form a thick boundary layer at the film–substrate interface. This "weak boundary layer," which causes major adhesion problems, was attributed to a lesser degree of compatibility with the copolymer. Surfactant enrichment was attributed to three factors: (1) initial enrichment at both interfaces in order to lower interfacial free energy, (2) enrichment at the film–air interface due to the transport of nonadsorbed surfactant by the water flux out of the film, and (3) longer-term migration to both interfaces due to surfactant incompatibility.

Although these studies have provided some insight into the factors governing surfactant compatibility and surfactant exudation behavior, they have generally been confined to the study of either a specific series of latices or a specific series of surfactants, and, as such, it is difficult to assess the ultimate chemical factors that

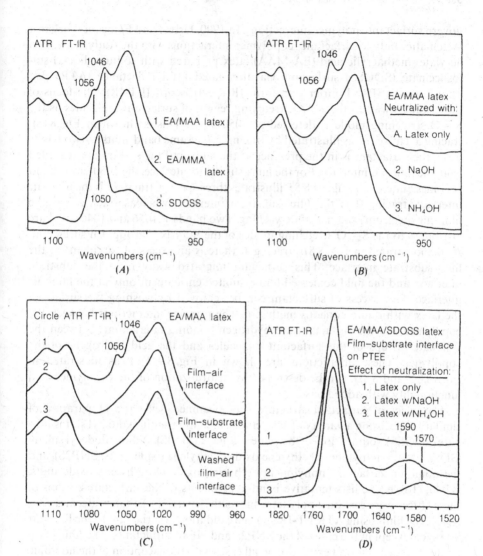

Figure 8.8 (*A*) ATR FT-IR spectra in the 1100–950 cm^{-1} region: tracing 1—film–air interface of the EA/MAA latex; tracing 2—film–substrate interface of the EA/MAA latex; tracing 3—SDOSS surfactant; (*B*) ATR FT-IR spectra of EA/MAA latex: tracing 1—film–air interface; tracing 2—film–substrate interface; tracing 3—film–air interface washed with Meoh/DDI H$_2$O solution; (*C*) ATR FT-IR spectra in the 1100–950-cm^{-1} region of EA/MAA latex (tracing 1), EA/MAA latex neutralized with NaOH (tracing 2), EA/MAA latex neutralized with NH$_4$OH (tracing 3); (*D*) ATR FT-IR spectra in the 1800–1520-cm^{-1} region of EA/MAA latex (tracing 1) EA/MAA latex neutralized with NaOH (tracing 2) EA/MAA latex neutralized with NH$_4$OH (tracing 3). Characteristic C—O stretching bands due to the COO$^-$Na$^+$ or COO$^-$NH$_4^+$ salt formation are detected. (Reproduced with permission from ref. 74, copyright 1990 Butterworth-Heinemann, and refs. 75–77, copyright 1991, John Wiley & Sons).

govern surfactant–polymer interactions. In 1990, Urban and Evanson[74] began a systematic study of surfactant–copolymer interactions with the study of an ethyl acrylate–methacrylic acid (EA-MAA) latex prepared with sodium dioctyl sulfosuccinate (SDOSS) surfactant via attenuated total reflectance (ATR) and photoacoustic (PA) Fourier transform IR spectroscopy (FT-IR). The focus of these studies was on the S—O stretching region of surfactant molecules as the S—O stretching mode is detected at $1050 \, cm^{-1}$. This is shown in Fig. 8.8A, tracing 3. However, as illustrated by tracing 1, the same band splits to two bands when the surfactant is in the presence of the latex and the spectra are recorded from the film–air interface. For the film–substrate interface, the $1050\text{-}cm^{-1}$ band becomes quite weak. Figure 8.8B illustrates three spectra: tracing 1 is the film–air interface, tracing 2 is the film–substrate interface, whereas tracing 3 is the film–air spectrum obtained after washing. Two bands at 1056 and $1046 \, cm^{-1}$ are attributed to the S—O stretching modes of the SO_3^- Na^+ groups on surfactant. As demonstrated in Fig. 8.8B, tracing 1, there is an excess of surfactant at the film–substrate interface. This surfactant migrated away from the substrate interface, and the final coalesced film exhibited enrichment only at the film–air interface. This excess of surfactant can be removed by washing the surface of the latex with dilute aqueous methanol. The primary interactions between the copolymer and the surfactant are hydrogen-bonding interactions between the sulfonate groups of the surfactant molecules and the acid hydrogens of the copolymer. These interactions are shown in Fig. 8.9 and, as illustrated in Figs. 8.8C and 8.8D, can be destroyed by neutralization of the copolymer acid functionality with added base.

The effect of surfactant structure was examined[75] via the preparation of similar latices using a variety of surfactants, including sodium dioctyl sulfosuccinate (SDOSS), sodium dodecylbenzene sulfonate (SDBS), sodium dodecyl sulfate (SDS), sodium nonylphenol ethylene oxide (2 ethylene oxide units; SNP2S), and the nonionic surfactant nonylphenol ethylene oxide (40 ethylene oxide units; NP40). Table 8.12 lists tentative band assignments of the surfactants. Films of these latices were cast on a poly(tetrafluoroethylene) (PTFE) substrate, and analysis of the films using ATR FT-IR showed exudation of surfactant to the film–air interface, except in the cases of the SNP2S and NP40 surfactants. The film–substrate interfaces showed exudation in all cases with the exception of the nonionic

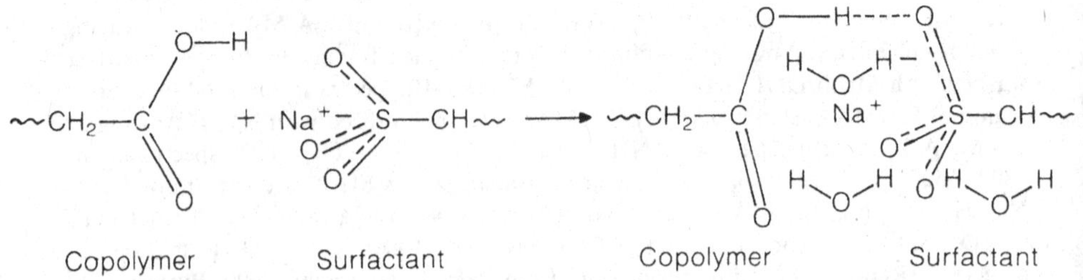

Figure 8.9 Schematic representation of surfactant–weak acid interactions.

TABLE 8.12 Infrared Bands and Their Tentative Assignments for SDBS, SDOSS, SNP2S, SDS, and NP-40, Along with the Bands Observed for EA/MAA Copolymer

A

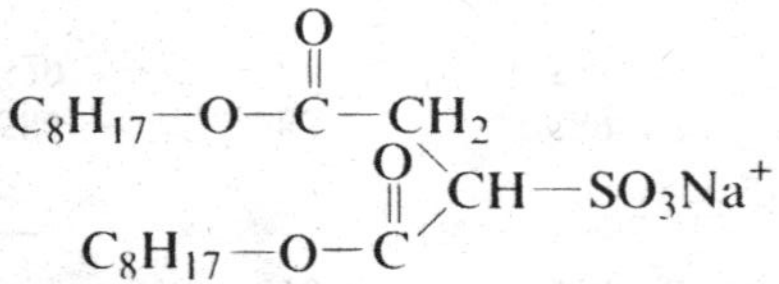

C_8H_{17}—O—C(=O)—CH_2
C_8H_{17}—O—C(=O)—CH—SO_3Na^+

Sodium dioctylsulfosuccinate (SDOSS)

B

$C_{12}H_{25}$—⟨benzene⟩—SO_3Na^+

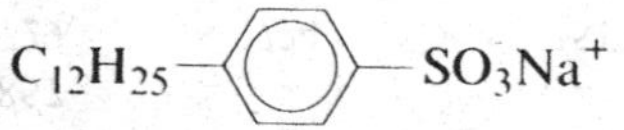

Sodium dodecylbenzenesulfonate (SDBS)

Sulfonate

C

C_9H_{19}—⟨benzene⟩—$OCH_2CH_2OCH_2CH_2$—SO_3Na^+

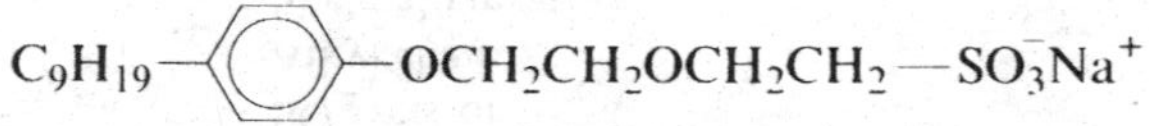

Sodium nonylphenol ethylene oxide (SNP2S)

D

$C_{12}H_{25}$—O—SO_3Na^+

Sodium dodecylsulfate (SDS)

E

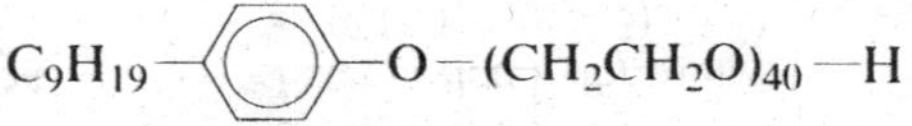

C_9H_{19}—⟨benzene⟩—O—$(CH_2CH_2O)_{40}$—H

Nonylphenol ethylene oxide (40 units) (NP-40)

SDOSS	SDBS	SNP2S	SDS	NP	Copolymer	Assignment
—	—	—	—	—	2981	Asymmetric C—H stretch (CH_3)
2960	2958	2954	2956	2950	2960	Asymmetric C—H stretch (CH_3)
2934	2925	2930	2919	2930	2934	Asymmetric C—H stretch (CH_2)
2879	2871	2877	2875	2884	2879	Symmetric C—H stretch (CH_3)
1735	—	—	—	—	1735	C=O stretch
—	1625	—	—	—	1700	H-bonded—COOH
—	1603	1611	—	1609	—	Ar—S stretch
—	—	1582	—	1580	—	p-Substituted aromatic
—	—	1513	—	1513	—	C=C aromatic
—	1493	—	—	—	—	C=C aromatic
1464	1463	1466	1468	1466	1466	C—H deformatic
1416	—	—	—	1455	1447	CH_2 scissor
1393	—	1395	—	—	—	HC—S deformation
1360	1378	1364	1380	1360	1382	C—(CH_3) symmetric deformation
1314	—	1295	—	1279	1299	CH_2 wagging

TABLE 8.12 (*Continued*)

SDOSS	SDBS	SNP2S	SDS	NP	Copolymer	Assignment
—	—	—	1248	—	—	S—O stretch (SO_4)
1241	—	1245	1221	1241	1252	C—O stretch
1216	1192	1185	—	—	—	Asymmetric S—O stretch (SO_3)
1175	—	—	—	1146	1173	Asymmetric C—O—C stretch
1094	—	—	—	1108	1098	Symmetric C—O—C stretch
—	—	—	1071	—	—	Symmetric S—O stretch (SO_4)
1050	1046	1056	—	—	—	Symmetric S—O stretch (SO_3)
1025	—	—	—	—	1025	C—C—O (ester)
—	1013	1013	—	—	—	=C—H in-plane deformation
—	—	942	—	947	—	CH_2—O (ether)
—	832	828	—	843	—	=C—H out-of-plane
—	—	—	830	—	—	S—O—C stretch
857	—	—	—	—	854	Ester skeletal vibration
729	724	751	722	—	—	—$(CH_2)_n$— ($n > 3$)
—	691	683	—	—	—	CH out-of-plane
652	616	614	—	—	—	S—O bending (SO_3)
—	—	—	631	—	—	S—O bending (SO_4)
581	583	589	585	—	—	SO_2 scissor
529	—	531	—	529	—	Alkyl chain
—	—	—	—	511	—	Skeletal vibrations

Source: Adopted from ref. 75.

NP40. With the aforementioned effects of polarity on surfactant–copolymer compatibility in mind, aqueous base was added to the latices to neutralize the acid functionality of the copolymer and thus increase the polarity of the copolymer environment. Films of the neutralized latices were then cast on PTFE and analyzed in a similar manner. Films of the neutralized latices were found to exhibit no surfactant exudation, except for the case of the SDS surfactant, which exhibits some exudation to the film–air interface. This observation was attributed to a lesser degree of compatibility of the SDS with the copolymer.

It is well established that the term *surfactant* is derived from its function as a surface-active molecule that has hydrophylic and hydrophobic ends. Hence, it is capable of modifying surface tension. In fact, surfactant molecules tend to minimize an excess of the surface tension, and this is why they often concentrate on surfaces. However, what if the substrate on which latex was deposited has an excess of surface energy? The effects of substrate surface tension and elongational strain in latices were also examined[76] and showed that when latex films were cast directly on the KRS-5 ATR element, which has a surface tension similar to that of glass (70 mN/m), it was found that SDOSS, SNP2S, and the nonionic NP40 exhibited no exudation to the film–substrate interface. On the other hand, SDS and SDBS did exhibit some exudation, but it was found to be substantially smaller than that for the films deposited on the PTFE substrate. On a mercury substrate, all anionic surfactants showed exhudation to the film–substrate interface while the nonionic NP40 showed no enrichment. When films of the various latices were subjected to mechanical elongations of 10, 30, and 50%, the anionic surfactants all showed enrichment that increased with the degree of elongation. The NP40 once again appeared to stay uniformly distributed across the film thickness, with no spectral changes being observed for elongations of up to 50%.

The effect of the *substrate* on surfactant *exhudation* behavior as illustrated in Fig. 8.10 can be explained in terms of the relative surface free energies of the copolymer and the substrate.[77] In the case of PTFE, which has a very low surface

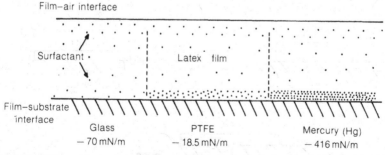

Figure 8.10 Schematic representation of the substrate effect on the distribution of surfactant across the latex films deposited on glass, polyterafluoroethylene (Teflon), and liquid mercury. (Reproduced with permission from ref. 77, copyright 1991, John Wiley & Sons).

free energy of 18.5 mN/m, there exists a considerable driving force for the migration of surfactant to the film–substrate interface in order to lower the high interfacial surface tension present there. In the case of the glass substrate, with a surface free energy of approximately 70 mN/m, the polymer (surface free energy approximately 30 mN/m) may readily wet the glass substrate; thus the driving force for surfactant migration to this interface is significantly reduced. In the case of the mercury substrate, its very high surface tension (416 mN/m) allows the polymer initially to wet the surface, but once coalescence has occurred, there exists a solid latex film–liquid mercury interface. This leads to a high interfacial tension, giving rise to a higher driving force for surfactant enrichment. Enhanced anionic surfactant enrichment on mechanical elongation was also explained in terms of surface tension. When the film is elongated, surface surfactant concentration decreases, resulting in a higher surface tension. This increased surface tension provides a driving force for further migration of surfactant to the film interfaces.

8.7.3 Thermosetting-Thermoplastic Interactions

Typically, coatings for plastics have poor adhesion to the substrate and interfacial failure may occur when relatively low stress is applied. Since a problem commonly encountered in the film technology and coatings industry is adhesion to substrates, a novel technique, called *rheophotoacoustic FT-IR* technique, has been developed.[78-80] In essence, it is a stress–strain device built into the photoacoustic sample compartment that, when combined to an FT-IR spectrometer, allows one to monitor molecular changes on applying stress to the substrate. If the sample consists of two layers, the coating and the substrate, such as shown in Fig. 8.11, and stress is applied to the substrate only, the coating will follow elongation of the substrate if interfacial forces are high. On the other hand, failure of the interfacial bonding will result in the formation of voids at the interface, followed by separation of the coating from the substrate. To demonstrate the feasibility of rheophotoacoustic measurements, Fig. 8.12 illustrates the spectra of polyethylene coated with a siloxane adhesive. Tracings *A*, *B*, and *C* of Fig. 8.12

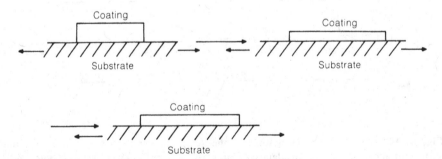

Figure 8.11 Schematic representation of the elongational process of the silane/polyethylene double layer.

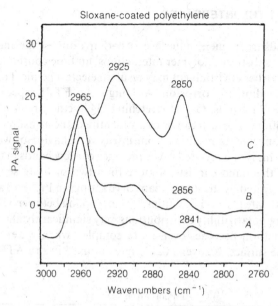

Figure 8.12   PA FT-IR spectra in the 3000–2700-cm^{-1} region of siloxane/polyethylene at various stages of elongation: (A) 0%; (B) 8.4%; (C) 16.8%. (Reproduced with permission from ref. 80).

represent the spectra recorded at 0.0, 8.4, and 16.8% elongations, respectively. While IR bands at 2965, 2925, and 2850 cm^{-1} serve as a reference for further analysis of the stretched double-layer films, tracings B and C in Fig. 8.12 indicate the increase of polyethylene substrate bands as the film is elongated. The initial intensity increase of the 2965-cm^{-1} band is attributed to the thinning of siloxane coating during the stretching process. Specifically, as the substrate is stretched, the siloxane coating follows the elongation of the polyethylene substrate and its thickness diminishes while the bands due to polyethylene increase. If the interfacial bonding between the coating and the substrate were strong, a gradual increase of the polyethylene bands with elongation would be expected simply because of the thinner siloxane layer on the surface of polyethylene. Surprisingly, when the sample is elongated to 16.8%, a strong increase of the polyethylene bands at 2925 and 2850 cm^{-1} is observed. Because thinning of the siloxane coating cannot be responsible for such drastic changes, the preceding observation indicates interfacial failure resulting in the formation of microvoids between the coating and the substrate. Subsequently, the acoustic waves generated at the interface can escape through microvoids without passing through the top siloxane layer. The microvoid formation was independently confirmed by taking the electron microscopy pictures at exactly the same elongations as the recorded photoacoustic spectra. To the author's best knowledge, the rheophotoacoustic FT-IR approach is the first study demonstrating the feasibility of spectroscopic detection of interfacial failure.

8.8 WATER AT THE INTERFACES

Water at the adsorber–metal interface is perhaps one of the most significant factors affecting substrate–polymer interactions, and the consequences of water entering the adsorber environment may result in electrochemical reactions at the metal surface leading to corrosion. Although the FT-IR spectrum of water contains two major bands, O—H stretching (3400-cm^{-1} region) and bending modes at $\sim 1640\,cm^{-1}$, which can be easily identified, the measurements of orientation or rearrangements resulting from intrusive behavior of water are not a simple matter. This is often related to the presence of a few bands in the 3400-cm^{-1} region. In fact, this band consists of more than two or more separate spectral features. As illustrated by the Lorentzian curve fittings in Fig. 8.13 and supported by theoretical Monte Carlo simulations,[81] the coupling between water molecules and the resulting polarizibility contributions may significantly affect this spectral region. The situation becomes even more complex when water molecules are adsorbed on the surface. Nguyen et al.[82] have utilized in situ ATR FT-IR setup

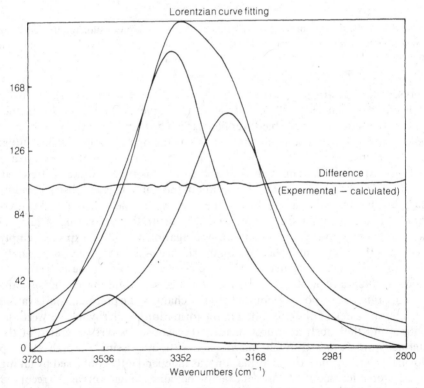

Figure 8.13 FT-IR spectrum of water. The boxed spectrum represents the Lorentzian decomposition of the O—H stretching region; the difference spectrum is the result of substraction of the experimental from the summed Lorentzian curves.

for monitoring diffusion of water through the pigmented and alkyd coatings, followed by subsequent polymer and interfacial degradation. Besides the expected intensity increase of water bands and diminishing bands due to epoxy or alkyds, no other changes were detected.

REFERENCES

1. H. Leidheiser and W. Funke, *J. Oil Colour Chem. Assoc.*, 1987, **70**(5), 121.
2. H. Leidheiser, *Corrosion*, 1983, **39**, 189.
3. E. P. Plueddemann, *Silane Coupling Agents*, Plenum Press, New York, 1982.
4. C. H. Chiang and J. L. Koenig, *Polymer Composites*, 1980, **1**, 88.
5. S. R. Culler, H. Ishida, and J. L. Koenig, *Polymer Composites*, 1986, **7**, 231.
6. A. Garton, *Polymer Composites*, 1984, **5**, 258.
7. A. Garton, *J. Polym. Sci., Polym. Chem. Ed.*, 1984, **22**, 1495.
8. N. H. Sung, A. Kaul, I. Chin, and C. S. P. Sung, *Polym. Eng. Sci.*, 1982, **22**, 637.
9. A. Kaul, N. H. Sung, I. Chin, and C. S. P. Sung, *Polym. Eng. Sci.*, 1984, **24**, 493.
10. S. Navoroj, S. R. Culler, J. L. Koenig, and H. Ishida, *J. Colloid. Interface Sci.*, 1984, **97**, 308.
11. P. Dreyfus and Y. Eckstein *J. Adhes.*, 1983, **15**, 163.
12. F. J. Boerio and D. J. Ondrus, in *Surface and Colloid Science in Computer Technology*, K. L. Mittel, Ed., Plenum Press, New York, 1987.
13. K. P. Battjes, A. M. Barolo, and P. Dreyfuss, *J. Adhes. Sci. Techn.*, 1991, **5**(10), 785.
14. D. J. Ondrus and F. J. Boerio, *J. Colloid Interface Sci.*, 1988, **124**, 349.
15. F. J. Boerio and D. J. Ondrus, *J. Colloid Interface Sc.*, 1990, **139**, 446.
16. R. G. Dillingham and F. J. Boerio, *J. Adhes. Sci. Techn.*, 1992, **6** (1), 207.
17. F. J. Boerio, P. P. Hong, P. J. Clark, and Y. Okamoto, *Langmuir*, 1990, **6**, 721.
18. W. H. Tsai, J. T. Young, and F. J. Boerio, *J. Adhes.*, 1991, **33**, 211.
19. E. A. Bekturov, *Polym. Commun.*, 1985, **26**(3), 81.
20. J. L. Lippert, J. A. Robertson, J. R. Havens, and J. S. Tan, *Macromolecules*, 1985, **63**, 18.
21. I. Cabasso, *J. Appl. Polym. Sci.*, 1989, **38**, 1653.
22. P. H. McCluskey, R. L. Snyder, and R. A. Condrate, *J. Solid State Chem.*, 1989, **83** (2), 332.
23. R. A. Condrate et al., *J. Solid State Chem.*, 1989, **26**, 1485.
24. C. F. Cullis and H. S. Berr, *Eur. Polym. J.*, 1978, **14**, 575.
25. D. L. Allara, *Org. Coat. Plast. Chem.*, 1976, **36**, 399.
26. H. G. Linde and R. T. Gleason, *J. Polym. Sci., Polym. Phys. Ed.*, 1989, **26**, 1485.
27. H. G. Linde, *J. Appl. Polym. Sci.*, 1990, **40**, 2049.
28. H. Leidheiser and P. D. Deck, *Science*, 1988, **241**, 1176.
29. H. G. Linde and R. T. Gleason, *J. Polym. Sci., Polym. Phys.*, 1989, **26**, 1485.
30. J. T. Young, W. H. Tsai, and F. J. Boerio, *Macromolecules*, 1992, **25**, 887.
31. H. G. Linde, *J. Appl. Polym. Sci.*, 1990, **40**, 2049.

32. R. Chen and F. J. Boerio, *J. Adhes. Sci. Techn.*, 1990, **4**(6), 453.

33. D. G. Kim, S. E. Molis, T. S. Oh, S. P. Kowalczyk, and J. Kim, *J. Adhes. Sci. Techn.*, 1991, **5** (7), 509.

34. G. N. Lewis, *J. Franklin Inst.*, 1938, **226**, 293.

35. R. S. Mulliken, *J. Phys. Chem.*, 1952, **56**, 801.

36. R. G. Pearson, *Hard and Soft Acids and Bases*, Dowen, Hutchington and Ross, Stroudsburg, Pa., 1973.

37. R. S. Drago, G. C. Vogel, and T. E. Needham, *J. Am. Chem. Soc.*, 1971, **93**, 6014.

38. F. M. Fowkes, D. O. Tischler, J. A. Wolfe, L. A. Lannigan, C. M. Ademu-John, and M. J. Halliwell, *J. Polym. Sci., Chem. Ed.*, 1984, **22**, 547.

39. M. N. Panel and B. N. Jani, *J. Macromol. Sci.*, 1985, **A22**(11), 1517.

40. A. Lehtinen, S. Purokoski, and J. J. Lindberg, *Makromol. Chem.*, 1975, **176**, 1553.

41. N. Hagihara, K. Sonogashira, and S. Takahashi, *Adv. Polym. Sci.*, 1981, **41**, 151.

42. M. G. Chan and D. L. Allara, *Polym. Eng. Sci.*, 1974, **4**, 12.

43. C. P. Cullis and H. S. Laver, *Eur. Polym. J.*, 1978, **14**, 575.

44. R. A. Dickie, J. S. Hammond, and J. Holubka, *Ind. Eng. Chem.*, 1981, **20**, 339.

45. R. O. Carter, VI, R. A. Dickie, J. W. Holubka, and N. E. Lindsay, *Ind. Eng. Chem. Res.*, 1989, **28**, 48.

46. D. L. DuBois and J. A. Turner, *J. Am. Chem. Soc.*, 1982, **104**(18), 4989.

47. J. Eschman, J. Strasser, M. Xu, Y. Okamoto, and E. Erying, *J. Phys. Chem.*, 1990, **94** (10), 3908.

48. E. P. Pleuddemann, *Silane Coupling Agents*, Plenum Press, New York, 1982.

49. G. Xue, G. Shi, J. Ding, W. Chang, and R. Chen, *J. Adhes. Sci. Techn.*, 1990, **4** (9), 723.

50. A. Lehtinen, S. Purokoski, and J. J. Lindberg, *Markromol. Chem.*, 1975, **176**, 1553.

51. R. J. Sunderberg and R. B. Martin, *Chem. Rev.*, 1974, **74**, 471.

52. G. W. Poling, *Corros. Sci.*, 1970, **10**, 359.

53. G. Xue, X. Y. Huang, J. Dong, and J. Zhang, *J. Electroanal. Chem.*, 1991, **310**, 139.

54. A. M. Tiefenthaler and M. W. Urban, *Composites*, 1989, **20**, 145.

55. I. W. Serfaty, in *Polyimides: Synthesis, Characterization, and Applications*, K. Mittel, Ed., Plenum Press, New York, 1984.

56. W. McDonald and M. W. Urban, *Composites*, 1991, **22** (4), 307.

57. C. Sellitti, J. L. Koenig, and H. Ishida, *Mater. Sci. Eng.*, 1990, **A126**, 235.

58. Y. Ishino and H. Ishida, *Anal. Chem.*, 1986, **58**, 2448.

59. Y. Ishino and H. Ishida, *Appl. Spectrosc.*, 1988, **42**, 1296.

60. C. Sellitti, J. L. Koenig, and H. Ishida, *Appl. Spectrosc.*, 1990, **44**(5), 830.

61. W. W. Hart, P. C. Painter, J. L. Koenig, and M. M. Coleman, *Appl. Spectrosc.*, 1977, **31**, 220.

62. M. Claybourn, P. Colombel, and J. Chalmers, *Appl. Spectrosc.*, 1991, **45**(2), 279.

63. F. Rositani, P. L. Antonucci, M. Minutoli, and N. Giordano, *Carbon*, 1987, **25**(3), 325.

64. T. Hirayama and M. W. Urban, *Prog. Org. Coat.*, 1992, **20**, 81.

65. M. W. Urban and E. M. Salazar-Rojas, *J. Polym. Sci., Chem. Ed.*, 1990, **28**, 1593.

66. J. W. Vanderhoff, *Br. Polym. J.*, 1970, **2**, 161–173.

67. B. R. Vijayendran, T. L. Bone, and C. Gajria, *J. Appl. Polym. Sci.*, 1981, **26**, 1351.

68. H. Arai and S. Horin, *J. Colloid Interface Sci.*, 1969, **30**, 372.

69. C. O. Timmons and W. A. Zisman, *J. Colloid Interface Sci.*, 1968, **28**, 106.

70. B. R. Vijayendran, *J. Appl. Polym. Sci.*, 1979, **23**, 733–742.

71. E. B. Bradford and J. W. Vanderhoff, *J. Macromol. Sci. phys.*, 1972, **B6**(4), 671–694.

72. C. L. Zhao, Y. Holl, T. Pith, and M. Lambla *Coll. Polym. Sci.*, 1987, **265**, 823–829.

73. C. L. Zhao, Y. Holl, T. Pith, and M. Lambla, *Br. Polym. J.*, 1989, **21**, 155–160.

74. M. W. Urban and K. W. Evanson, *Polym. Commun.*, 1990, **31**, 279.

75. M. W. Urban and K. W. Evanson, *J. Appl. Polym. Sci.*, 1991, **42**, 2287.

76. M. W. Urban and K. W. Evanson, *J. Appl. Polym. Sci.*, 1991, **42**, 2297.

77. M. W. Urban and K. W. Evanson, *J. Appl. Polym. Sci.*, 1991, **42**, 2309.

78. M. W. Urban and W. F. McDonald, *Sound and Vibration Damping in Polymers*, ACS Symposium Series, No.424, R. D. Corsaro and L. H. Sperling, Eds., American Chemical Society, Washington, D.C., 1990, Chapter 9, p. 151.

79. W. F. McDonald, H. Geottler, and M. W. Urban, *Appl. Spectrosc.*, 1989, **43** (8), 1387.

80. W. F. McDonald and M. W. Urban, *J. Adhes. Sci. Techn.*, 1990, **4**(9), 751.

81. J. R. Reimers and R. O. Watts, *Chem. Phys.*, 1984, **91**, 201.

82. T. Nguyen, E. Byrd, and C. Lin, *J. Adhes. Sci. Techn.*, 1991, **5** (9), 697.

CHAPTER 9

SURFACTANTS, COLLOIDAL INTERFACES, AND THIN FILMS ON SURFACES

9.1 CHALLENGES AND DIFFICULTIES

Although this book is devoted to surface vibrational spectroscopy, and early chapters elucidated on the fact that vibrational characteristics of molecules or macromolecules will be altered as a result of adsorption, very little has been said about the state from which species are being adsorbed. In other words, it was assumed that the adsorption process may occur from gaseous or liquid states. Such generalizations may have some value, but if one begins to consider molecules solubilized in a solvent, it is quickly realized that molecules (1) can be completely solubilized or dissolved, (2) may not dissolve at all and may precipitate out of the solvent and (3) may be only partially dissolved. While thermodynamic principles will dictate the occurence of one of thesest, the magnitude of solubilization will be determined not only by a chemical nature of the solute-solvent system but also by molecular weight. Furthermore, if one considers that conformational features of dissolved species may depend on not only chemical structures and molecular weight but also concentration, the issue becomes a bit complex. To make the entire problem even more challenging, the next question to be addressed is whether the aformentioned factors can affect adsorption processes on surfaces. With this in mind, let us consider vibrational features of a specific class of molecules that may have a dual character: one end of the molecule can be solubilized in water, whereas the other end can dissolve in hydrocarbon solvents. Such molecules are called *surfactants* (surface-active agents) because this dual feature allows them to alter surface tension in water.

9.2 SURFACTANTS IN AQUEOUS MEDIA

A dual property within one molecular entity composed of a polar head that is soluble in water and a nonpolar (hydrophobic) tail that is soluble in oil makes surfactants unique for various interfacial interactions. Because they may self-assemble in water to form aggregates called *micelles*, various phases having distinct architectural features can be obtained and their architectural makeup usually depends on structural composition, concentration, ionic strength, pressure, and temperature. The formed aggregates can be spherical, globular, spherical bilayers, rod-like, or similar. Figure 9.1 depicts schematically several selected structures of surfactant aggregates. Another distinguished feature of surfactant solutions is their ability to enhance solubility of hydrophobic solute molecules. With an aid of surfactants, usually the poor solubility of hydrocarbons can be enhanced by several orders of magnitude. As a result of their practical importance and interest in surfactant behavior, many experimental[1,2] and theoretical[3] studies documented in many symposia proceedings were conducted.

Although spectroscopic attempts have been made to identify sphere to rod transitions, the structural changes are relatively small, making detection of these

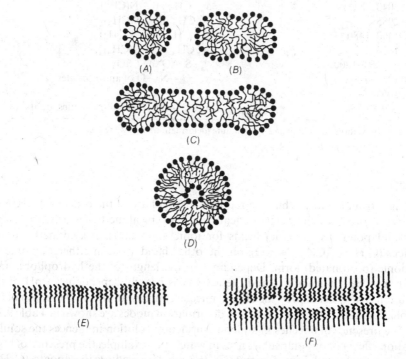

Figure 9.1 Various structural arrangements of surfactant aggregates in dilute aqueous solutions: (*A*) spherical; (*B*) globular; (*C*) spherocylindrical; (*D*) spherical bilayer vesicles; (*E*) single-layered; (*F*) double-layered.

TABLE 9.1   Typical IR Bands Observed in Surfactant Molecules and Their Sensitivity to Various Factors[a]

Wavenumber (cm^{-1})	Assignment
	Methylene Tails
2956	v_{as}, CH_3—R
2920	v_{as}, CH_2
2870	v_s, CH_3—R
2850	v_s, CH_2
1472 (singlet)	δ, CH_2 (triclinic subcell)
1472 + 1462 (doubler)	δ, CH_2 (orthorhombic subcell)
1468 (singlet)	δ, CH_2 (hexagonal or rotator phase
1468–1466	δ, CH_2("liquid-like" disorder)
1465–1456	δ_{as}, CH_3—R
1420	δ, α-CH_2
1380	δ_s, CH_3—R
1367, 1300	w, (CH_2 (-gtg'-, "kink" defect)
1354	w, CH_2 (-$tggt$-, "double gauche" defect)
1341	w, CH_2 (-tg, "end-gauche" defect)
	Head Groups
3040	v_{as}, CH_3 of —$N(CH_3)_3$
2985	v_s, CH_3 of —$N(CH_3)_3$
1490, 1480	δ_{as}, CH_3 of —$N(CH_3)_3$
1405	δ_s, CH_3 of —$N(CH_3)_3$
~1250–1200	v_{as}, S—O of —SO_4
1250–1200	v_{as}, C—N—O of amine oxides
1065	v_s, S—O of —SO_4
~970–950	v_{as}, C—N—C of amine oxides, quats

[a] Abbreviations: v = stretching, δ = bending (deformation), w = wagging.

Source: Adopted from ref. 4.

changes troublesome. While the reported changes are of the order of $1-3\,cm^{-1}$, Table 9.1 provides characteristic features of a typical methylene (hydrophobic) tail and polar (hydrophilic) heads for ionic surfactants. Alkyldimethylamine oxides ($C_nH_{2n+1}(CH_3)_2N \Rightarrow O$), on the other hand, exist in either nonionic or cationic (protonated) form. Depending on the length of the hydrophobic tail, both may be surface active and the activity is heavily dependent on pH of the solution. The effect of pK_a, the critical micelle concentration and the length of the aliphatic chain on the stretching and deformation modes are shown in Table 9.2.[4]

The presence of micelles in aqueous surfactant solution influences the solubilization effect of nonpolar substances in water. For example, the presence of both phenyl acetate and benzyl acetate in the aqueous sodium decanoate (C_9H_{18} COO^-Na^+) micelle solution affects the environment of the nonpolar phases, which is reflected in the C=O and C—O modes in acetate to split and broaden,[5]

TABLE 9.2 Effect of pK_a, Critical Micelle Concentration on the Stretching and Deformation Modes in $C_nH_{2n+1}(CH_3)_2N—O$

$C_nAO \equiv C_nH_{2n+1}(CH_3)_2N \rightarrow O$	pK_a	Critical Micelle Concentration (M)	Stretching Modes, ν_a					Deformation Modes[a]			
			C—N—O	C—N—O	C—N—O	C—N—O	H$_3$C—N—CH$_3$	δ_a CH$_3$—N	δ_s CH$_3$—N	δ α—CH$_2$	δ CH$_2$
C$_1$AO	4.5	—	—	1240	—	—	951	1466, 1482	1404	—	—
			—	1240	1254	—	952	1465, 1483	1407	—	—
			—	—	1254	—	953	1462, 1486	1410	—	—
C$_6$AO (monomer)	4.6	—	1196	—	1225	1245	Weak	1460, 1476	1403	1431	1470
			1197	—	1227	1245	"	1460, 1475	1404	1432 (w)	1470
			1198	1205	1230	1245	"	1460, 1479	1410, 1388	1433 (w)	1469
C$_8$AO (monomer)	4.6	—	1194	1206	1226	1250	"	1460, 1475	1403	1430	1471
			1194	1208	1228	1248	"	1460, 1477	1404	1431 (w)	1470
			1194	1213	1230	1244	"	1459, 1479	1410, 1348	1433 (w)	1469
C$_8$AO (micelle)	4.6	0.140	1194	1205	1227	1249	972	1459, 1475	1403	1430 (w)	1470
		0.110	1192	1203	1229	1243	973	1458, 1477	1404	1434 (w)	1469
		0.150	1193	1212	1232	1244	974	1458, 1479	1410, 1388	1437 (w)	1469
C$_{12}$AO (micelle)	4.9	0.0020	1196	1223	1235	1249	969	1459, 1475	1402	1430	1468
		0.0024	1190	1220	1233	1242	969	1458, 1477	1404	1437 (w)	1468
		0.0060	1199	1209	weak	1242	974	1457, 1480	1409, 1388	1436 (w)	1468

[a] δ_s = symmetric deformation; δ_a = antisymmetric deformation; (w) = band intensity is relatively weak.

Source: Adopted from ref. 4.

suggesting that acetates are solubilized into the surface polar layer of micelles and may diffuse into the core of micelle.

The C—H stretching region is also sensitive to the concentration changes. Determination of partitioning of hydrocarbons between micelle interior and aqueous phase has been a subject of numerous studies. In the case of Raman spectroscopy, it is based on the observation that the C—H stretching bands display a shift when parameters of the solvent environment change. Larson and Rand[6] were the first to observe that the C—H stretching band in the hydrocarbon solvent shifts by $7 \, cm^{-1}$ when a solute is dissolved in water. These observations were confirmed by others[7,8] for surfactant solutions and surfactant–polymer systems.[9] For example, the C—H stretching band of the sodium dodecyl sulfate (SDS) and the OH stretching of H_2O are sensitive to the changes in the SDS concentration, and the C—H stretching bands shift to lower frequency with the increased concentration. Furthermore, a mutual presence of the neutralized styrene/acrylic acid (S/AA) random copolymer and surfactant in the aqueous media leads to a splitting of the asymmetric and symmetric C—O stretching bands of the COO^- groups, from 1547 (asymmetric) to 1557 and $1542 \, cm^{-1}$ and from 1407 (symmetric) to 1401 and $1407 \, cm^{-1}$. It was suggested that these spectral changes are attributed to the SDS micelle interactions with the carboxylic groups of the S/AA in the following manner: the nonpolar CH_3 groups occupy the interior of the micelle, whereas polar SO_3^- exterior groups can interact with negatively charged COO^- groups through counterions of Li^+ or Na^+. These interactions destroy a local symmetry of the COO^- species, and give rise to the splitting. It should be noted that the use of numerical methods may be desired to enhance the spectral information. The presence of immiscible aqueous and hydrocarbon phases in the S/AA solution was also examined.[10]

Another issue is the effect of surfactant concentration in aqueous solution. According to the studies conducted on n-hexanoate solutions,[11] the C—H stretching of the IR spectrum can be affected such the asymmetric C—H stretching modes in CH_3 and CH_2 groups shift down by approximately $6 \, cm^{-1}$ while changing concentration from $0 \, M$ to $3 \, M$. In contrast, C—O symmetric stretch would exhibit no shift, and the C—O asymmetric C—O modes would shift upfield about $3 \, cm^{-1}$. It appears that to sample the effect of concentration in aqueous solutions is to use D_2O as the D—O stretching and bending modes are shifted from approximately 3500 to $2600 \, cm^{-1}$ for stretching and from 1640 to $1400 \, cm^{-1}$ for D—O bending.[12]

> One issue should be kept in mind: the resolution of instrument should be set such that the claimed wavenumber shift should exceed the resolution limit. Unfortunately, this is not the case in many studies.

Another interesting feature of the frequency shifts observed for ionic surfactants is the effect of their counterion environment. For example, in contrast to the upfield frequency shifts for asymmetric COO^- stretching modes, the asymmetric S—O stretching modes in SO_3^- appear to shift to lower wavenumber. These differences were interpreted as being due to the position of ions with respect to their hydro-phylic counterpart; namely, sodium ion is located between two COO^- groups of the alkanoate micelle, whereas the same ion seats infront of the SO_3^- groups in the alkanesulfonate micelle.[13]

The thermotropic nature of phase transitions in water-surfactant systems has also been a topic of many studies. In spite of supposively simple binary system, surfactant–water systems may display fairly complex and often difficult to detect structural changes that depend on temperature and concentration. Among other studies, sodium palmitate,[14] potassium laurate,[15] sodium stearate,[16,17] octadecyl-trimethylammonium chloride and dioctadecyldimethylammonium chloride,[18,19] sodium dodacyl sulfate[20] are perhaps the most representative ones. Micelle formation was investigated by IR and Raman spectroscopy.[21-33]

9.3 AMPHIPHILIC MONOLAYERS

There have been numerous studies dealing with thin films on various surfaces. In principle, the main interest is in monolayers spread on liquid and solid surfaces. Many water-insoluble amphiphiles can be spread at the air–water interface to form monomolecular layers,[34] and under suitable conditions, can be transferred from water to a solid support.[35] If this transfer is successful, these monolayers are called Langmuir–Blodgett (LB) films.

Although amphiphilic monolayers at the air–water interface have been exten-sively studied as models for a variety of interfacial processes, the information pertaining to their detailed physical structures has been limited for numerous reasons.[36,37] The primary reason was because most spectroscopic methods were unable to provide information about flat, low-surface-area interfaces. In recent years the situation has improved after utilization of cynclotron x-ray diffrac-tometers[38,39] or in situ external FT-IR measurements.[40,41] Perhaps the most useful source for spectroscopy of amphiphilic compounds are alkanes, which have been studied in great depth.[42,43] A particular emphasis was given to the width, intensity, and integrated intensities of the C—H asymmetric and symmet-ric bands near 2850 and 2925 cm^{-1}. These are sensitive to the gauche/trans conformer ratios and ordering of the aliphatic chains resulting from pressure, temperature, and composition changes.

Molecular structures of adsorbate molecules and their arrangements on surfaces may create an extremely suitable environment for a specific surface orientation. In addition, the surface dynamics, that is, the response of the adsorbed layer on the surface to the external condition changes, is of great importance. These are conditions such as temperature changes, electrical current, and chemical environ-ment. Of course, in order for the molecular layer to respond to an external condition change, it is necessary to design thin-layer surface chemistry such that the chemical groups will "jump as high as we can pull up a string." As an example, let us consider several molecular arrangements such as that depicted in Fig. 9.2. These molecules contain either amine or acid groups that will result in the mutual

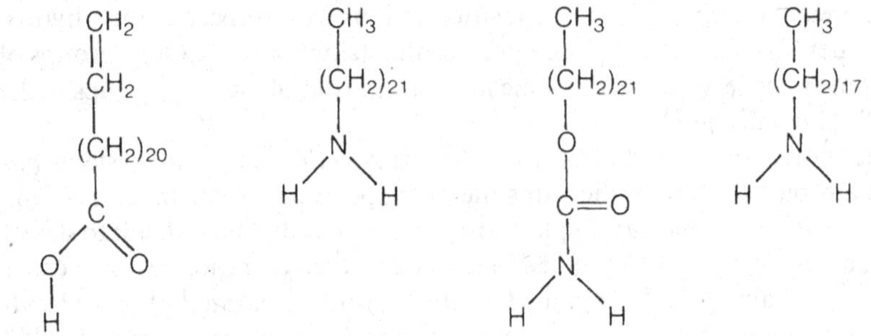

22-Tricosenoic acid   1-Docosyalamine   Docosyl 4-aminobenzoate   4-Octadecylaminine

A   B   C   D

Figure 9.2 Functionalities of molecules used to produce pyroelectric effects using Langmuir–Blodgett deposition.

interactions between them. Although it is known that amine and acid groups react with each other, the simultaneous presence of each group will result in different behavior on the surface. In other words, depending on the position of each species on the multilayer surface, one can modify the surface characteristics.

9.4 THE NATURE OF INTERACTION IN LAYERED STRUCTURES

Spreading liquid monomolecular films of 1,2-dipalmitoyl-*sn*-glycero-3-phosphocholine (DPPC) on water surfaces may lead to liquid-expanded (LE) to liquid-condensed (LC) phase transitions.[44] One interesting feature of these monolayers is the decreasing of the CH_2 stretching with the decrease of a molecular area occupied by the monolayer molecules. This decrease detected by the single-reflection FT-IR method was attributed to the conformational changes of the hydrocarbon chains.

When proper structural features are assembled on the surface, pyroelectric response,[45,46] can be induced. This can be achieved, for example, if one would alternatively dip a substrate in fatty acid and amine molecules so that the molecular dipoles would line up head-to-head, and a large residual temperature dependent permanent dipole would be induced. However, any orientation changes or rearrangements would lead to dipole changes. This is illustrated below by arrows pointing in the direction of dipole changes upon structural rearrangements.

$$\mu_{acid} \quad \downarrow \qquad \mu_{acid} \quad \downarrow \quad \uparrow \quad \mu_{amine}$$

$$\mu_{amine} \quad \uparrow \qquad \mu_{acid} \quad \uparrow \quad \uparrow \quad \mu_{amine}$$

The molecules that can produce these changes are shown in Fig. 9.2. As an example, let us consider structures A and B being assembled to produce and LB film, such as that shown in Fig. 9.3*A*. When infrared spectra of such prepared species show the presence of C—O stretching modes at a $\sim$ 1420 cm^{-1} due to

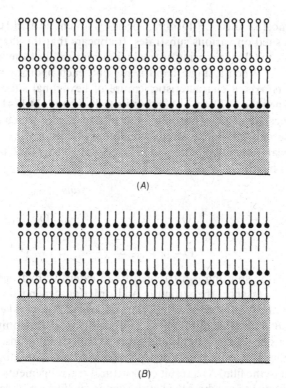

Figure 9.3 Y (*A*) and Y' (*B*) structural arrangements of Lagmuir–Blodgett deposition.

COO^- on COOH neutralization, and the NH bending modes at $1550–1600 \, cm^{-1}$ due to NH_3^+ on protonation, there will most likely be proton transfer between structures **A** and **B**. On the other hand, if the $C{=}O$ stretch at $\sim 1730–1700 \, cm^{-1}$ is detected instead of the $C{-}O$ stretch, most likely two acid groups of **A** would formed a cyclic dimer. Such structures would most likely give no pyroelectric response, and the presence of **A** and **C** would most likely result in these properties. When structures **A** and **D** are involved in the LB formation, it was suggested[47,48] that sideways (lateral) chains are formed to which NH_2 groups are attached. The basis for these assessments was lack of $C{-}O$ stretching bands and the presence of $1730 \, cm^{-1}$ due to nonbonded or weakly bonded $C{=}O$ groups, leading to formation od species given below:

Perhaps the most studied LB-forming species is arachidic acid $[CH_3 (CH_2)_{18}\text{-}COOH]$ deposited on various substrates. It appears that the shape of the 1470 cm^{-1} band due to CH_2 scissoring normal vibrations results from intermolecular interactions.[49,50] When a single band is present, each hydrocarbon can freely rotate along its axis, a characteristic feature of hexagonal subcell packing in n-paraffins. When the 1470-cm^{-1} band splits to two bands at 1741 and and 1465 cm^{-1}, it is indicative of an orthorombic subcell packing for which the chains can pack in an alternative fashion. Another region of interest is the carbonyl region where the appearance of two bands could be attributed to cis and trans groups,[51] such as shown below

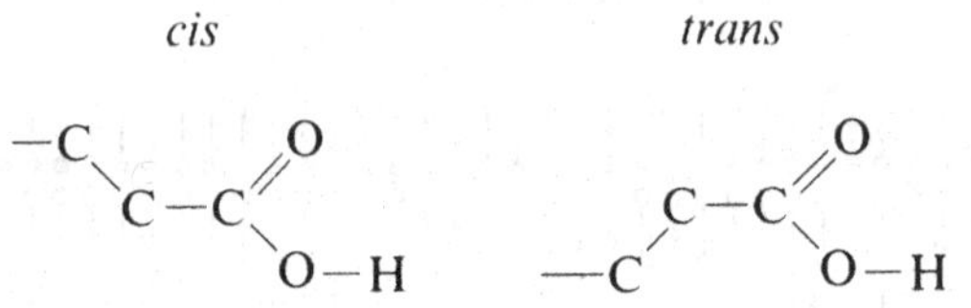

but most likely the splitting is attributed to an extensive hydrogen bonding of both isomers. In essence, the structure and ordering depend on the substrate characteristics and a number of monolayers deposited. Yarwood et al.[52,53] have studied the LB multilayer films of 12/1 arachidic acid/valinomycin (AA/VA) deposited on a silica ATR crystals and exposed to KCl solutions. These studies showed that the increasing bands at 1653 and 1744 cm^{-1} are due to incorporation of KCl into the film. As a result of structural rearrangements within the LB films, the bands due to arachdic acid cyclic dimers at 1705 cm^{-1} become weaker, whereas the increase of the C—H stretching bands at 2856 and 2924 cm^{-1} were detected.

Another group of LB film materials are cyanine dyes, such as 3-octadecyl-2-[3-3-octadecyl-2-benzothiazolinylidene)-1-propenyl] benzothiaazolium iodine.[54] These species are capable of forming "head-on" (or Y) and "tail-on" (or Y') configurations, which are depicted in Figs. 9.3A and 9.3B, respectively. One of the claimed features of the hydrocarbon chains attached to an aromatic moiety is a significant amount of gauche conformation in one monolayer of Y' type.

Tetracyanoquinodimethylene (TCNQ) with n-alkyl substitution or other derivatives is another group of interest for their electrical conductivity. Whereas the presence of the CH_2 antisymmetric and symmetric bands at 2918 and 2848 cm^{-1} and the CH_2 scissoring modes at 1471 and 1462 cm^{-1} suggested the presence of trans-zigzag conformation in orthorhombic packing, the band at 2222 cm^{-1} is due to the CN stretching modes of the cyanogroups.[55] Except for the intensity differences, a single monolayer LB film relfection-absorption spectrum of octadecyl TCNQ with the tail-on configuration (Y' type) appears to be similar to that of three monolayers with the head-on configuration (Y type).

The LB films have been particularly appealing for the nonlinear optical materials. Such species as the LB multilayers of dyes, which have acceptor and donor groups, may exhibit enhanced second-harmonic signal generation (SHG) properties. Hemicyanine (HC) and nitrostibene (4HANS) dyes are only representative examples,[56-59] but, in general, noncentrosymmetric positioning of the donor (D) and accepter (A) groups is the key feature. It appears that the presence of multiple bands in the 1600–1700-cm^{-1} region is sensitive to ordering of the monolayers, and the bands at 1710–1750 cm^{-1} are due to hydrogen bonding between the layers. These are the primary interactions between monolayers.[60] Although to detect conclusively the nature of these interactions more work is needed.

9.5 ORDERED MONOLAYERS ON SOLID SURFACES

The pioneering studies of ordered barium stearate deposited on metal surfaces were reported by Francis and Ellison.[61] Since that time there have been numerous studies of formic acid on copper,[62-64] aluminum,[3] and nickel.[4] In these studies, isoamyl nitrate and nitric oxide monolayer adsorptions were also investigated. The first spectra of 15-Å-thick epoxy monolayers on iron and copper surfaces were reported by Boerio and Chen.[65] When acitic acid is deposited from the gas phase, proton dissociation occurs, as all studies reported the presence of acetate ion on the surface.[66,67] However, when acidic acid was deposited on the Al_2O_3 surface from the solution, the bands at 1590, 1405, and 1335 cm^{-1} were observed.[68] Based on the analysis of the Raman intensities, it was concluded that numerous and complicated surface species having various orientations can be present. For the adsorption of 2,4-pentandione from the solution,[8] the bands at 1610, 1535, 1460, and 1405 cm^{-1} were detected, indicating the presence of Al(AcAc)$_3$ complexes.

The determination of conformational disorder of phospholipid acyl chains of dipalmitoylphosphatidylcholine (DPPC), dipalmitoylphosphatidylethanolamine (DPPE), cholesterol, gramicidin D, and their mixtures[69] can be accomplished by taking advantage of the sensitivity of the CD_2 rocking region. While Table 9.3 provides the effect of trans/gauche conformers on vibrational energy of the CD_2 rocking modes,[70] the relative amounts of trans/gauche conformers are sensitive to the temperature changes.

Figure 9.4 illustrates a series of PA FT-IR spectra of LB layers deposited on

TABLE 9.3   Effect of Trans and Gauche Conformations on CD$_2$ Rocking Modes

Conformer Class	(Conformation)	v_{obs} (cm^{-1})
et	(end–trans)	622
egt	(end–gauche)	657
egg	(end–gauche)	650
tttt	(trans)	622
ttgtt + tg'tgt	(single gauche bend + kink)	652
ttggt + tg'tggt	(multiple gauche)	646

Source: M. L. Maroncelli, H. L. Strauss, and R. G. Snyder, J. Phys. Chem., 1985, **89**, 4390.

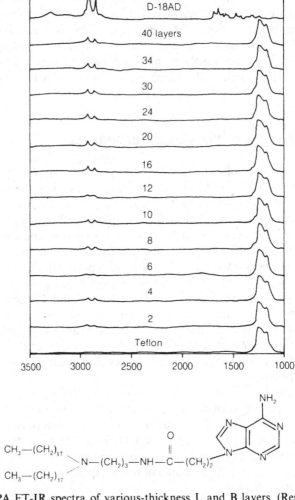

Figure 9.4   PA FT-IR spectra of various-thickness L and B layers. (Reproduced with permission from ref. 71, copyright 1988, American Chemical Society).

poly(tetrafluoroethylene) along with the structure of an individual molecule.[71] It appears that photoacoustic FT-IR spectroscopy is not only sensitive to a two monolayer thin coverage, but also orientation of the surface species can be estimated using He and Xe coupling gases in the photoacoustic experiment.[72] A classic experimental setup, however, is to use polarized IR source assembled with ATR or grazing angle spectroscopy. The latter approach was pioneered by Allara and Swalen,[73] who related the band positions and intensities to film structures. A useful comparison of the C—H stretching modes for crystalline, liquid, and ad-sorbtion on gold $CH_3(CH_2)_nSH$ with n ranging from 5 to 21 is given in Table 9.4. Figure 9.5 illustrates alkyl thiol adsorbed on a metal surface and the orientations of the symmetric and asymmetric CH stretching vibrations.

TABLE 9.4 Frequency Shifts in CH Stretching Region for Crystalline, Liquid, and Adsorbtion on Gold $CH_3(CH_2)_n$ SH with n Ranging from 5 to 21[a]

Structural Group	Stretching Mode	Crystalline	Liquid	$n = 21$	$n = 15$	$n = 19$	$n = 5$
$-CH_2-$	v_a	2918	2924	2918	2918	2920	2921
	v_s	2851	2855	2850	2850	2851	2852
CH_3-	$v_a(ip)$	—	—	2965	2965	2966	2966
	$v_a(op)$	2956	2957	—	—	—	—
	$v_s(FR)$	—	—	2937	2938	2938	2939
	$v_s(FR)$	—	—	2879	2879	2879	2879

[a] Abbreviations: ip, in-plane; op, out-of-plane; FR, contributions from fermi resonance—wavenumber not determined because of overlap. (Adopted from ref. 73).

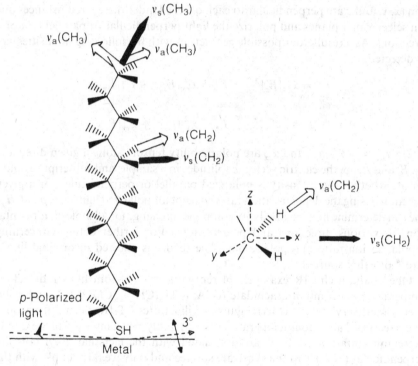

Figure 9.5 Alkyl thiol adsorbed on gold surface. CH_2 groups are parallel to the surface, whereas symmetric and asymmetric CH stretching modes are perpendicular to each other. (Adopted from ref. 73).

9.6 DEGREE OF MOLECULAR ORIENTATION AND PACKING

Thin films of organic and polymeric materials deposited on substrates can undergo rearrangements on the surface. External factors affecting dynamics of molecular reorganization of surface segments may be of physical (temperature or pressure changes, electrostatic charges) or chemical origin. However, the orientational information provided by IR studies of thin films depends on the nature of interactions between the electric vector of the electromagnetic radiation and the dipole moment of a given bond. In a standard transmission arrangement, the electric field vector is polarized in perpendicular or parallel directions, and only those dipole moments that can interact with the polarized electric vector can be detected. In the case of surface species, the situation is similar; if, for example, any molecular group having a component with a transition dipole moment parallel to the surface, parallel polarized light will be absorbed at the energy corresponding to this transition. In contrast, when parallel polarized light is used, no bands will be detected. This is illustrated in Chapter 3 along with further insights of grazing angle reflection FT-IR spectroscopy. As far as Raman spectroscopy is concerned, the equations given below will define intensities of inelastically scattered light.

The laser light is incident in the xy direction, and z is the direction of observation (x, y, and z are perpendicular to each other). Under these circumstances one can select x or y planes and polarize the light perpendicular or parallel to x or y directions. As a result, four possible geometries with the following intensities can be detected:

$$I_{\perp x} = \alpha_{zx}^2 |E_z|^2 \qquad I_x = \alpha_{xx}^2 |E_x|^2 + \alpha_{yx}^2 |E_y|^2$$

$$I_{\perp y} = \alpha_{zy}^2 |E_z|^2 \qquad I_y = \alpha_{xy}^2 |E_x|^2 + \alpha_{yy}^2 |E_y|^2$$

where α_{xx}, α_{xy}, α_{yy}, α_{zy}, and α_{zx} are polarizability tensors along a given direction; E_x, E_y and E_z are the electric vector amplitudes in a sample; and subscripts $\perp$ and $\parallel$ denote whether the intensity is polarized parallel or perpendicular for a given direction. Using the depolarization ratio concept for polarized light (e.g., $\rho = I_x/I_y$), one can determine how polarizibility components change for all molecular orientations in various directions and polarizations. For further details concerning depolarization ratio and polarizibility the reader is referred to original literature[74] or other sources.[75,76]

One of the useful IR examples of reorganization of thin films induced by temperature is cadmium arachidate (CdA; $(CH_3(CH_2)_{18}COO)_2Cd$) and poly-(methylmethacrylate) (PMMA). Figure 9.6 illustrates FT-IR spectra of 6 monolayers recorded at various temperatures.[77] Evidently, weak intensity of the C—H stretching vibrations at 25°C are associated with the fact that alkyl tails are perpendicular ($\sim 8°$)[78] to the substrate surface and only weakly couple with the perpendicular electric vector of the electromagnetic radiation. However, as temperature increases, a gradual steady increase of the band intensity is detected and is attributed to the conformational disordering of alkyl tails resulting from the presence of gauche defects and disturbance of the crystalline order.[79] At 110°C the film melts and above that temperature, random structures are present.

Another example of the effect of temperature on ordering and packing of thin films is PMMA. In this case, we are dealing not only with a melting point but also with a glass transition temperature (T_g). Apparently, the C=O stretching and

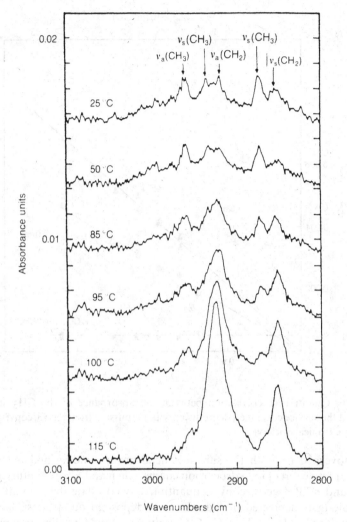

Figure 9.6 RA FT-IR spectra in the C—H stretching region of six L-monolayers of CdA as a function of temperature. (Reproduced with permission from ref. 77, copyright 1985, Society for Applied Spectroscopy).

CH_2 bending vibrations are not affected by the heating film from 25 to 187°C. However, the relative intensity of the 1240–1270- and 1150–1190-cm^{-1} change appreciably above T_g. Because these bands are attributed to combined C—O and C—C stretching modes in ester (—C—O—CH$_3$) side chains, these observations were attributed to two rotational isomeric states for ester groups.[80] Long-chain diacetylene in a monolayer form have also been studied, and a significant amount of information concerning factors affecting orientation has emerged.[81,82]

Finally, organic molecular monolayers formed by spontaneous adsorption from solution can self-assemble forming organized films. Kinetics of adsorption is

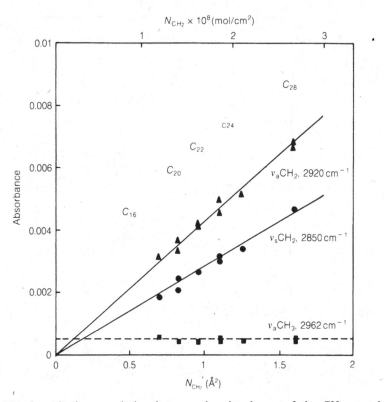

Figure 9.7 Quantitative correlation between the absorbance of the CH_2 stretching modes and the surface coverage. (Reproduced with permission from ref. 83, copyright 1983, American Chemical Society).

clearly governed by both the substrate-head group binding and the van der Waals attractive forces in hydrocarbon chains. Chen and Frank[83] utilized transmission and ATR spectroscopy to quantitatively correlate the amount of Langmuir–Blodgett deposited n-alkanoic acids deposited on polar surfaces. The CH_2 stretching modes of C_{16}—$C_{28}Cd^{2+}$ salts were used for calibration purposes, and the absolute surface concentration of the CH_2 was determined. Figure 9.7 illustrates a quantitative correlation between the number of moles (N) per square centimeter and absorbance due to CH_2 stretching modes.

9.7 ORIENTATION AND INTERACTIONS OF SURFACTANTS ON LATEX SURFACES

Based on the previous considerations it is obvious that surfactants play an important role not only in latex technology, but also in the interfacial and thin-film areas. A lack of compatibility with the polymer or copolymer network and surface tension may result in exudation of surfactants from a polymer matrix.

Surfactant enrichment was attributed to three factors: (1) initial enrichment at both interfaces in order to lower interfacial free energy, (2) enrichment at the film–air interface due to the transport of non-adsorbed surfactant by the water flux out of the film, and (3) longer-term migration to both interfaces due to surfactant incompatibility.

Let take as an example ATR FT-IR spectra of ethyl acrylate/methacryclic acid (EA/MMA) latex prepared using sodium dioctyl sulphosuccinate (SDOSS). Figure 9.8 illustrates three spectra: tracing A is the film–air interface, tracing B is the film–substrate interface, and tracing C is the film–air interface (trace A) washed with methanol/water solution.[84] It appears that tracing A contains two additional bands due to S—O stretching vibrations of SO_3^- groups at 1046 and $1056\,cm^{-1}$. The splitting of the original $1050\text{-}cm^{-1}$ band in pure surfactant is attributed to the acid groups and water environments in latex. The terminal

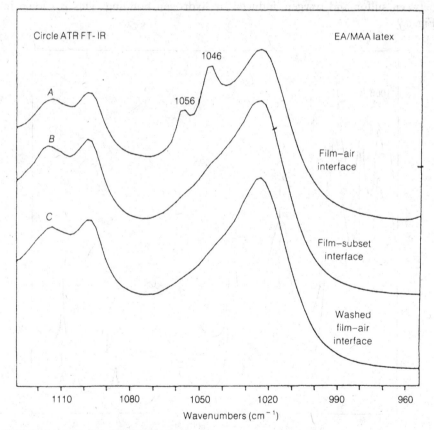

Figure 9.8   ATR FT-IR spectra of ethyl acrylate/methacrylic acid latex in the 960–1110-cm^{-1} region: (tracing A) film–air interface; (tracing B) film–substrate interface; (tracing C) washed film–air interface. (Reproduced by permission from ref. 84, copyright 1990, Butterworth-Heinemann).

—$SO_3^- Na^+$ ion pair of surfactant is associated with the COOH acid groups through hydrogen. This was schematically depicted in Fig. 8.9. When COOH groups are neutralized, the $1056 \, cm^{-1}$ is eliminated, breaking apart —S—O $\cdots$ H—O—C=O associations. With a coordination number of 4, there are four possible local symmetries of the —SO_3^- group: tetrahedral arrangement (T_d,) tetragonal plane (D_{4h}), tetragonal piramid (C_{4v}), and irregular tetrahedral arrangement (C_{3v}). According to the reduction tables, and keeping in mind that the heterogeneous atoms surround the central sulfur atom (three oxygens and carbon), only tetragonal pyramidal structure with the dp^3- and d^3-orbital configurations or irregular tetrahedral arrangement with the dp^3-, d^3p-, and d^2sp-orbital configurations are possible. While the tetragonal pyramid structure exists in a nonhydrated form of surfactant, manifested by the presence of the S—O stretching band at $1050 \, cm^{-1}$, the presence of weak acid and water disturbs the $SO_3^- Na^+$] environment by the formation of partial double bonds between sulfur and oxygen, induced by hydrogen bonding. This is shown in Fig. 8.9.

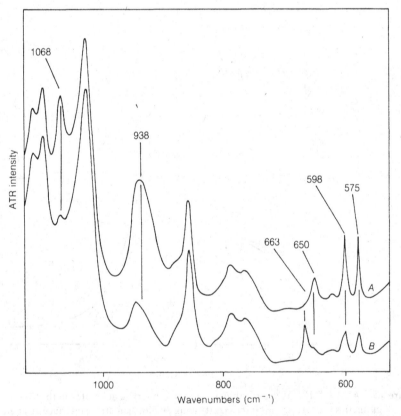

Figure 9.9 Polarized ATR FT-IR spectra of ethyl acrylate/methacrylic acid latex in the $1100–550$-cm^{-1} region: (tracing A) TE polarization; (tracing B) TM polarization.

In addition to the issue of compatibility, the nature of interactions and spatial distribution and orientation of surfactant molecules and other species at the film–air and film–substrate interfaces is important. Using TE and TM polarized light in an ATR mode of detection, formation and orientation of acid dimers at the film–substrate interface can be deduced. From the analysis of the band intensities of the OH $\cdots$ O out-of-plane deformation normal vibrations attributed to the acid dimer groups at 938 cm^{-1} and the bands at 663, 598, and 575 cm^{-1} shown in Fig. 9.9, local acid dimer orientation can be estimated. Semiquantitative analysis of the band intensities indicates that of the two nonplanar structures illustrated in Fig. 9.10 vertical orientation depicted in the lower diagram (dimer 2) appears to be the most probable orientation.[85]

ATR FT-IR spectroscopy and a particle size analysis can be utilized to elucidate the effects of flocculation and coalescence on the surfactant mobility in latex films as well as orientation of hydrophilic surfactant ends near or at the film–substrate or film–air interfaces.[86] Using polarization ATR FT-IR experimental setup and monitoring intensity changes of the 1046 and 1056 cm^{-1} bands, among at least three structural arrangements of the SO$_3^-$Na$^+$ ends at the interfaces that is random, S—O bond predominantly in plane and out of plane, the intensity changes favor out-of-plane arrangements, such as that depicted in Fig 9.11. It should be kept in mind that the analysis of the latex literature clearly indicates that depending on polymer and surfactant environments and coales-

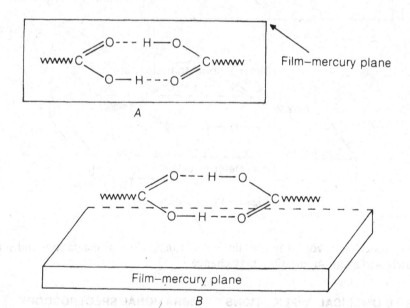

Figure 9.10 Possible arrangements of acid dimers on metal surfaces: (A) ring in film plane; (B) ring perpendicular to film plane (C—C axis in plane); (C) ring perpendicular to film plane (C—C axis perpendicular to plane).

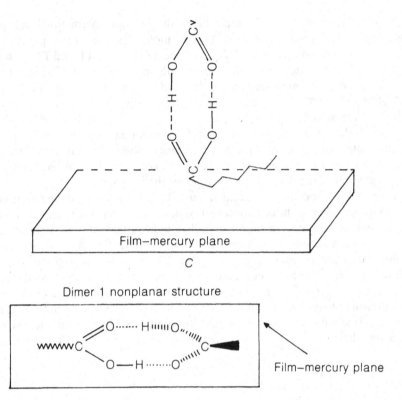

Dimer 1 nonplanar structure

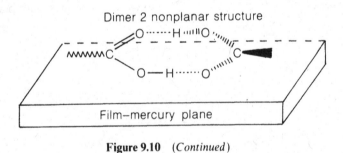

Dimer 2 nonplanar structure

Figure 9.10   (*Continued*)

cence processes involved in the film formation, surfactant exudation, and ultimately surfactant orientation, may change.

9.8   BIOMEDICAL APPLICATIONS OF VIBRATIONAL SPECTROSCOPY

Although there are numerous applications of vibrational spectroscopy in biology and medicine, and several good monographs cover this area, here we provide

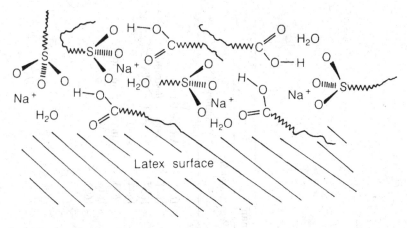

Figure 9.11 Possible out-of-plane orientation of $SO_3^- Na^+$ hydrophilic surfactant ends with respect to the latex surface.

only selected examples related to adsorption, bands asssignments and a few examples. The adsorption of proteins onto various surfaces is of primary interest because molecule-level understanding of bonding characteristics provides further insights concerning biocompatibility. In the late 1970s, Jakobsen et al.[87,88] demonstrated the use of ATR methodology as a mean for direct protein adsorption measurements. Other studies focused on quantitative[89,90] aspects of adsorption or adsorption of individual components.[91] Inspite of the complexity of the protein structures, the band assignments for α-helix and β-sheet structures have been proposed, but no agreement was reached on the amide I spectral region. The issue was whether the bands in the $1600-1700\ cm^{-1}$ region are due to turn structures, disorder, or other side-chain vibrations. This issue was resolved by Kirsch and Koenig[92] in the temperature studies of helix proteins. The summary of these and other studies, along with an elegant discussion provided by Jakobsen and Wasacz,[93] have led to the IR and Raman bands assignments for albumin, which are summarized in Table 9.5.

The analysis of the vibrational spectra of proteins is particularly difficult because amino acid residues and protein chains can form various conformers. As in many other spectroscopic studies, a commonly accepted approach is to study model species[94] and, on that basis, deduce structural features in the actual systems. Further difficulty arises from the environmental effects, such as solvents, pH, or concentration. In fact, the presence of D_2O versus H_2O will effect the vibrational band energies.[95] Table 9.6 provides IR and Raman bands along with their assignments for poly-L-(Asn, Asp) system.

Although the late T. B. Hirschfeld was the first person to admit that FT Raman spectroscopy cannot exist in practice, he was also the first one to demonstrate and develop the FT–Raman spectrometer in the near IR region.[96] As a result, the late 1980s have been saturated with numerous applications of near-IR Fourier transform Raman spectroscopy, which eventually will replace

TABLE 9.5 Infrared and Raman Spectra of Albumin and Their Assignments

Infrared[a]				Raman[b]	
Wavenumber [cm⁻¹]	Assignment[c]	Wavenumber [cm⁻¹]	Assignment[d]	Wavenumber [cm⁻¹]	Assignment[d]
1678	Amide I, turns Arg, [Asn]			1683	Amide I, turns
				1671	Amide I, β-strand
1655	Amide I, α-helix	1652	Amide I, α-helix	1657	Amide I, α-helix
				1644	H_2O
1633	Amide I, turns			1633	Amide I, β-strand
1597	Phe	1618	Tyr	~1618	
		1602	Phe	~1602	
1575	$\nu(COO^-)_{asym}$				
1547	Amide II⁻	1551	Trp		
1518	Glu, Lys, Leu				
1499	PHE, [TYR]				
1470	$\delta(CH_3)$, Leu, Val, Aal				
1447	$\delta(CH_2)$, Lys, [Glu, Arg]	1449	$\delta(CH_2)$, $\delta(CH_3)$		
1420		1419			

IR[a]	Assignment	Raman (solid)[b]	Assignment	Raman (solution)	Assignment
1398	$\nu(COO^-)_{sym}$	1402	$\nu(COO^-)_{sym}$		$\nu(COO^-)_{sym}$
1363	$\delta(CH_3)$, Leu, Val	1337	$\delta(CH_3)$		$\delta(CH_3)$
1344	LYS, [Phe, Arg,Val]	1317	$\delta(CH_3)$		$\delta(CH_3)$
1315	Amide III, α-helix				
1300	Amide III, α-helix	1268	Amide III		Amide III
	Amide III, turns				$\delta(CH_3)$
1270	TYR	1252	Amide III		Amide III
1244	Amide III, disorded	1206	TYR, PHE		TYR, PHE
1213	Tyr	1175	TYR		TYR
1172	Leu, Lys, Tyr, Val	1158	C—N		C—N
1151	Glu[Asp, Phe]	1126	C—N		C—N
1126		1102	C—N		C—N
1106	Ala, Leu, Lys				

[a] The IR frequencies are taken from deconvoluted spectra of albumin in solution.

[b] The first column of Raman frequencies is taken from nondeconvoluted spectra of albumin in solution. The second column of Raman frequencies (1600–1700 cm⁻¹) is taken from deconvoluted spectra of solid albumin.

[c] Key: Arg—arginine; Asn—asparagine; Glu—glutamic acide; Lys—lysine; Leu—leucine; Tyr—tyrosine; Ala—alanine; Val—valine; Phe—phenylalanine.

[d] V. J. C. Lin and J. L. Koenig, *Biopolymers*, 1976, **15**, 203.

Source: Adopted from ref. 93.

TABLE 9.6 Infrared and Raman Bands of L-Asparagine/L-Aspartate Random Copolypeptide[a]

IR Band (cm⁻¹)		Raman Band (cm⁻¹)		Assignment
H₂O Solution	D₂ Solution	H₂O Solution	D₂O Solution	
1677 (sh)		1670 (sh)		νC=O (—CONH₂)
(1676) (s)		(1670) (s)		
1652 (vs)		1655 (vs)		Amide I, δH₂O
	1652 (sh)		1653 (vs)	Amide I', random coil
	1646 (vs)		1639 (sh)	νC=O (—COND₂)
	1639 (sh)		(1638) (s)	
	(1637) (vs)			
1590 (s)	1596 (s)	1599 (sh)	1603 (w)	ν_{as}COO⁻, δNH₂
(1608) (m)		(1603) (m)		
1557 (sh)				Amide II, turns
1537 (sh)				
1521 (sh)				Amide II, random coil
1448 (w)		1447 (sh)		δCH₂
	1450 (m)		1450 (sh)	νC—ND₂, δCH₂
	1433 (sh)		1432 (s)	
	1421 (sh)			
1419 (w)		1415 (sh)		νC—NH₂

				Assignment
1394 (m)	1395 (m)	1395 (s)	1403 (sh)	$\nu_{as}COO^-$
	1347 (w)		1342 (sh)	} $twCH_2 + \omega CH_2 + \delta C_xH$
			1328 (w)	
1319 (w)		1315 (vw) }		
1282 (w)	1208 (s)	1274 (w)		
1247 (w)		1248 (w)	1239 (sh)	Amide III
			1204 (m)	δND_2 / δD_2O
1197 (w)		1186 9w		$\nu C_xN + \nu CC$
		1107 (m)		rNH_2
		(1130) (s)		
		1069 (m)	1059 (vw)	νCC
		1041 (vw)		$\nu C_xN + \nu CC$
		992 (m)	975 (vs)	$\nu C_xN + \nu CC$ + amide III'
		938 (vw)	932 (m)	νCC
			(931) (m)	rND_2
879 (m)		879 (m)	874 (m)	νCC
830 (m)		830 (m)	836 (m)	νCC

[a] Values in parentheses correspond to acetamide in aqueous solution. Abbreviations: s, strong; vs, very strong; m, medium; w, weak; vw, very weak; sh, shoulder; ν, stretching; δ, in-plane-deformation; ω, wagging; r, rocking; tw, twisting.

Source: Adopted from ref. 95.

and dominate the field of Raman vibrational spectroscopy. The principles governing FT–Raman detection impose the use of a near-IR excitation source. As a result, species that exhibit strong fluorescence in the traditional visible or ultraviolet Raman excitation region will not fluoresce, making the scattering process fairly effective. This section focuses on selected practical aspects of the FT–Raman use for biological systems that could not be detected without a

TABLE 9.7 FT–Raman Bands Observed in (*A*) Pigmented Rodent and Human Eye Lens; (*B*) Chicken Leg Bone and Human Tooth

		A Eye Lens		
Pigmented Rodent Lenses		Normal Human Lenses		
Chipmuk	Squirrel	82-Year-Old	Lyophilized	Band Assigment
2917	2924	2928	2919	C—H stretch
2870	2877	2900	2888	C—H stretch
1673	1671	1669	1669	Amide I
1613	1607	1611	1607	
1549	1551	1557	1547	Tryptophan
1447	1449	1447	1444	CH$_2$ deformation
1341	1338		1341	
	1331		1331	
1254	1247	1243	1237	Amide III
1209	1210	1210	1207	Tyrosine
1129	1125			
1004	1006	1006	1003	
855	877		879	Tryptophan
	857		853	Tyrosine
830	834	832	830	Tyrosine
757	762		759	Tryptophan

	A *B* Bones	
Chicken Leg Bone	Human Tooth	Band Assignment
2918	2932	C—H stretch
2882	2880	C—H stretch
1670	1670	Amide I
1448	1451	CH$_2$ deformation
1260	1266	Amide III
1243	1243	Amide III
1210		Hydroxyproline modes
1072	1069	P—O asymmetric stretch of (PO$_4$) (OH)
960	961	P—O symmetric stretch of (PO$_4$) (OH)$_2$
856	855	Proline modes
451	464	O—P—O symmetric bend of (PO$_4$)$_6$ (OH)$_2$

Source: Adopted from ref. 97.

near-IR excitation source. Although there are unlimited applications of the technique and so-called good quality spectra of fluorescing compounds along with the "shake and bake" research on all sorts of polymers were presented, perhaps the most entertaining and appealing area is that of a biomedical nature. In fact, it is not only the issue of the real-world samples where the shake-and-bake applications would find their home, but the issue is of recording the real-world sample spectra that can be interpreted.

Table 9.7 provides a list of FT–Raman bands of various biological and biomedical materials.[97] Surface-enhanced Raman spectroscopy was also used to elucidate the differences between normal human lens, a senile cataractous human lens, and a chipmunk lens.[98] Considering the intensity and wavenumber analysis, the chipmunk and normal human lens contain ademine-5'-monophosphate, whereas intense bands due to tyrosine and tryptophan are found in the cataractous human lens. Model compounds contained in eye lens pigments were also examined using SERS.[99]

REFERENCES

1. K. L. Mittel, Ed., *Micelliczation, Solubilization, and Microemulsions*, Plenum Press, New York, 1977.

2. K. L. Mittel and B. Lindman, Eds., *Surfactants in Solution; Theoretical and Applied Aspects*, Plenum Press, New York, 1974.

3. R. Nagarajan and E. Ruckenstein, *Langmuir*, 1991, **7**, 2934.

4. J. F. Rathman and D. R. Scheuing, in *Fourier Transform Infrared Spectroscopy in Colloid and Interface Science*, ACS Symp. Series, D. R. Scheuing, Ed., American Chemical Society, Washington, D.C., 1990.

5. S. Shinjiro, T. Matsui, and S. Tanaka, *Appl. Spectrosc.*, 1987, **41**(8), 1438.

6. U. Larson and R. P. Rand, *Biochem. Biphys. Acta*, 1973, **326**, 245.

7. M. H. Brooker, D. J. Jobe, and V. C. Reinsborough, *J. Chem. Soc., Faraday Trans. 2*, 1984, **80**, 73.

8. T. B. Hierschfeld, *Anal. Chem.*, 1981, **53**, 2232.

9. M. W. Urban and J. L. Koenig, *Appl. Spectrosc.*, 1987, **41**(6), 1028.

10. M. W. Urban, J. L. Koenig, L. B. Shih, and J. Allaway, *Appl. Spectrosc.*, 1987, **41**(4), 590.

11. J. Umemura, D. G. Cameron, and H. H. Mantsch, *J. Phys. Chem.*, 1980, **84**, 2272.

12. J. Umemura, H. H. Mantsch, and D. G. Cameron, *J. Colloid Interface Sci.*, 1981, **83**, 558.

13. H. Wennerstrom and B. Lindman, *Phys. Rept.*, 1979, **52**, 1.

14. D. Chapman, *J. Chem. Soc.*, 1958, **1958**, 784.

15. R. Faiman and D. A. Long, *J. Raman Spectrosc.*, 1975, **3**, 371.

16. N. Trzebowski and E. Langholf, *Z. Chem.*, 1967, **7**, 245.

17. N. Trzebowski and E. Langholf, *Z. Chem.*, 1967, **7**, 282.

18. T. Kawai, J. Umemura, T. Takenaka, M. Kodoma, and S. Seki, *J. Colloid Interface Sci.*, 1985, **103**, 56.

19. J. Umemura, T. Kawai, T. Takenaka, M. Kodama, Y. Ogawa, and S. Seki, *Mol. Cryst. Liq. Cryst.*, 1984, **112**, 293.

20. T. Kawai, J. Umemura, and T. Takenaka, *Bull. Inst. Chem. Res., Kyoto Univ.*, 1983, **61**, 314.

21. P. Mukerjee and K. J. Mysels, *Critical Micelle Concentration of Aqueous Surfactant Systems*, NSRDS-NBS36, U.S. Government Printing Office, Washington, D. C., 1971.

22. H. Okabayashi, M. Okuyama, and T. Kitagawa, *Bull. Chem. Soc. Jpn.*, 1975, **48**, 2264.

23. H. Okabayashi and M. Abe, *J. Phys. Chem.*, 1980, **84**, 999.

24. J. B. Rosenholm, P. Stenius, and I. Danielsson, *J. Colloid Interface Sci.*, 1976, **57**, 551.

25. J. B. Rosenholm, K. Larson, and N. Dinh-Nguyen, *Colloid Polym. Sci.*, 1977, **255**, 1098.

26. D. J. Gardiner, J. Barker, and J. W. Rasburn, *Adv. Mol. Relax. Interact. Proc.*, 1982, **24**, 7.

27. H. Okabayashi, *Z. Naturforsch.*, 1977, **32A**, 1569.

28. K. Kamogawa, K. Tajima, K. Hayakawa, and T. Kitagawa, *J. Phys. Chem.*, 1984, **88**, 2494.

29. H. Takahashi, J. Umemura, and T. Takenaka, *J. Phys. Chem.*, 1982, **86**, 4660.

30. K. Machida, K. Lee, and A. Kuwae, *J. Raman Spectrosc.*, 1980, **9**, 198.

31. T. Imae, S. Ikeda, and K. Itoh, *J. Chem. Soc., Faraday Trans. 1*, 1983, **79**, 2843.

32. J. Umemura, D. G. Cameron, and H. H. Mantsch, *J. Phys. Chem.*, 1980, **84**, 2272.

33. T. Kawai, J. Umemura, and T. Takenaka, *Colloid Polymer Sci.*, 1984, **262**, 61.

34. I. J. Langmuir, *J. Am. Chem. Soc.*, 1917, **39**, 1848.

35. I. J. Langmuir, *Trans. Faraday. Soc.*, 1920, **15**, 62.

36. J. A. Mann, *Langmuir*, 1985, **1**, 10.

37. J. D. Swalen, D. L. Allara, J. D. Andrade, E. A. Chandross, S. Garoff, J. Israelachvilli, T. J. McCarthy, R. Murray, R. P. Pease, J. F. Rabolt, K. J. Wynne, and H. Yu, *Langmuir*, 1987, **3**, 932.

38. J. P. Matousek, B. J. Orr, and M. Selby, *Prog. Anal. Atom. Spectrosc.*, 1984, **7**, 275.

39. D. L. Haas and J. A. Caruso, *Anal. Chem.*, 1984, **56**, 2014.

40. K. J. Milligan, M. Zerezhgiand, and J. A. Caroso, *Spectrochim. Acta*, 1983, **38B**, 369.

41. D. J. Douglas and J. B. French, *Anal. Chem.*, 1981, **53**, 37.

42. M. Maroncelli, S. P. Qi, H. L. Strauss, and R. G. Snyder, *J. Am. Chem. Soc.*, 1982, **104**, 6237.

43. M. Maroncelli, S. P. Qi, H. L. Strauss, and R. G. Snyder, *Science*, 1981, **214**, 188.

44. M. L. Mitchell and R. A. Dluhy, *Mikrochimica Acta*, 1988, **11**, 349.

45. C. A. Jones, M. C. Petty, G. G. Roberts, G. Davies, J. Yarwood, N. M. Ratcliffe, and J. W. Barton, *Thin Solid Films*, 1987, **155**, 187.

46. C. A. Jones, M. C. Petty, G. G. Roberts, G. Davies, J. Yarwood, N. M. Ratcliffe, and J. W. Barton, *Thin Solid Films*, 1988, **159**, 461.

47. L. J. Bellamy, *Infrared Spectra of Complex Molecules*, Chapman and Hall, London, 1975.

48. G. H. Davies and J. Yarwood, *Spectrochim. Acta*, 1987, **43**, 1619.

49. R. G. Snyder, *J. Mol. Spectrosc.*, 1961, **7**, 116–132.

50. M. Tasumi, T. Shimanouchi, and T. Miyazawa, *J. Mol. Spectrosc.*, 1962, **9**, 261.

51. Y. Ozaki, Y. Fujimoto, S. Terashita, and N. Katayama, *Spectroscopy*, 1993, **8**(1), 36.

52. V. A. Howarth, M. C. Petty, G. H. Davies, and J. Yarwood, *Thin Solid Films*, 1988, **160**, 483.

53. V. A. Howarth, M. C. Petty, G. H. Davies, and J. Yarwood, *Langmuir*, 1989, **5**, 330.

54. N. Katayama, Y. Ozaki, T. Araki, and K. Iriyama, *J. Mol. Struct.*, 1991, **242**, 27–37.

55. M. Kubota, Y. Ozaki, T. Araki, S. Ohki, and K. Iriyama, *Langmuir*, 1991, **7**, 772.

56. D. B. Neal, M. C. Petty, G. G. Roberts, M. M. Ahmad, W. J. Feast, I. R. Gerling, N. A. Cade, P. V. Kalinsky, and I. R. Peterson, *Electron. Lett.*, 1986, **22**, 40.

57. D. B. Neal, M. C. Petty, G. G. Roberts, M. M. Ahmad, W. J. Feast, I. R. Gerling, N. A. Cade, P. V. Kalinsky, and I. R. Peterson, *Opt. Commun.*, 1985, **22**, 40.

58. P. Stroeve, M. P. Srinwasan, B. G. Higgens, and S. T. Kowel, *Thin Solid Films*, 1987, **146**, 209.

59. P. Stroeve, M. P. Srinwasan, B. G. Higgens, and S. T. Kowel, *Opt. Commun.*, 1987, **61**, 351.

60. J. Yarwood, *Spectroscopy*, 1990, **5**(6), 34.

61. S. A. Francis and A. H. Ellison, *J. Opt. Soc. Am.*, 1959, **49**, 131.

62. R. W. Strobie and M. J. Dignam, *Can. J. Chem.*, 1978, **56**, 1088.

63. M. Ito and W. Suetaka, *J. Phys. Chem.*, 1975, **79**, 1190.

64. M. Ito and W. Suetaka, *J. Catal.*, 1978, **54**, 13.

65. F. J. Boerio and S. L. Chen, *Appl. Spectrosc.*, 1979, **33**, 121.

66. H. G. Tompkins and D. L. Allara, *J. Colloid Interface Sci.*, 1974, **47**, 697.

67. A. F. Hebard, J. R. Arthur, and D. L. Allara, *J. Appl. Phys.*, 1978, **49**, 6039.

68. D. L. Allara, *Advances in Chemistry Series*, American Chemical Society, Washington, D.C., 1980.

69. R. Mendelsohn and M. A. Davies, in *Fourier Transform Infrared Spectroscopy in Colloids and Interface Science*, ACS Symp. Series No. 447, American Chemical Society, Washington, D.C., 1990.

70. M. Maroncelli, H. L. Strauss, and R. G. Snyder, *J. Chem. Phys.*, 1985, **82**, 2811.

71. E. G. Chatzi, M. W. Urban, H. Ishida, and J. L. Koenig, *Langmuir*, 1988, **4**, 846.

72. M. W. Urban and J. L. Koenig, *Appl. Spectrosc.*, 1986, **39**, 1051.

73. D. L. Allara and J. D. Swallen, *J. Phys. Chem.*, 1982, **86**, 2700.

74. P. Pulay, *Mol. Phys.*, 1969, **17**, 197; 1970, **18**, 473; P. Pulay and W. Meyer, *J. Mol. Spectrosc.*, 1971, **40**, 59.

75. N. Nakamoto, *Infrared and Raman Spectra of Inorganic and Coordination Compounds*, 4th ed., Wiley-Interscience, New York, 1986, p. 72.

76. T. Takenaka and J. Umemura, in *Vibrational Spectra and Structure*, J. R. Durig, Ed., Elsevier, New York, 1991, Chapter 5.

77. N. E. Schlotter and J. F. Rabolt, *Appl. Spectrosc.*, 1985, **39**(6), 994.

78. D. Allara and J. D. Swallen, *J. Phys. Chem.*, 1982, **86**, 2700.

79. C. Naselli, J. F. Rabolt, and J. D. Swallen, *J. Chem. Phys.*, 1985, **82**, 2136.

80. S. Havriliak and N. Roman, *Polymer*, 1966, **7**, 3887.

81. B. Tieke, G. Wegner, D. Naegele, and H. Ringsdorf, *Angew. Chem.*, 1976, **15**, 764.

82. B. Tieke, H. J. Graf, G. Wegner, D. Naegele, H. Ringsdorf, A. Banerjie, D. Day, and J. Lando, *Colloid Polym. Sci.*, 1977, **255**, 521.

83. C. Chen and C. Frank, in *Advances in Chemistry Series*, No. 203, C. D. Craver, Ed., American Chemical Society, Washington, D. C., 1983.

84. M. W. Urban and K. W. Evanson, *Polym. Commun.*, 1990, **31**, 279.

85. T. Thorstenson and M. W. Urban, *J. Appl. Polym. Sci.*, 1993, **47**, 1387.

86. T. A. Thorstenson, L. K. Tebelius, and M. W. Urban, *Appl. Polym. Sci.*, 1993, **50**, 1207.

87. R. M. Gendreau and R. J. Jakobsen, *J. Biomed. Mater. Res.*, 1979, **13**, 893.

88. S. Winters, R. M. Gendreau, I. Leininger, and R. J. Jakobsen, *Appl. Spectrosc.*, 1982, **36**, 404.

89. D. J. Fink, T. B. Hutson, K. K. Chitter, and R. M. Gendreau, *Anal. Biochem.*, 1987, **15**, 57.

90. K. K. Chittur, D. J. Fink, R. I. Leininger, and T. B. Hutson, *J. Coll. Interface Sci.*, 1986, **111**, 419.

91. E. J. Castillo, J. L. Koenig, and J. M. Anderson, *Biomaterials*, 1984, **5**, 319.

92. J. L. Kirsch and J. L. Koenig, *Appl. Spectrosc.*, 1989, **43**, 445.

93. R. J. Jakobsen and F. M. Wasacz, *Appl. Spectrosc.*, 1990, **44**(9), 1478.

94. B. G. Frushour and J. L. Koenig, in *Advances in Infrared and Raman Spectroscopy*, Vol. 1, G. J. H. Clark and R. E. Hester, Eds., Heyden, London, 1975.

95. P. Carmona, M. Molina, P. Martinez, and Z. B. Altabef, *Appl. Spectrosc.*, 1991, **45**(6), 977.

96. T. Hirschfeld and B. Chase., *Appl. Spectrosc.*, 1986, **40**(2), 133.

97. S. Nie, K. L. Bergbauer, J. J. Ho, J. F. R. Kuck, and N. T. Yu, *Spectroscopy*, 1990, **5**(7), 24.

98. K. V. Sokolov, S. V. Lutsenko, I. R. Nabiev, S. Nie, and N. T. Yu, *Appl. Spectrosc.*, 1991, **45**(7), 1143.

99. S. Nie, C. G. Castillo, K. L. Bergbauer, J. F. R. Kuck, I. R. Nabiev, and N. T. Yu, *Appl. Spectrosc.*, 1990, **44**(4), 571.

APPENDIX A

POINTS GROUPS AND BAND PREDICTIONS

The enclosed tables list all characteristic symmetry elements for various molecular structures. The following ground rules are used in assigning a given species proper symmetry: Knowing a structure of a given species find all symmetry elements, for example, each species has an identity operator i along with other symmetry elements. In the case of water, C_{2v} point group is assigned because water structure is such that the molecule can be converted to itself with an identity operator i, rotation along the z axes by $180°$ (C_2), and two vertical xz (σ_{zx}) and yz (σ_{yz}) plane reflections. Each character table contains normal vibrations listed according to the following principles: A and B represent nondegenerate normal vibrations, symmetric and nonsymmetric, respectively. This is represented by 1 or -1 in the character table. E and F are reserved for doubly and triply degenerated species. In an effort to identify which species are IR or Raman active, one needs to take a look at the character table and realize that wherever symbol α is shown it indicates that a given vibration is Raman active because this normal vibration exhibits a nonzero polarizibility tensor α. In a similar way, T_x, T_y, or T_z, symbols indicate IR active modes as they designate translations along or around a given axis. R_x, R_y, and R_z are designated to rotations. In a band prediction process, the first step is to identify a number of IR and Raman active bands ($3n - 6$ for nonlinear or $3n - 5$ for linear molecules, $n -$ number of atoms), then based on a structural analysis identify a point group, followed by calculating an irreversible representation. The latter is beyond a scope of this book and can be found in the references listed at the end of Chapter 1. Knowing a number of IR and Raman active bands allows us to compare experimental and theoretical predictions and utilize a normal coordinate analysis to calculate vibrational energies.

C_s	I	$\sigma(xy)$		
A'	$+1$	$+1$	T_x, T_y, R_z	$\alpha_{xx}, \alpha_{yy}, \alpha_{zz}, \alpha_{xy}$
A''	$+1$	-1	T_z, R_x, R_y	α_{yz}, α_{xz}

C_2	I	$C_2(z)$		
A	$+1$	$+1$	T_z, R_z	$\alpha_{xx}, \alpha_{yy}, \alpha_{zz}, \alpha_{xy}$
B	$+1$	-1	T_x, T_y, R_x, R_y	α_{yz}, α_{xz}

C_i	I	i		
A_g	$+1$	$+1$	R_x, R_y, R_z	all components of α
A_u	$+1$	$+1$	T_x, T_y, T_z	

C_{2v}	I	$C_2(z)$	$\sigma_v(xz)$	$\sigma_v(yz)$		
A_1	$+1$	$+1$	$+1$	$+1$	T_z	$\alpha_{xx}, \alpha_{yy}, \alpha_{zz}$
A_2	$+1$	$+1$	-1	-1	R_z	α_{xy}
B_1	$+1$	-1	$+1$	-1	T_x, R_y	α_{xz}
B_2	$+1$	-1	-1	$+1$	T_y, R_x	α_{yz}

C_{3v}	I	$2C_3(z)$	$3\sigma_v$		
A_1	$+1$	$+1$	$+1$	T_z	$\alpha_{xx}+\alpha_{yy}, \alpha_{zz}$
A_2	$+1$	$+1$	-1	R_z	
E	$+2$	-1	0	$(T_x, T_y), (R_x, R_y)$	$(\alpha_{xx}-\alpha_{yy}, \alpha_{xy}), (\alpha_{yz}, \alpha_{xz})$

C_{4v}	I	$2C_4(z)$	$C_4{}^2 \equiv C_2'$	$2\sigma_v$	$2\sigma_d$		
A_1	$+1$	$+1$	$+1$	$+1$	$+1$	T_z	$\alpha_{xx}+\alpha_{yy},\ \alpha_{zz}$
A_2	$+1$	$+1$	$+1$	-1	-1	R_z	
B_1	$+1$	-1	$+1$	$+1$	-1		$\alpha_{xx}-\alpha_{yy}$
B_2	$+1$	-1	$+1$	-1	$+1$		α_{xy}
E	$+2$	0	-2	0	0	$(T_x,T_y),(R_x,R_y)$	$(\alpha_{yz},\alpha_{xz})$

$C_p{}^n$ (or $S_p{}^n$) denotes that C_p (or S_p) operation is carried out successively n times.

$C_{\infty v}$	I	$2C_\infty{}^\phi$	$2C_\infty{}^{2\phi}$	$2C_\infty{}^{3\phi}$	$\cdots$	$\infty\sigma_v$		
Σ^+	$+1$	$+1$	$+1$	$+1$	$\cdots$	$+1$	T_z	$\alpha_{xx}+\alpha_{yy},\ \alpha_{zz}$
Σ^-	$+1$	$+1$	$+1$	$+1$	$\cdots$	-1	R_z	
Π	$+2$	$2\cos\phi$	$2\cos 2\phi$	$2\cos 3\phi$	$\cdots$	0	$(T_x,T_y),(R_x,R_y)$	$(\alpha_{yz},\alpha_{xz})$
Δ	$+2$	$2\cos 2\phi$	$2\cos 2\cdot 2\phi$	$2\cos 3\cdot 2\phi$	$\cdots$	0		$(\alpha_{xx}-\alpha_{yy},\alpha_{xy})$
Φ	$+2$	$2\cos 3\phi$	$2\cos 2\cdot 3\phi$	$2\cos 3\cdot 3\phi$	$\cdots$	0		
$\cdots$	$\cdots$	$\cdots$	$\cdots$	$\cdots$	$\cdots$	$\cdots$		

C_{2h}	I	$C_2(z)$	$\sigma_h(xy)$	i		
A_g	$+1$	$+1$	$+1$	$+1$	R_z	$\alpha_{xx},\alpha_{yy},\alpha_{zz},\alpha_{xy}$
A_u	$+1$	$+1$	-1	-1	T_z	
B_g	$+1$	-1	-1	$+1$	R_x,R_y	α_{yz},α_{xz}
B_u	$+1$	-1	$+1$	-1	T_x,T_y	

D₃	I	$2C_3(z)$	$3C_2$		
A_1	+1	+1	+1		$\alpha_{xx}+\alpha_{yy},\ \alpha_{zz}$
A_2	+1	+1	−1	$T_z,\ R_z$	
E	+2	−1	0	$(T_x,T_y),(R_x,R_y)$	$(\alpha_{xx}-\alpha_{yy},\ \alpha_{xy}),(\alpha_{yz},\alpha_{xz})$

$D_{2d}\equiv V_d$	I	$2S_4(z)$	$S_4^2\equiv C_2''$	$2C_2$	$2\sigma_d$		
A_1	+1	+1	+1	+1	+1		$\alpha_{xx}+\alpha_{yy},\ \alpha_{zz}$
A_2	+1	+1	+1	−1	−1	R_z	
B_1	+1	−1	+1	+1	−1		$\alpha_{xx}-\alpha_{yy}$
B_2	+1	−1	+1	−1	+1	T_z	α_{xy}
E	+2	0	−2	0	0	$(T_x,T_y),(R_x,R_y)$	$(\alpha_{yz},\alpha_{xz})$

D_{3d}	I	$2S_6(z)$	$2S_6^2\equiv 2C_3$	$S_6^3\equiv S_2\equiv i$	$3C_2$	$3\sigma_d$		
A_{1g}	+1	+1	+1	+1	+1	+1		$\alpha_{xx}+\alpha_{yy},\ \alpha_{zz}$
A_{1u}	+1	−1	+1	−1	+1	−1		
A_{2g}	+1	+1	+1	+1	−1	−1	R_z	
A_{2u}	+1	−1	+1	−1	−1	+1	T_z	
E_g	+2	−1	−1	+2	0	0	(R_x,R_y)	$(\alpha_{xx}-\alpha_{yy},\ \alpha_{xy}),(\alpha_{yz},\alpha_{xz})$
E_u	+2	+1	−1	−2	0	0	(T_x,T_y)	

D_{4d}

D_{4d}	I	$2S_8(z)$	$2S_8^2 \equiv 2C_4$	$2S_8^3$	$S_8^4 \equiv C_2''$	$4C_2$	$4\sigma_d$		
A_1	$+1$	$+1$	$+1$	$+1$	$+1$	$+1$	$+1$		$\alpha_{xx}+\alpha_{yy},\ \alpha_{zz}$
A_2	$+1$	$+1$	$+1$	$+1$	$+1$	-1	-1	R_z	
B_1	$+1$	-1	$+1$	-1	$+1$	$+1$	-1		
B_2	$+1$	-1	$+1$	-1	$+1$	-1	$+1$	T_z	
E_1	$+2$	$+\sqrt{2}$	0	$-\sqrt{2}$	-2	0	0	(T_x, T_y)	
E_2	$+2$	0	-2	0	$+2$	0	0		$(\alpha_{xx}-\alpha_{yy},\ \alpha_{xy})$
E_3	$+2$	$-\sqrt{2}$	0	$+\sqrt{2}$	-2	0	0	(R_x, R_y)	$(\alpha_{yz},\ \alpha_{xz})$

$D_{2h} \equiv V_h$

$D_{2h} \equiv V_h$	I	$\sigma(xy)$	$\sigma(xz)$	$\sigma(yz)$	i	$C_2(z)$	$C_2(y)$	$C_2(x)$		
A_g	$+1$	$+1$	$+1$	$+1$	$+1$	$+1$	$+1$	$+1$		$\alpha_{xx}, \alpha_{yy}, \alpha_{zz}$
A_u	$+1$	-1	-1	-1	-1	$+1$	$+1$	$+1$		
B_{1g}	$+1$	$+1$	-1	-1	$+1$	$+1$	-1	-1	R_z	α_{xy}
B_{1u}	$+1$	-1	$+1$	$+1$	-1	$+1$	-1	-1	T_z	
B_{2g}	$+1$	-1	$+1$	-1	$+1$	-1	$+1$	-1	R_y	α_{xz}
B_{2u}	$+1$	$+1$	-1	$+1$	-1	-1	$+1$	-1	T_y	
B_{3g}	$+1$	-1	-1	$+1$	$+1$	-1	-1	$+1$	R_x	α_{yz}
B_{3u}	$+1$	$+1$	$+1$	-1	-1	-1	-1	$+1$	T_x	

D_{3h}

D_{3h}	I	$2C_3(z)$	$3C_2$	σ_h	$2S_3$	$3\sigma_v$		
A_1'	$+1$	$+1$	$+1$	$+1$	$+1$	$+1$		$\alpha_{xx}+\alpha_{yy},\ \alpha_{zz}$
A_1''	$+1$	$+1$	$+1$	-1	-1	-1		
A_2'	$+1$	$+1$	-1	$+1$	$+1$	-1	R_z	
A_2''	$+1$	$+1$	-1	-1	-1	$+1$	T_z	
E'	$+2$	-1	0	$+2$	-1	0	(T_x, T_y)	$(\alpha_{xx}-\alpha_{yy},\ \alpha_{xy})$
E''	$+2$	-1	0	-2	$+1$	0	(R_x, R_y)	$(\alpha_{yz},\ \alpha_{xz})$

D4h

D4h	I	2C4(z)	C4²≡C2''	2C2	2C2'	σh	2σv	2σd	2S4	S2≡i		
A1g	+1	+1	+1	+1	+1	+1	+1	+1	+1	+1		$\alpha_{xx}+\alpha_{yy},\ \alpha_{zz}$
A1u	+1	+1	+1	+1	+1	-1	-1	-1	-1	-1		
A2g	+1	+1	+1	-1	-1	+1	-1	-1	-1	+1	R_z	
A2u	+1	+1	+1	-1	-1	-1	+1	+1	+1	-1	T_z	
B1g	+1	-1	+1	+1	-1	+1	+1	-1	-1	+1		$\alpha_{xx}-\alpha_{yy}$
B1u	+1	-1	+1	+1	-1	-1	-1	+1	+1	-1		
B2g	+1	-1	+1	-1	+1	+1	-1	+1	-1	+1		α_{xy}
B2u	+1	-1	+1	-1	+1	-1	+1	-1	+1	-1		
Eg	+2	0	-2	0	0	-2	0	0	0	+2	(R_x, R_y)	$(\alpha_{yz}, \alpha_{xz})$
Eu	+2	0	-2	0	0	+2	0	0	0	-2	(T_x, T_y)	

D5h

D5h	I	2C5(z)	2C5²	σh	5C2	5σv	2S5	2S5³		
A1'	+1	+1	+1	+1	+1	+1	+1	+1		$\alpha_{xx}+\alpha_{yy},\ \alpha_{zz}$
A1''	+1	+1	+1	-1	+1	-1	-1	-1		
A2'	+1	+1	+1	+1	-1	-1	+1	+1	R_z	
A2''	+1	+1	+1	-1	-1	+1	-1	-1	T_z	
E1'	+2	2 cos 72°	2 cos 144°	+2	0	0	+2 cos 72°	+2 cos 144°	(T_x, T_y)	
E1''	+2	2 cos 72°	2 cos 144°	-2	0	0	-2 cos 72°	-2 cos 144°	(R_x, R_y)	$(\alpha_{yz}, \alpha_{xz})$
E2'	+2	2 cos 144°	2 cos 72°	+2	0	0	+2 cos 144°	+2 cos 72°		$(\alpha_{yy}, \alpha_{xy})$ $(\alpha_{xx}-\alpha_{yy}, \alpha_{xy})$
E2''	+2	2 cos 144°	2 cos 72°	-2	0	0	-2 cos 144°	-2 cos 72°		

D_{6h}	I	$2C_6(z)$	$2C_6^2\equiv 2C_3$	$C_6^3\equiv C_2''$	$3C_2$	$3C_2'$	σ_h	$3\sigma_v$	$3\sigma_d$	$2S_6$	$2S_3$	$S_6^3\equiv S_2\equiv i$		
A_{1g}	$+1$	$+1$	$+1$	$+1$	$+1$	$+1$	$+1$	$+1$	$+1$	$+1$	$+1$	$+1$		$\alpha_{xx}+\alpha_{yy},\ \alpha_{zz}$
A_{1u}	$+1$	$+1$	$+1$	$+1$	$+1$	$+1$	-1	-1	-1	-1	-1	-1		
A_{2g}	$+1$	$+1$	$+1$	$+1$	-1	-1	$+1$	-1	-1	$+1$	$+1$	$+1$	R_z	
A_{2u}	$+1$	$+1$	$+1$	$+1$	-1	-1	-1	$+1$	$+1$	-1	-1	-1	T_z	
B_{1g}	$+1$	-1	$+1$	-1	$+1$	-1	-1	$+1$	-1	$+1$	-1	$+1$		
B_{1u}	$+1$	-1	$+1$	-1	$+1$	-1	$+1$	-1	$+1$	-1	$+1$	-1		
B_{2g}	$+1$	-1	$+1$	-1	-1	$+1$	-1	-1	$+1$	$+1$	-1	$+1$		
B_{2u}	$+1$	-1	$+1$	-1	-1	$+1$	$+1$	$+1$	-1	-1	$+1$	-1		
E_{1g}	$+2$	$+1$	-1	-2	0	0	-2	0	0	-1	$+1$	$+2$	(R_x,R_y)	$(\alpha_{yz},\alpha_{xz})$
E_{1u}	$+2$	$+1$	-1	-2	0	0	$+2$	0	0	$+1$	-1	-2	(T_x,T_y)	
E_{2g}	$+2$	-1	-1	$+2$	0	0	$+2$	0	0	-1	-1	$+2$		$(\alpha_{xx}-\alpha_{yy},\alpha_{xy})$
E_{2u}	$+2$	-1	-1	$+2$	0	0	-2	0	0	$+1$	$+1$	-2		

$D_{\infty h}$	I	$2C_\infty^\phi$	$2C_\infty^{2\phi}$	$2C_\infty^{3\phi}$	$\cdots$	σ_h	∞C_2	$\infty\sigma_v$	$2S_\infty^\phi$	$2S_\infty^{2\phi}$	$\cdots$	$S_2\equiv i$		
Σ_g^+	$+1$	$+1$	$+1$	$+1$	$\cdots$	$+1$	$+1$	$+1$	$+1$	$+1$	$\cdots$	$+1$		$\alpha_{xx}+\alpha_{yy},\ \alpha_{zz}$
Σ_u^+	$+1$	$+1$	$+1$	$+1$	$\cdots$	-1	-1	$+1$	-1	-1	$\cdots$	-1	T_z	
Σ_g^-	$+1$	$+1$	$+1$	$+1$	$\cdots$	$+1$	-1	-1	$+1$	$+1$	$\cdots$	$+1$	R_z	
Σ_u^-	$+1$	$+1$	$+1$	$+1$	$\cdots$	-1	$+1$	-1	-1	-1	$\cdots$	-1		
Π_g	$+2$	$2\cos\phi$	$2\cos 2\phi$	$2\cos 3\phi$	$\cdots$	-2	0	0	$-2\cos\phi$	$-2\cos 2\phi$	$\cdots$	$+2$	(R_x,R_y)	$(\alpha_{yz},\alpha_{xz})$
Π_u	$+2$	$2\cos\phi$	$2\cos 2\phi$	$2\cos 3\phi$	$\cdots$	$+2$	0	0	$+2\cos\phi$	$+2\cos 2\phi$	$\cdots$	-2	(T_x,T_y)	
Δ_g	$+2$	$2\cos 2\phi$	$2\cos 4\phi$	$2\cos 6\phi$	$\cdots$	$+2$	0	0	$+2\cos 2\phi$	$+2\cos 4\phi$	$\cdots$	$+2$		$(\alpha_{xx}-\alpha_{yy},\alpha_{xy})$
Δ_u	$+2$	$2\cos 2\phi$	$2\cos 4\phi$	$2\cos 6\phi$	$\cdots$	-2	0	0	$-2\cos 2\phi$	$-2\cos 4\phi$	$\cdots$	-2		
Φ_g	$+2$	$2\cos 3\phi$	$2\cos 6\phi$	$2\cos 9\phi$	$\cdots$	-2	0	0	$-2\cos 3\phi$	$-2\cos 6\phi$	$\cdots$	$+2$		
Φ_u	$+2$	$2\cos 3\phi$	$2\cos 6\phi$	$2\cos 9\phi$	$\cdots$	$+2$	0	0	$+2\cos 3\phi$	$+2\cos 6\phi$	$\cdots$	-2		
$\cdots$	$\cdots$	$\cdots$	$\cdots$	$\cdots$	$\cdots$	$\cdots$	$\cdots$	$\cdots$	$\cdots$	$\cdots$	$\cdots$	$\cdots$		

T_d

T_d	I	$8C_3$	$6\sigma_d$	$6S_4$	$3S_4^2 \equiv 3C_2$		
A_1	$+1$	$+1$	$+1$	$+1$	$+1$		$\alpha_{xx}+\alpha_{yy}+\alpha_{zz}$
A_2	$+1$	$+1$	-1	-1	$+1$		
E	$+2$	-1	0	0	$+2$		$(\alpha_{xx}+\alpha_{yy}-2\alpha_{zz},\ \alpha_{xx}-\alpha_{yy})$
F_1	$+3$	0	-1	$+1$	-1	(R_x, R_y, R_z)	
F_2	$+3$	0	$+1$	-1	-1	(T_x, T_y, T_z)	$(\alpha_{xy}, \alpha_{yz}, \alpha_{zx})$

O_h

O_h	I	$8C_3$	$6C_2$	$6C_4$	$3C_4^2 \equiv 3C_2''$	$S_2 \equiv i$	$6S_4$	$8S_6$	$3\sigma_h$	$6\sigma_d$		
A_{1g}	$+1$	$+1$	$+1$	$+1$	$+1$	$+1$	$+1$	$+1$	$+1$	$+1$		$\alpha_{xx}+\alpha_{yy}+\alpha_{zz}$
A_{1u}	$+1$	$+1$	$+1$	$+1$	$+1$	-1	-1	-1	-1	-1		
A_{2g}	$+1$	$+1$	-1	-1	$+1$	$+1$	-1	$+1$	$+1$	-1		
A_{2u}	$+1$	$+1$	-1	-1	$+1$	-1	$+1$	-1	-1	$+1$		
E_g	$+2$	-1	0	0	$+2$	$+2$	0	-1	$+2$	0		$(\alpha_{xx}+\alpha_{yy}-2\alpha_{zz},\ \alpha_{xx}-\alpha_{yy})$
E_u	$+2$	-1	0	0	$+2$	-2	0	$+1$	-2	0		
F_{1g}	$+3$	0	-1	$+1$	-1	$+3$	$+1$	0	-1	-1	(R_x, R_y, R_z)	
F_{1u}	$+3$	0	-1	$+1$	-1	-3	-1	0	$+1$	$+1$	(T_x, T_y, T_z)	
F_{2g}	$+3$	0	$+1$	-1	-1	$+3$	-1	0	-1	$+1$		$(\alpha_{xy}, \alpha_{yz}, \alpha_{zx})$
F_{2u}	$+3$	0	$+1$	-1	-1	-3	$+1$	0	$+1$	-1		

INDEX